SOUND AND SOURCES OF SOUND

ELLIS HORWOOD SERIES IN ENGINEERING SCIENCE

STRENGTH OF MATERIALS
J. M. ALEXANDER, University College of Swansea.
TECHNOLOGY OF ENGINEERING MANUFACTURE
J. M. ALEXANDER, R. C. BREWER, Imperial College of Science and Technology, University of London, J. R. CROOKALL, Cranfield Institute of Technology.
VIBRATION ANALYSIS AND CONTROL SYSTEM DYNAMICS
CHRISTOPHER BEARDS, Imperial College of Science and Technology, University of London.
COMPUTER AIDED DESIGN AND MANUFACTURE 2nd Edition
C. B. BESANT, Imperial College of Science and Technology, University of London.
STRUCTURAL DESIGN AND SAFETY
D. I. BLOCKLEY, University of Bristol.
BASIC LUBRICATION THEORY 3rd Edition
ALASTAIR CAMERON, Imperial College of Science and Technology, University of London.
STRUCTURAL MODELLING AND OPTIMIZATION
D. G. CARMICHAEL, University of Western Australia
ADVANCED MECHANICS OF MATERIALS 2nd Edition
Sir HUGH FORD, F.R.S., Imperial College of Science and Technology, University of London and J. M. ALEXANDER, University College of Swansea.
ELASTICITY AND PLASTICITY IN ENGINEERING
Sir HUGH FORD, F.R.S. and R. T. FENNER, Imperial College of Science and Technology, University of London.
INTRODUCTION TO LOADBEARING BRICKWORK
A. W. HENDRY, B. A. SINHA and S. R. DAVIES, University of Edinburgh
ANALYSIS AND DESIGN OF CONNECTIONS BETWEEN STRUCTURAL JOINTS
M. HOLMES and L. H. MARTIN, University of Aston in Birmingham
TECHNIQUES OF FINITE ELEMENTS
BRUCE M. IRONS, University of Calgary, and S. AHMAD, Bangladesh University of Engineering and Technology, Dacca.
FINITE ELEMENT PRIMER
BRUCE IRONS and N. SHRIVE, University of Calgary
PROBABILITY FOR ENGINEERING DECISIONS: A Bayesian Approach
I. J. JORDAAN, University of Calgary
STRUCTURAL DESIGN OF CABLE-SUSPENDED ROOFS
L. KOLLAR, City Planning Office, Budapest and K. SZABO, Budapest Technical University.
CONTROL OF FLUID POWER, 2nd Edition
D. McCLOY, The Northern Ireland Polytechnic and H. R. MARTIN, University of Waterloo, Ontario, Canada.
TUNNELS: Planning, Design, Construction
T. M. MEGAW and JOHN BARTLETT, Mott, Hay and Anderson, International Consulting Engineers
UNSTEADY FLUID FLOW
R. PARKER, University College, Swansea
DYNAMICS OF MECHANICAL SYSTEMS 2nd Edition
J. M. PRENTIS, University of Cambridge.
ENERGY METHODS IN VIBRATION ANALYSIS
T. H. RICHARDS, University of Aston, Birmingham.
ENERGY METHODS IN STRESS ANALYSIS: With an Introduction to Finite Element Techniques
T. H. RICHARDS, University of Aston, Birmingham.
ROBOTICS AND TELECHIRICS
M. W. THRING, Queen Mary College, University of London
STRESS ANALYSIS OF POLYMERS 2nd Edition
J. G. WILLIAMS, Imperial College of Science and Technology, University of London.

SOUND AND SOURCES OF SOUND

A. P. DOWLING, B.A., Ph.D.
Fellow of Sidney Sussex College, Cambridge

and

J. E. FFOWCS WILLIAMS, B.Sc., Ph.D., M.A.
Fellow of Emmanuel College, Cambridge
Rank Professor of Engineering (Acoustics)

both of Department of Engineering
University of Cambridge

ELLIS HORWOOD LIMITED
Publishers · Chichester

Halsted Press: a division of
JOHN WILEY & SONS
New York · Brisbane · Chichester · Toronto

First published in 1983 by

ELLIS HORWOOD LIMITED

Market Cross House, Cooper Street, Chichester, West Sussex, PO19 1EB, England

The publisher's colophon is reproduced from James Gillison's drawing of the ancient Market Cross, Chichester.

Distributors:

Australia, New Zealand, South-east Asia:
Jacaranda-Wiley Ltd., Jacaranda Press,
JOHN WILEY & SONS INC.,
G.P.O. Box 859, Brisbane, Queensland 40001, Australia

Canada:
JOHN WILEY & SONS CANADA LIMITED
22 Worcester Road, Rexdale, Ontario, Canada.

Europe, Africa:
JOHN WILEY & SONS LIMITED
Baffins Lane, Chichester, West Sussex, England.

North and South America and the rest of the world:
Halsted Press: a division of
JOHN WILEY & SONS
605 Third Avenue, New York, N.Y. 10016, U.S.A.

British Library Cataloguing in Publication Data
Dowling, A.P.
Sound and sources of sound
1. Sound
I. Title II. Ffowcs Williams, John E.
534 QC225.15

Library of Congress Card No. 82-15687

ISBN 0-85312-400-0 (Ellis Horwood Limited – Library Edn.)
ISBN 0-85312-527-9 (Ellis Horwood Limited – Student Edn.)
ISBN 0-470-27371-2 (Halsted Press) (Library Edn.)
ISBN 0-470-27388-7 (Halsted Press) (Student Edn.)

Printed in Great Britain by Unwin Brothers of Woking.

Contents

Preface

Sound has an impact on engineering because powerful machines are noisy and the need to limit their noise raises questions of where and how the noise is generated. Sound scattering off material inhomogeneities or being emitted by transient stresses during the non-destructive testing of mechanical parts, or people, is useful only when some interpretation is made of what the source of the observed wave might be. Sonar, the powerful navigation and detection aid of bats and submarines, again directs attention on the source of scattered waves and how those sources might be modified to avoid detection. The whole science of noise generated by unsteady turbulent flows, both underwater and in the air, rests on an adequate modelling of the source process.

Though practical concern invariably centres on the sources of sound it is a subject rarely treated in acoustical texts. The science of sound itself has been superbly documented for over a century but concern with sources comes from the relatively recent engineering interest in the subject. This book is intended to put that right in a development from first principles of the things we practising engineers and researchers in the subject most need to know. Sound and sources of sound are described, analysed and illustrated with material selected for its interest to us. Our choice is determined by those things we have most enjoyed studying and found most useful.

Throughout the book we concentrate on the precise description of useful idealisations of the real world and develop the mathematical tools needed for their study as the need arises. This is the technique and material we have used in teaching a final-year undergraduate course to engineers at Cambridge University and we would not change the material significantly in addressing mathematicians or physicists. We hope that the development from first principles of the subject material and techniques of analysis makes the book self-contained but stress that only good will come from additional reference to the works of Rayleigh and Lighthill who, for us, provide the subject's main foundation.

The length of the book is determined by what we think can, with reasonable selectivity, be put over in a thirty-six-hour lecture course. There are many omissions in both depth of treatment and material, made inevitable by our wish to get quickly within reach of the ground we find most interesting. We think that worked examples are useful and have therefore given several questions and worked answers, but these are separated in the text to encourage readers to learn for themselves the level of effort needed for a solution. We hope that these examples contain non-trivial illustrations of the practical subject.

At the end of the book we gather together a brief listing of facts, definitions and formulae in a ready reference 'handbook' that will help remind the reader of some commonly recurring points.

We are grateful to those Cambridge undergraduates who have worked from the early versions of this text and checked our examples, and to colleagues, particularly Tom Hynes and David Crighton for helpful comments on the manuscript and choice of material. We much appreciate the careful typing (from sometimes execrable handwriting) of various versions of the text by Marion Childs, Ginny Morrow and Jennifer Blows and the artwork of Elaine Avis. Finally, we are grateful to the University of Cambridge for allowing publication of course material and examples which have been used in formal examinations.

Ann P. Dowling
John E. Ffowcs Williams
Cambridge, May 1982

Chapter 1

Characteristics of Sound

1.1 INTRODUCTION

The science of acoustics was thoroughly developed in the nineteenth century, Stokes and Rayleigh being the subject's greatest figures. The sounds they were interested in were generally pleasant. For example they studied the sound generated by vibrating strings and organ pipes. Nowadays most of the sounds of engineering interest are unpleasant, there being a large research effort concerned with the sound generated by powerful machinery and jet engines. But not all modern-day acoustics is concerned with unwanted noise. Architects need to be able to design a room or concert hall with good acoustical properties. Work goes on to improve the performance of 'hi-fi', and to understand exactly how musical instruments work. Sonar is used widely as an underwater navigation and detection aid, and is continually being refined. In medicine ultrasonics are used to produce 'photographs' of tissue transparent to X-rays, and can show, for example, the malfunction of a valve in a patient's heart. Ultrasonics are widely used as a means of testing for structural integrity and sound is a principal means of monitoring the performance of machines. Nuclear power plants are continually monitored for the tell-tale sounds of malfunction and pressure vessels routinely undergo ultrasonic testing for cracks. Modern acoustics is a subject with many diverse applications and most industries now have teams of engineers specialising in the various aspects of noise and vibration control.

1.2 PROPAGATION OF SOUND ENERGY

Sound propagates as a wave. In air it travels at a speed of some 340 metres per second. Its speed in water is a lot faster, around 1500 metres per second. The sound wave evolves as it travels away and far from the source region its curvature has so diminished that it resembles a one-dimensional or plane wave field propagating radially outwards with constant speed c.

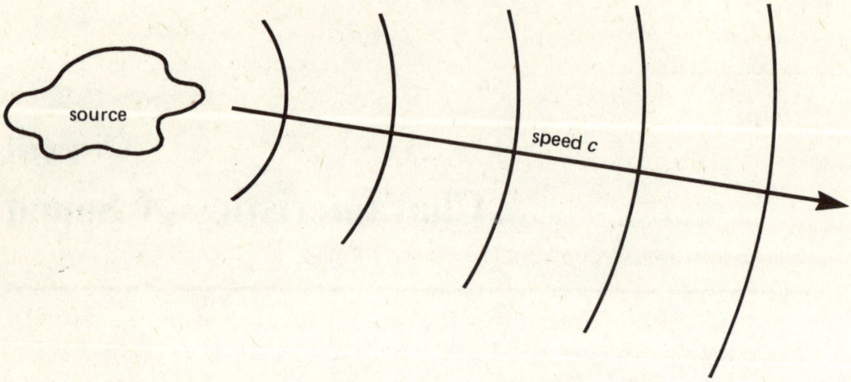

Fig. 1.1 — Sound waves propagating away from a source.

As sound travels it transports energy with it. Sound energises our eardrums to vibrate; that is how we hear the waves. (Tones of frequency ranging between some 20 and 20,000 cycles per second, or Hertz (Hz), are audible, the human ear being most sensitive to sound in the 1 to 5 kHz range.) Since sound waves carry energy their source must supply that energy to satisfy the power demand of the wave field. The quiet human whisper involves an acoustic power of some 10^{-10} watts and the human shout about 10^{-5} watts. A large jet transport at take-off emits about 10^5 watts while a large rocket launch vehicle generates some 10^7 watts of acoustical energy. Some amusing comparisons can be constructed from these figures. The large jet aircraft makes as much acoustic power as all the world's population shouting at once! But sounds are usually of low energy, the total energy radiated by the combined shouts of the Wembley cup final crowd during an exciting game being about that required to fry one egg! Because of the enormous numerical range involved in measuring these sound powers the logarithmic scale is used the power level being given in decibels (PWL).

$$\text{PWL} = 10\log_{10}\left(\frac{\text{sound power output}}{10^{-12}\text{ watts}}\right)$$

$$= 10\log_{10}(\text{sound power in watts}) + 120\text{ dB}. \qquad (1.1)$$

The human shout has a PWL of 70 dB and the large rocket motor 190 dB.

1.3 SOUND IS A LINEAR MOTION

As a sound wave propagates it disturbs the fluid from its mean state. These disturbances are nearly always small. We will consider departures from a state in which the fluid is at rest with a uniform pressure p_0 and density ρ_0. When this is perturbed by a sound wave the pressure at position **x** and time t changes to $p_0 + p'(\mathbf{x}, t)$, the density to $\rho_0 + \rho'(\mathbf{x}, t)$ and the fluid particles move, very slowly, with a velocity $\mathbf{v}(\mathbf{x}, t)$. The ratios $|p'/p_0|$ and $|\rho'/\rho_0|$ are in most cases very much less than unity.

Though always weak, the range of amplitudes commonly experienced in sound waves is very great and it is convenient to express the pressure perturbation on the numerically more compact logarithmic scale. Again, somewhat confusingly the unit is termed the decibel, or dB. The sound pressure level (SPL) is a measure of the mean square level of fluctuation and is by convention defined as

$$\text{SPL in dB} = 20 \log_{10}\left(\frac{p'_{\text{rms}}}{0.0002\ \mu\text{bar}}\right)$$

$$= 20 \log_{10}\left(\frac{p'_{\text{rms}}}{2 \times 10^{-5}\ \text{N/m}^2}\right) \tag{1.2}$$

where $p'^2_{\text{rms}} = \overline{p'^2}$, the overbar denoting the mean value.

On this scale a fluctuation of one atmosphere in pressure corresponds to 194 dB. The threshold of pain is between 130 and 140 dB corresponding to a pressure variation of only one thousandth of an atmosphere, i.e. $p'/p_0 = 10^{-3}$. The threshold of hearing is around the zero of the SPL decibel scale at which point the sound pressure amplitude is only 10^{-10} atmospheres.

We will consider a 1 kHz wave in order to illustrate the order of magnitude of the disturbance in an acoustic wave. At the threshold of pain fluid particles in a 1 kHz wave vibrate with a velocity of about 0.1 m/s, which is only a fraction of about 10^{-4} of the speed at which the sound wave is travelling. The displacement amplitude of the particles is then between 10^{-4} and 10^{-5} metres, while the wavelength, the distance the wave travels in one period, is about a third of a metre, so that the displacement amplitude is only about 10^{-4} of the wavelength of sound. At the threshold of hearing the vibration amplitude is an incredibly small 10^{-11} metres at this frequency, that is, only some 10^{-3} of the mean free molecular path length! (Those very weak sounds may not be described very well in terms of continuum mechanics, but we shall not acknowledge that fact any further in this book. Most interesting sounds are well modelled when the air is regarded as a continuum.) These numbers illustrate that the flow perturbations involved in acoustic waves are very small indeed. It therefore follows that all products of the perturbed quantities are negligible and the response of the acoustic field is linear, i.e. doubling the excitation doubles the sound field. This has two main consequences. Firstly, there is no interaction between different acoustic waves, and two or more waves can propagate without distorting one another. Their sound fields just add linearly. It is well known that it is possible to hear at least two conversations at once and that the two sounds just add, neither being distorted by the presence of the other. The second consequence is that the flow variables satisfy the linearised equations of fluid motion and each flow variable is linearly related to any other. This leads to a great simplification in the mathematics. A knowledge of one variable, the pressure say, provides a basis for the simple evaluation of all the others like, for example, the density or the particle velocity.

Example

Two sound sources radiate sound waves of different frequencies. If their sound pressure levels at some point are85 and 80 dB respectively, find the total sound pressure level due to the two sources together. What happens if the two sources radiate sound waves of the same frequency?

We express the pressure perturbation p' as the sum of two elements, $p_1 \cos \omega_1 t$ due to the first source and $p_2 \cos(\omega_2 t + \varphi)$ due to the second source. The radian frequencies of the two sounds are ω_1 and ω_2 respectively and φ is a phase angle.

$$p' = p_1 \cos \omega_1 t + p_2 \cos(\omega_2 t + \varphi).$$

Squaring this we find

$$\begin{aligned} p'^2 &= p_1^2 \cos^2 \omega_1 t + 2p_1 p_2 \cos \omega_1 t \cos(\omega_2 t + \varphi) + p_2^2 \cos^2(\omega_2 t + \varphi) \\ &= p_1^2 \cos^2 \omega_1 t + p_1 p_2[\cos\{(\omega_1 + \omega_2)t + \varphi\} + \cos\{(\omega_1 - \omega_2)t - \varphi\}] \\ &\quad + p_2^2 \cos^2(\omega_2 t + \varphi). \end{aligned}$$

The two terms within the square brackets average to zero if the two sounds are of different frequencies and then

$$\overline{p'^2} = \tfrac{1}{2}(p_1^2 + p_2^2).$$

But if the two sounds are of the same frequency

$$\overline{p'^2} = \tfrac{1}{2}p_1^2 + \tfrac{1}{2}p_2^2 + p_1 p_2 \cos \varphi$$

and the resultant sound pressure levels depends on the phase difference between the two sound waves. For example if they are 'in phase', $\varphi = 0$ and

$$\overline{p'^2} = \tfrac{1}{2}(p_1 + p_2)^2.$$

By definition the SPL is $20 \log_{10}(p'_{\text{rms}}/2 \times 10^{-5})$ so that for the two individual sounds,

$$85 = 20 \log_{10}\left(\frac{p_1/\sqrt{2}}{2 \times 10^{-5}}\right); \quad \text{i.e.} \quad p_1 = 0.503 \text{ N/m}^2$$

and

$$80 = 20 \log_{10}\left(\frac{p_2/\sqrt{2}}{2 \times 10^{-5}}\right) \quad \text{so that} \quad p_2 = 0.283 \text{ N/m}^2.$$

The SPL of the combined sounds of different frequencies is therefore

$$20 \log_{10}\left[\frac{\frac{1}{\sqrt{2}}(p_1^2 + p_2^2)^{1/2}}{2 \times 10^{-5}}\right] = 86.2 \text{ dB}.$$

but if they were both of the same frequency and had identical phase the SPL of the combined sounds would be

$$20 \log_{10} \frac{\frac{1}{\sqrt{2}}(p_1 + p_2)}{2 \times 10^{-5}} = 88.9 \text{ dB}.$$

1.4 VISCOSITY IS UNIMPORTANT IN SOUND WAVES

Viscous effects are usually negligible in a sound field because the pressure represents a far greater stress field than that induced by viscosity at frequencies of most practical interest. The ratio of the two stresses is the Reynolds number. The relevant Reynolds number is $2\pi c\lambda/\nu = \omega\lambda^2/\nu$ where ω is the angular frequency of the sound, λ the wavelength and ν the kinematic viscosity. This Reynolds number is usually very large, being of the order of 10^8 for sound in air at the most audible frequency, and for that reason we will regard sound as being essentially a weak motion of an inviscid fluid. Of course if very long-distance propagation is involved, viscous effects can be important but sound has to travel for about $\omega\lambda^2/\nu$ wavelengths before these effects are felt. Actually non-equilibrium effects in the molecular structure of moist air are much more important than viscosity, but even they are quite negligible in a great many practically significant sound fields. These effects are more easily understood when they are viewed as small modifying influences on a sound field that is essentially that established in a perfect inviscid fluid.

1.5 THE ONE-DIMENSIONAL WAVE EQUATION

Sound involves a weak motion of an inviscid fluid from an initial state of rest. We will begin by considering one-dimensional or plane waves, that is waves in which the flow parameters p, ρ, $\mathbf{v}$ are functions of a single space variable, x say, and t.

$$p = p(x, t).$$

Such waves often occur in nature. We have already noted that far from their source all sound waves become virtually plane. One-dimensional waves also occur when sound travels through narrow pipes which are the essence of many musical instruments.

Since there is no pressure gradient in either the y or the z direction there is no mechanism to provide an acceleration in those directions and the fluid velocity is directed entirely in the x direction. We can write

$$\mathbf{v}(x, t) = (u(x, t), 0, 0).$$

Consider a control volume of unit length in the y and z directions and of width δx. Since mass must be conserved, the change of mass within the control

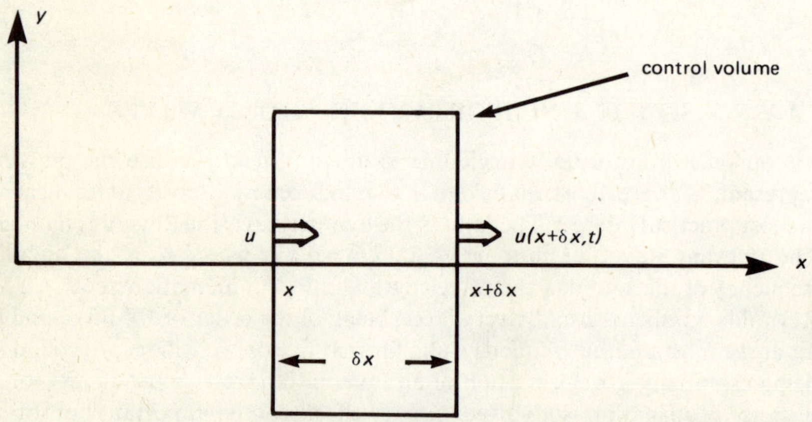

Fig. 1.2 — The control volume.

volume in unit time is related to the difference between the rate of inflow and outflow of the fluid crossing the end faces of the element, and we can write:

$$\frac{\partial \rho'}{\partial t}\,\delta x = \{(\rho_0 + \rho')u\}(x, t) - \{(\rho_0 + \rho')u\}(x + \delta x, t)$$

$\rho' u$ involves the product of two small quantities and is therefore negligible. Hence,

$$\frac{\partial \rho'}{\partial t} + \rho_0 \frac{\partial u}{\partial x} = 0, \qquad (1.3)$$

is the linearised equation of mass conservation. Similarly a linearised form of the equation of conservation of momentum, when all products of the small quantities ρ', u and p' have been neglected, is

$$\rho_0 \frac{\partial u}{\partial t} \delta x = p'(x, t) - p'(x + \delta x, t).$$

So that

$$\rho_0 \frac{\partial u}{\partial t} + \frac{\partial p'}{\partial x} = 0. \tag{1.4}$$

If we differentiate (1.3) with respect to t, and subtract from it (1.4) differentiated with respect to x, we obtain

$$\frac{\partial^2 \rho'}{\partial t^2} - \frac{\partial^2 p'}{\partial x^2} = 0. \tag{1.5}$$

Sound is a disturbance in which the pressure p can be determined from a knowledge of the density alone, $p = p(\rho)$. The particular form of the relation between the pressure and density is a property of the fluid and whether or not heat exchange takes place during the compressions and rarefactions.

We can expand the relationship between pressure and density and write

$$p = p_0 + (\rho - \rho_0) \frac{\mathrm{d}p}{\mathrm{d}\rho}(\rho_0) + \cdots,$$

hence to the accuracy of linearisation

$$p' = \rho' \frac{\mathrm{d}p}{\mathrm{d}\rho}(\rho_0). \tag{1.6}$$

In general an increase of pressure tends to increase the density and so the gradient $(\partial p/\partial \rho)(\rho_0)$ is a positive constant. We will now introduce a new variable c defined by

$$c^2 = \frac{\mathrm{d}p}{\mathrm{d}\rho}(\rho_0),$$

and then (1.6) becomes

$$p' = c^2 \rho'. \tag{1.7}$$

When we substitute for p' in equation (1.5) we obtain

$$\frac{1}{c^2} \frac{\partial^2 p'}{\partial t^2} - \frac{\partial^2 p'}{\partial x^2} = 0. \tag{1.8}$$

This is a one-dimensional equation for waves travelling at speed c. The small perturbations of pressure which we call sound are therefore organised into waves—sound waves. The same equation describes the propagation of waves in a tensioned wire or the propagation of electromagnetic waves in free-space.

A general solution of a wave equation like (1.8) is

$$p'(x,t) = f(x - ct) + g(x + ct). \tag{1.9}$$

(It is easy to check by differentiating that this solution does in fact satisfy (1.8).) $f(X)$ and $g(X)$ can be any functions. $p'(x,t) = f(x - ct)$ describes a pressure disturbance travelling in the direction of increasing x. The pressure perturbation maintains the same form, but travels at speed c. This travelling pattern is called a wave, and c is called the wave or sound speed. Similarly g describes a wave travelling to the left in the direction of $-x$.

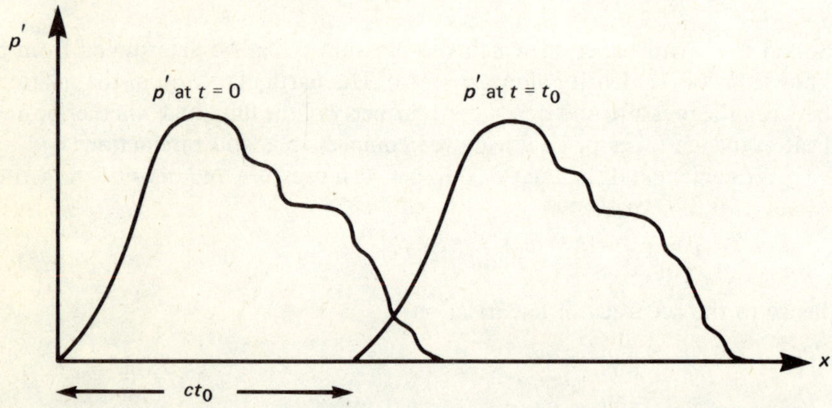

Fig. 1.3 — Development of pressure perturbations in the travelling wave $p'(x,t) = f(x - ct)$.

Harmonic waves

One particular form of waves is described by the solution (1.9) when f and g are harmonic functions. Then we can write

$$p'(x,t) = Ae^{i\omega(t - x/c)} + Be^{i\omega(t + x/c)}; \tag{1.10}$$

ω is the frequency in radians per second. There are $\omega/2\pi$ cycles per second and the period of the wave is $2\pi/\omega$ seconds. As is usual when writing p' as a complex function we in fact mean that p' is equal to the Real part of the right-hand side. The wavelength, λ, the distance between adjacent crests (or troughs), is given by

$$\lambda = 2\pi c/\omega.$$

p' can also be written as $Ae^{i(\omega t - kx)} + Be^{i(\omega t + kx)}$, where $k = \omega/c$. $k = 2\pi/\lambda$ is called the wavenumber. $\omega/k(=c)$ is called the phase speed.

The particle velocity

Once we have found the pressure perturbation in a one-dimensional wave we can easily evaluate u or ρ' because they are simple multiples of p'. Consider first a wave travelling to the right

$$p' = f(x - ct).$$

From (1.7)

$$\rho' = p'/c^2 = c^{-2}f(x - ct).$$

If we write $X = x - ct$, then

$$\frac{\partial \rho'}{\partial t} = \frac{1}{c^2}\frac{\partial f}{\partial t}(X) = \frac{1}{c^2}\frac{\mathrm{d}f}{\mathrm{d}X}\frac{\partial X}{\partial t}$$

by application of the Chain Rule, and so

$$\frac{\partial \rho'}{\partial t} = -\frac{1}{c}\frac{\mathrm{d}f}{\mathrm{d}X}.$$

Equation (1.3) then shows

$$\frac{\partial u}{\partial x} = \frac{1}{\rho_0 c}\frac{\mathrm{d}f}{\mathrm{d}X}(X) = \frac{1}{\rho_0 c}\frac{\partial f}{\partial x}(x - ct).$$

Hence after integration (and noting that u has a zero mean value)

$$u(x, t) = \frac{1}{\rho_0 c} f(x - ct).$$

We see that all the flow quantities are functions of x and t only in the combination $x - ct$, and that

$$p' = \rho_0 cu. \tag{1.11}$$

$p'/u = \rho_0 c$ is called the acoustic impedance. In a plane wave p' and u are in phase.

In a wave progressing in the negative direction, i.e. when $p' = g(x + ct)$, the same procedure as that described above shows

$$p' = -\rho_0 cu. \tag{1.12}$$

1.6 THE GENERATION OF ONE-DIMENSIONAL SOUND WAVES IN A PIPE

A simple way of generating one-dimensional sound waves in a pipe is by the vibration of a piston which fits tightly into one end of the pipe. Let us consider a piston which is initially at $x = 0$ and has a slow velocity $U(t)$. We see from equation (1.9) that any disturbances generated in the pipe by the motion of the

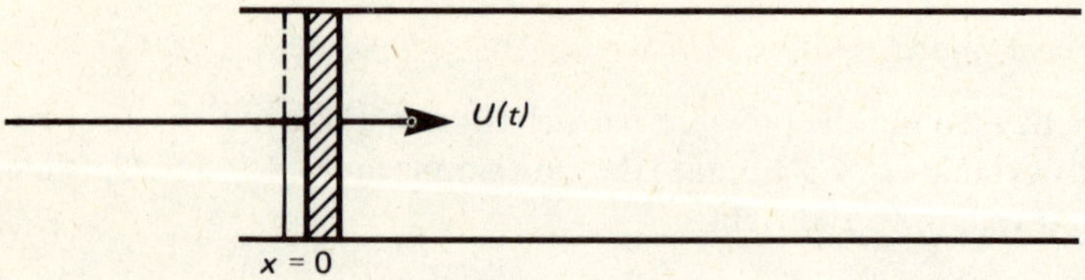

Fig. 1.4 — The tightly fitting piston moves with velocity $U(t)$.

piston must have the form of a wave propagating away from the piston with speed c,

$$p'(x, t) = f(t - x/c) \quad \text{say.}$$

Hence from (1.11)

$$u = \frac{1}{\rho_0 c} f(t - x/c).$$

The velocities of the fluid and piston must be equal on the front of the piston, and so

$$u = U(t) \quad \text{on the piston.} \tag{1.13}$$

But $U(t)$ is small in a sound wave, very much smaller than c, and the piston has only moved by a negligible amount from its initial position at $x = 0$, and so we may linearise (1.13) to obtain

$$u = U(t) \quad \text{on} \quad x = 0,$$

i.e. $f(t) = \rho_0 c U(t)$. The pressure perturbation in the pipe is therefore equal to $\rho_0 c U(t - x/c)$. The sound wave travels with speed c down the pipe and with no change in amplitude.

1.7 EVALUATION OF c

p may be related to ρ in several different ways according to the fluid motion examined. In the seventeenth century Newton used the newly discovered Boyle's law to say that

$$p/\rho = f(T)$$

where T is the temperature and so for an isothermal process

$$c^2 = \frac{dp}{d\rho} = \frac{p_0}{\rho_0} = 84{,}000 \text{ m}^2/\text{s}^2 \quad \text{at} \quad T = 293 \text{ K}.$$

This gives $c \approx 290$ m/s, 15% lower than the observed value. A correct evaluation was not done until 1816 when Laplace observed that the compression in a sound wave occurred too rapidly for an exchange of heat, and that the process is not, therefore, isothermal but isentropic.

For an isentropic expansion in a perfect gas,

$$p/\rho^{\gamma} = p_0/\rho_0^{\gamma},$$

$$c^2 = \frac{\mathrm{d}p}{\mathrm{d}\rho} = \gamma RT = \gamma p_0/\rho_0 = 117{,}700\ \mathrm{m^2/s^2}$$

where γ is the ratio of specific heat capacities and R is the gas constant. At $T = 293$ K, this leads to $c = 343$ m/s. This agrees well with the experimental results. A theoretical prediction of the speed of sounds in liquids is considerably more difficult than it is for gases. For example the sound speed in sea water depends on the pressure and salinity and on the amount of suspended gas as well as the water temperature. The speed of sound in a bubbly medium can be very low indeed, in fact much lower than in either of the two fluids that constitute the bubbly mixture.

1.8 THE ENERGETICS OF ONE-DIMENSIONAL ACOUSTIC MOTIONS

An important energy conservation principle can be derived from equations (1.3) and (1.4). First we multiply (1.3) by $c^2\rho'$

$$c^2\rho'\frac{\partial\rho'}{\partial t} + \rho_0 c^2\rho'\frac{\partial u}{\partial x} = 0,$$

both terms, being of second order in the small quantities, are extremely small and will be seen to represent the low energy levels in a sound wave. Since $\rho'\partial\rho'/\partial t$ can be rewritten as $\frac{1}{2}(\partial/\partial t)\rho'^2$ and $c^2\rho' = p'$, it follows that

$$\frac{c^2}{\rho_0}\frac{1}{2}\frac{\partial}{\partial t}\rho'^2 + p'\frac{\partial u}{\partial x} = 0 \tag{1.14}$$

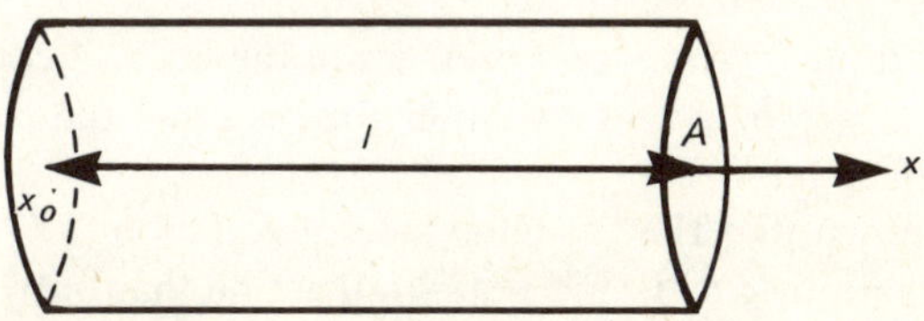

Fig. 1.5 — The control volume.

After multiplying (1.4) by u and rearranging the equation we obtain

$$\tfrac{1}{2}\rho_0 \frac{\partial u^2}{\partial t} + u \frac{\partial p'}{\partial x} = 0. \tag{1.15}$$

The addition of (1.14) and (1.15) leads to

$$\frac{\partial}{\partial t}\left(\tfrac{1}{2}\rho_0 u^2 + \frac{1}{2}\frac{c^2}{\rho_0}\rho'^2\right) + \frac{\partial}{\partial x}(p'u) = 0. \tag{1.16}$$

This equation describes energy conservation. To see this explicitly we integrate the equation over a control volume of length l in the x direction and of cross-sectional area A.

$$\frac{\partial}{\partial t}\int_{x_0}^{x_0+l} A\left(\tfrac{1}{2}\rho_0 u^2 + \frac{1}{2}\frac{c^2}{\rho_0}\rho'^2\right)\mathrm{d}x = A(p'u(x_0, t) - p'u(x_0 + l, t)) \tag{1.17}$$

Each term in this equation can be interpreted in terms of physical quantities. $\tfrac{1}{2}\rho_0 u^2$ is the kinetic energy density, e_k; it is the amount of kinetic energy in the acoustic fluctuations per unit volume of fluid. A fluid undergoing acoustic compression also has potential energy, it stores energy in much the same way as a compressed spring stores energy. This potential energy can be evaluated by calculating the work that has to be done to compress the gas to such a state. Consider a fixed mass of gas. As the pressure is increased the volume V occupied by the gas will decrease and the work done by the excess pressure in such a compression is $-\int p'\mathrm{d}V$.

For a fixed mass of gas ρV is constant, and changes in V can be related to changes in ρ by differentiation;

$$\mathrm{d}V = \frac{-V_0}{\rho_0}\mathrm{d}\rho$$

for small changes, where V_0 is the initial volume. When we substitute for $\mathrm{d}V$ and put $p' = c^2\rho'$ in the product $p'\mathrm{d}V$ we find:

$$-\int p'\mathrm{d}V = \frac{V_0}{\rho_0}c^2\int \rho'\mathrm{d}\rho' = \frac{V_0 c^2}{\rho_0}\tfrac{1}{2}\rho'^2.$$

Hence the work done by the excess pressure in the wave in compressing unit volume of fluid is $\tfrac{1}{2}(c^2/\rho_0)\rho'^2$, and we therefore say that the potential energy density, e_p, is $\tfrac{1}{2}(c^2/\rho_0)\rho'^2$.

We have shown that the left-hand side of equation (1.17) is the rate of change of energy (both kinetic and potential) within the control volume. The right-hand side of this equation is even easier to interpret. $Ap'u$ is the rate at which excess pressure in the fluid *outside* a control surface does work on the fluid inside. Therefore there is an energy flux of magnitude $p'u$ through unit area of the surface in unit time.

The product $p'u$ is called the intensity, I. It is the rate at which acoustic energy crosses unit area of space

$$I = p'u. \tag{1.18}$$

Equation (1.16) can be rewritten as

$$\frac{\partial}{\partial t}(e_k + e_p) = -\frac{\partial I}{\partial x}. \tag{1.19}$$

To summarise then, the energy in acoustic disturbances is composed of two distinct elements, one potential and the other kinetic. That energy increases only as a result of energy fluxing into the fluid elements owing to gradients in the acoustic intensity, I. Energy is conserved in acoustic waves.

The sound intensity level (IL) is measured in dB and concerns the mean value of the intensity $\bar{I}$, say

$$\text{IL in dB} = 10\log_{10}\left(\frac{\bar{I}}{10^{-12}\ \text{W/m}^2}\right).$$

For one-dimensional waves travelling to the right, from equation (1.11)

$$u = p'/\rho_0 c = \frac{c}{\rho_0}\rho'.$$

Hence

$$e_k = \frac{c^2}{2\rho_0}\rho'^2 = e_p. \tag{1.20}$$

In one-dimensional waves the potential energy density e_p is equal to the kinetic energy density e_k. This equipartition of energy between the potential and kinetic forms is characteristic of many vibration processes. Also in the one-dimensional sound wave it is easily checked that the intensity, $I = p'u = (c^3\rho'^2/\rho_0)$ and this is equal to ce, where e is the total energy density $e_p + e_k$.

$$I = ec.$$

Hence for a wave moving to the right (1.19) is just

$$\frac{\partial e}{\partial t} + c\frac{\partial e}{\partial x} = 0, \tag{1.21}$$

i.e. e is a function of $(x - ct)$.

e is a constant in a reference frame moving at speed c. This is an important result. It says that the energy propagates at the wave speed c, the energy travels at the same speed as the wave crests and all wave crests travel at the same speed regardless of their wavelength. Such a wave motion is called non-dispersive. A group of sound waves will not disperse as they would if some wavelets travelled faster than others.

Example

Suppose that at an open-air concert the sound is radiated from an arrangement of loudspeakers. At a distance of 20 m from the speakers the noise level is a stimulating 110 dB (*SPL*). What is the mean power output of the speakers? (Assume all sound waves incident on the ground are absorbed and that the sound field is omnidirectional.)

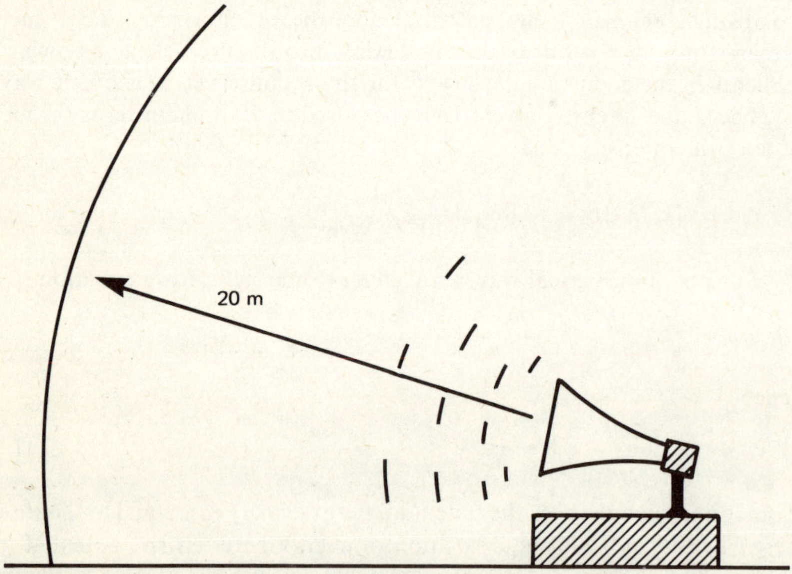

Fig. 1.6 — Waves from a loudspeaker.

$$SPL = 110 \text{ dB}, \quad \text{i.e.} \quad 20 \log_{10}\left(\frac{p'_{rms}}{2 \times 10^{-5} \text{ N/m}^2}\right) = 110,$$

so that the rms value of the pressure fluctuation is

$$p'_{rms} = 6.32 \text{ N/m}^2.$$

At 20 m the sound waves will be essentially plane and therefore

$$u_{rms} = \frac{p'_{rms}}{\rho_0 c} = 1.55 \times 10^{-2} \text{ m/s}.$$

The mean intensity at a radial distance of 20 m is

$$\bar{I} = \overline{p'u} = 9.8 \times 10^{-2} \text{ W/m}^2.$$

The energy flux through a hemisphere of radius 20 m is

$$2\pi(20)^2\bar{I} = 246.5 \text{ W}.$$

Since we are supposing that the sound field at the source is omnidirectional and that an equal amount of energy has been absorbed by the ground, the power output of the speakers is 493 watts.

1.9 SUBJECTIVE UNITS OF NOISE

The human ear responds more strongly to sound at some frequencies than it does to others. Sound of the same intensity but of different frequencies will not appear to be equally loud, and in fact sound at some frequencies is always inaudible. Pressure and intensity are physical quantities and can be measured by a meter but loudness is purely subjective since it depends on the response of the listener.

The loudness level is measured in phons. A tone will have a loudness level of ***N*** phons if it is judged by the average ear to be as loud as a pure tone of

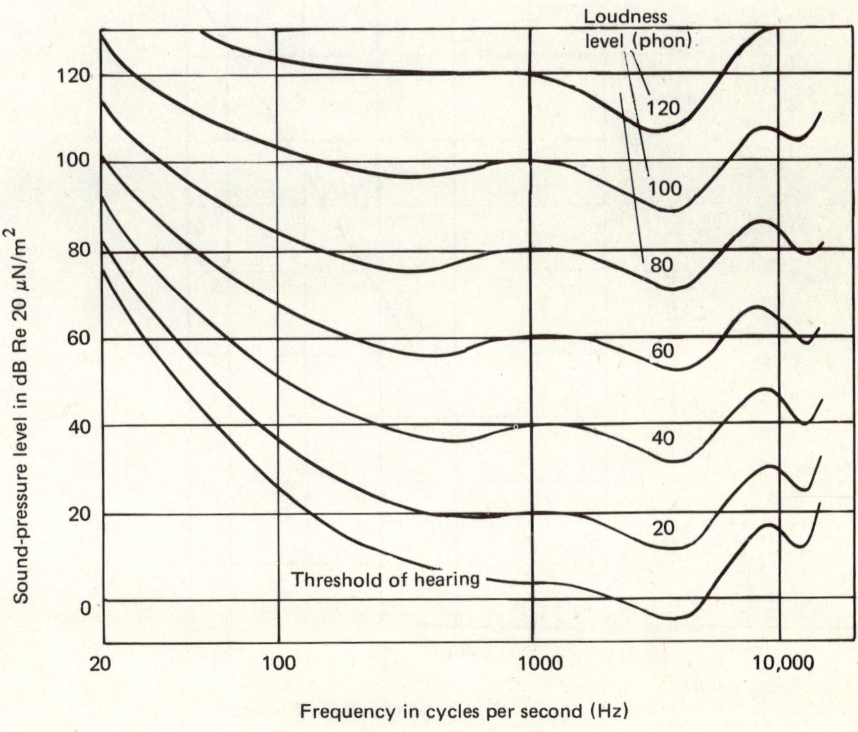

Fig. 1.7 — Loudness level of pure tones.

frequency 1 kHz at a sound pressure level of *N* dB. Many groups of listeners have taken part in subjective experiments to determine the loudness of sounds as perceived by the average ear and the results are plotted in Figure 1.7. Individuals will have different sensitivities and some hypersensitive people are very much more conscious of noise annoyance than the average on which the noise assessment scales are based.

Unfortunately, because a sound of 30 phons does not seem half as loud as a sound of 60 phons, a further measure is needed and that is the 'sone' which is a 'linear' measure of loudness. A sound of 10 sones is ten times as loud as a sound of 1 sone, and so on. The sone scale is chosen so that it has the property of being directly proportional to the loudness and it is normalised so that 1 sone is the loudness of a sound whose loudness level is 40 phons, 40 being chosen so that the audible sounds lie in a range of approximately 0 to 100 sones. *N* sones is the loudness of a sound which appears to the observer to be *N* times louder than the standard 40 phons level.

A graph showing how the average ear relates loudness in sones to the *SPL* of a 1 kHz tone (i.e. the loudness level in phons) is shown in Figure 1.8 from

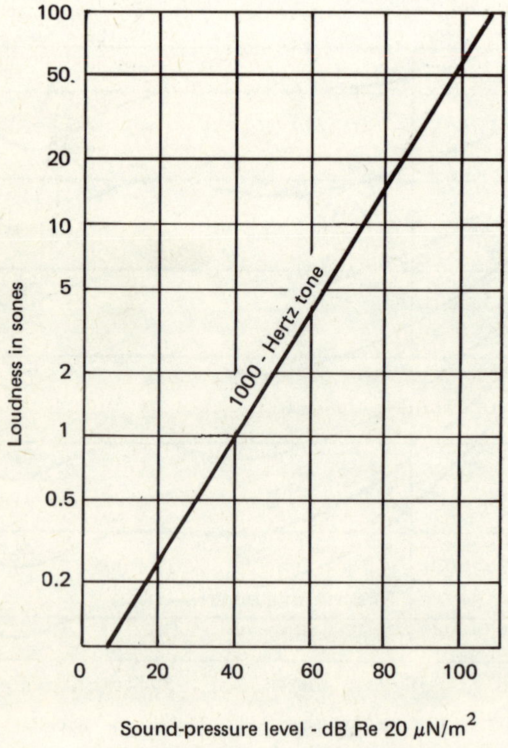

Fig. 1.8 — Loudness of a 1 kHz tone.

which it can be seen that the 50 dB tone seems only twice as loud as the 40 dB tone which contains only one-tenth of the energy. A level increase of 10 dB apparently doubles the perceived loudness.

Since the loudness of a sound, and therefore our awareness of it, depends on the sound's frequency content, it is useful to have decibel scales for measuring noise which take this into account. We have seen that the average ear's response to pure tones of differing frequencies can be expressed as a table, or in the form of a graph, and we can use these to estimate the human response to any single tone. But such pure tones rarely occur in practice and we need to be able to predict the response of the ear to less harmonic sounds. Fortunately by using spectral analysis a quite general sound field can be expressed in terms of pure tones essentially by Fourier decomposition of the signal.

1.10 SOUND SPECTRA

The tonal composition of a noise or musical sound can be investigated by spectral analysis. The heart of spectral analysis is the Fourier transform. Fourier's theorem states that a function $s(t)$ can be represented as an integral of its harmonic elements

$$s(t) = \frac{1}{2\pi}\int_{-\infty}^{\infty} F(\omega)\mathrm{e}^{i\omega t}\mathrm{d}\omega, \tag{1.22}$$

and the strength of each harmonic element is a similar integral of the signal

$$F(\omega) = \int_{-\infty}^{\infty} s(t)\mathrm{e}^{-i\omega t}\mathrm{d}t. \tag{1.23}$$

F and s are called Fourier transform pairs and F is the spectral decomposition of s. $s(t)$ may be, for example, the sound pressure perturbation. Equation (1.22) tells us that this sound can be considered to be made up of harmonic waves each with time dependence $\mathrm{e}^{i\omega t}$ and that the amplitude of the waves with frequencies in the band between ω and $\omega + \delta\omega$ is $F(\omega)\delta\omega/2\pi$.

We have already considered harmonic waves with a time dependence of the form $\mathrm{e}^{i\omega t}$ as a simple example of a wave and Fourier's theorem shows that no generality is lost by doing this; an arbitrary pulse or disturbance can be written in terms of such harmonic waves.

The sound pressure level in a frequency interval of width 1 Hz centred at ω is called the pressure spectrum level (PSL). Similarly the sound intensity at frequency ω is defined to be the combined intensity of wave elements within a frequency interval 1 Hz centred at ω.

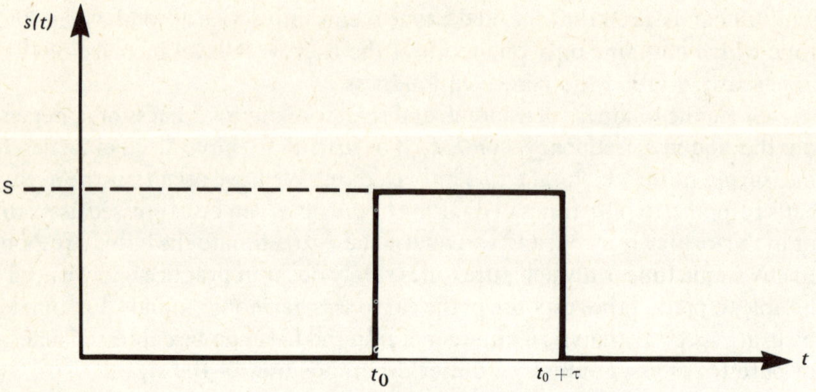

Fig. 1.9 — A square pulse.

Example[3]. *The harmonic decomposition of a single pulse*

We consider a signal $s(t)$ such that $s(t) = S$ for t between t_0 and $t_0 + \tau$ but is zero at all other times, see Figure 1.9. The Fourier transform of this signal is,

$$\boldsymbol{F}(\omega) = S\int_{t_0}^{t_0+\tau} \bar{e}^{i\omega t}\mathrm{d}t = \frac{-iS}{\omega}\bar{e}^{i\omega t_0}(1 - \bar{e}^{i\omega\tau})$$

$$= S\tau\bar{e}^{i\omega(t_0+\tau/2)}\frac{\sin\omega\tau/2}{\omega\tau/2}.$$

This single pulse has spectral components over most frequencies and the sharper the pulse the wider is the range of the frequency spectrum.

In the limit, when the pulse duration is zero the signal becomes an impulse. An impulse is a signal of negligible duration but of sufficiently large

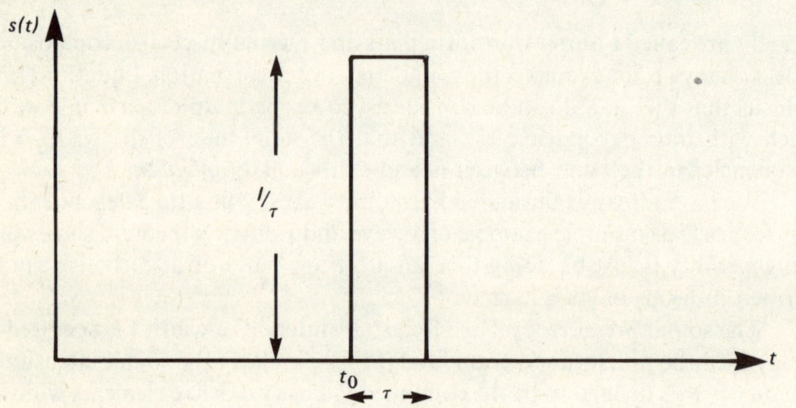

Fig. 1.10 — An impulse.

amplitude that $\int s(t)\mathrm{d}t$ has a non-zero value, i.e. it can be considered to be a pulse of magnitude $S = I/\tau$ and of a small time duration τ. Then $\int s(t)\mathrm{d}t = \tau I/\tau = I$.

An impulse of unit magnitude is called a delta-function. $\delta(t - t_0)$ is a unit impulse at $t = t_0$. It is a function defined by

$$\delta(t - t_0) = 0 \quad \text{for } t \neq t_0,$$

and in the vicinity of $t = t_0$ it is sufficiently large that $\int_{-\infty}^{\infty} \delta(t - t_0)\mathrm{d}t = 1$.

A δ-function has the property that

$$\int_{-\infty}^{\infty} g(t)\delta(t - t_0)\mathrm{d}t = g(t_0), \quad \text{for any function } g(t). \tag{1.24}$$

To understand this we note that, since $\delta(t - t_0) = 0$ unless $t = t_0$, we can write

$$\int_{-\infty}^{\infty} g(t)\delta(t - t_0)\mathrm{d}t = g(t_0) \int_{-\infty}^{\infty} \delta(t - t_0)\mathrm{d}t$$
$$= g(t_0)$$

because the integral of the δ-function is 1.

The Fourier transform or spectral decomposition of the impulsive signal

$$s(t) = I\delta(t - t_0)$$

is therefore

$$F(\omega) = I \int_{-\infty}^{\infty} \mathrm{e}^{-i\omega t}\delta(t - t_0)\mathrm{d}t = I\mathrm{e}^{-i\omega t_0}$$

(see Figure 1.11).

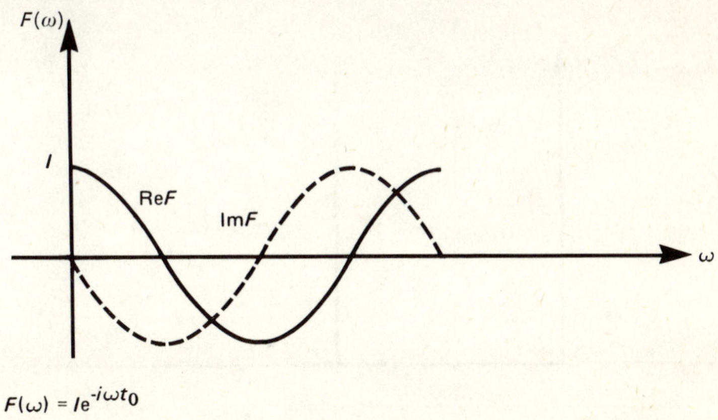

Fig. 1.11 — $F(\omega) = I\mathrm{e}^{-iwt_0}$ is the Fourier transform of an impulse at time t_0 of strength I.

When we substitute for F in equation (1.22) we find

$$\delta(t - t_0) = \frac{1}{2\pi}\int_{-\infty}^{\infty} e^{i\omega(t-t_0)}d\omega.$$

This is an important result and a simple change of the variables' names shows that it can be written as

$$\int_{-\infty}^{\infty} e^{ixy}dx = 2\pi\delta(y). \qquad (1.25)$$

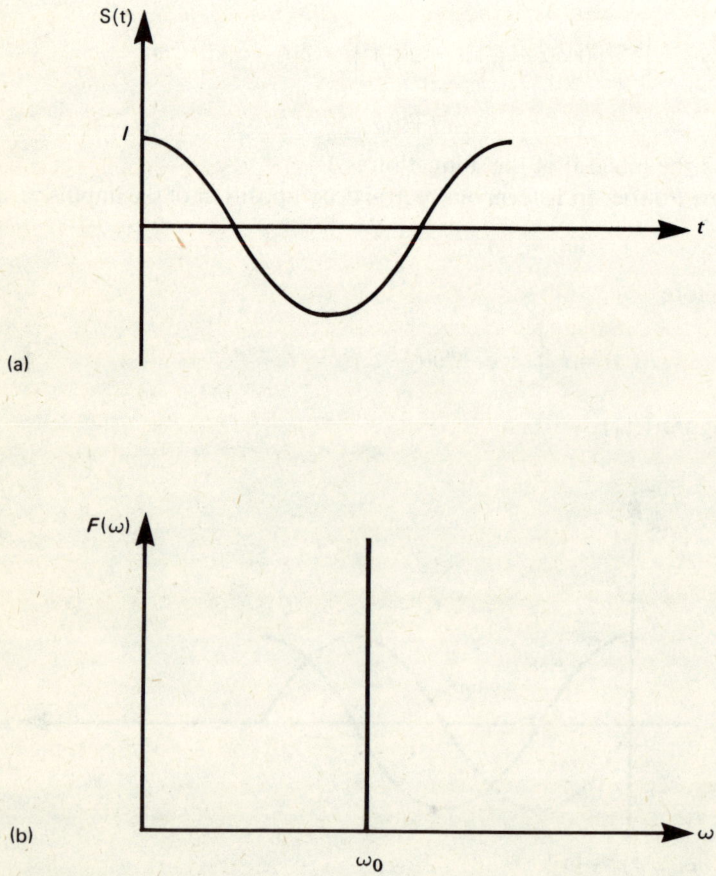

Fig. 1.12 — (b) shows the Fourier transform of the pure tone in (a), and vice versa.

Suppose now that $s(t)$ is a harmonic signal, $s(t) = I\mathrm{e}^{i\omega_0 t}$. Then its Fourier transform is seen to be

$$F(\omega) = I\int_{-\infty}^{\infty} \mathrm{e}^{i(\omega_0 - \omega)t}\mathrm{d}t = I2\pi\delta(\omega_0 - \omega), \tag{1.26}$$

after applying the useful integral relationship in (1.25).

This example shows that the spectrum of a function may be continuous, or discrete. If pure tones are present the spectrum will have spikes but most real noises have continuous spectra.

A continuous pressure spectrum level is generally found experimentally by measuring the sound pressure level in a specified frequency band. We could choose to measure, say, the sound pressure level in each 1 Hz frequency band over the frequencies of interest but most sound meters have a bandwidth wider than 1 Hz. Bandwidths of one octave or one-third-octave are the most common. An octave is a frequency range over which the frequency doubles. The commonly used octaves are 37.5–75 Hz, 75–150 Hz, etc. A one-third-octave is the frequency interval between two frequencies having a ratio $(2)^{1/3}:1$. The sound pressure level in decibels in a given band of width Δf is called the pressure band level (PBL). The interval in each octave, or one-third-octave, band is proportional to the frequency and so the bandwidth increases linearly with frequency. This frequency weighting must be born in mind when discussing the pressure band levels.

We can calculate the average sound pressure level in a frequency band by dividing the pressure band level by the bandwidth, or as we are using decibels by subtracting $10 \log_{10}(\Delta f)$. If the bandwidth is small enough for the pressure spectrum to be approximately constant within each frequency band, this mean value will be nearly equal to the pressure spectrum level per unit frequency interval (PSL) at the centre frequency in the band.

$$\mathrm{PSL} = \mathrm{PBL} - 10 \log_{10}(\Delta f). \tag{1.27}$$

[Note: The 'centre' frequency in an octave is defined to be the frequency which is a factor $\sqrt{2}$ bigger than the lowest frequency found in the octave. Similarly for a one-third-octave the 'centre' frequency is a factor $2^{1/6}$ bigger than the lowest frequency in the band.]

Noise data is often expressed in terms of pressure spectra; the pressure band level is measured in each frequency band, and an average pressure spectrum level is calculated from (1.27). The data is then presented with this average pressure spectrum level plotted as a function of the centre frequency in the band. When results are plotted in this way it is important to state the bandwidth because the given PSL represents an average over this frequency interval. The bandwidth is usually chosen to be the largest possible that will still give the resolution required for any particular application.

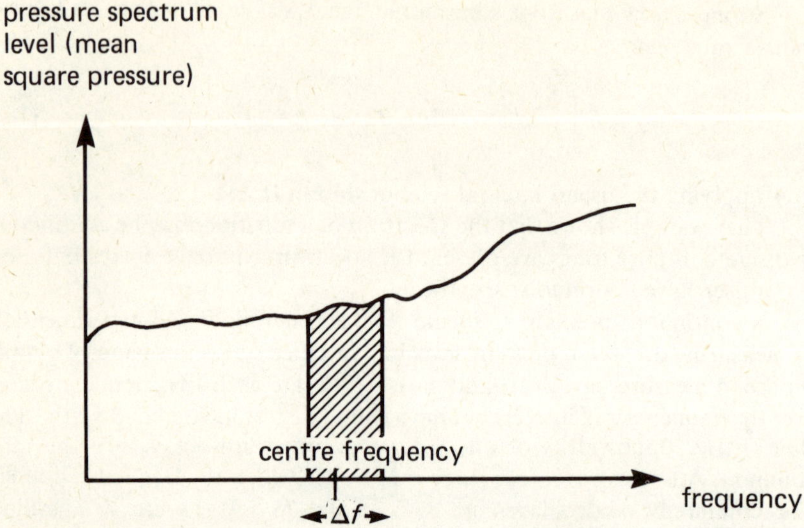

Fig. 1.13 — The hatched area indicates the pressure band level, *PBL*, in the frequency band of width Δf and the mean pressure spectrum level per unit frequency in the band is $= PBL - 10\log_{10}(\Delta f)$.

The intensity and power band levels can be defined similarly to the pressure band level. They are the levels in that particular frequency band.

1.11 VARIOUS SUBJECTIVE DECIBEL SCALES

Because the human ear responds more strongly to some frequencies than it does to others, if a sound meter is to assess the annoyance caused by a noise, it must take this into account. A method of doing this is to weight the sound pressure level in each frequency band by a factor which takes into account the ear's sensitivity to that frequency range. 'A-weighting' is one such biasing curve which has been internationally agreed. There are also B and C weightings but they are rarely used. An A-weighted sound level is measured by dBA. A-weighting is easy to use because the weighting can be done automatically by a sound meter and a reading in dBA can be obtained directly from the instruments.

The average A-weighted sound level over a sufficiently long sample of noise is denoted by L_{eq} and is often used as a means of quantifying noise. This measure does not, however, account for time variations in sound; many noises tend to fluctuate between intensely loud and quieter periods (for example, the noise produced by road traffic will peak with the passage of a heavy truck and will have quiet intervals when there is little or no noise). Humans find such

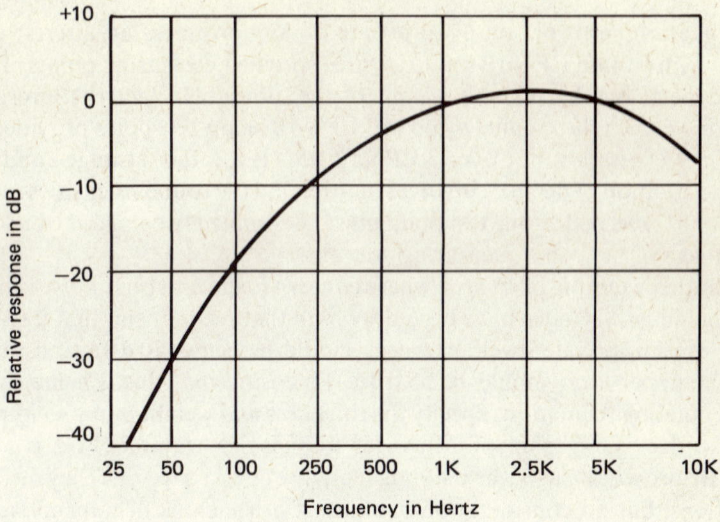

Fig. 1.14 — The 'A-weighting' bias for the dBA scale.

intermittent noises more annoying than those with a steady sound level even if the total sound energy over a long period of time is the same. The *noise pollution* scale has been devised to take this into account;

$$L_{NP} = L_{eq} + 2.56\sigma,$$

where L_{eq} is the mean A-weighted sound level over a long time-interval, and σ is the standard deviation of the same sample of A-weighted sound.

The list of different weightings and the decibel scales used to measure loudness is very long and not particularly instructive. It is perhaps worth mentioning just one more scale for sound pressure and that is the *perceived noise* decibel, PNdB. Perceived noise decibels are used instead of A-weighted sound to measure aircraft noise. They are based on a different weighting which takes account of the high frequency component of the noise of a jet engine which can be particularly irritating to the ear. The numerical difference between PNdB and dBA depends on the type of aircraft and the manoeuvre it is performing, but typically a PNdB reading is between 12 and 16 dB higher than a reading in dBA. An aircraft's noise is judged in *effective perceived noise* decibels. These are PNdB with a weighting for any pure tones in the noise signal, like the whine of turbomachinery, which can be particularly annoying. They also contain an adjustment for the time it takes an aircraft to fly over.

1.12 CODES OF PRACTICE

It is generally accepted that noise is an environmental health hazard and codes of practice have been drawn up to try and control noise.

For example to obtain a Certificate of Airworthiness an aircraft must comply with certain EPNdB limits at three specified measuring points. These limits depend on the take-off weight of the aircraft. In fact the aircraft is allowed to exceed these limits by up to 2 EPNdB at any test point provided the sum of the excesses is not over 3 EPNdB (this is a rather strange condition since the addition of decibels involves multiplying the sound intensities!). Also the amount exceeded at one test point must be completely balanced by a lower level at one of the other measuring stations.

Another example of an area where concern is shown about noise is in the building industry. Codes have been drawn up that recommend that day-time sound levels in private dwelling places should be below 50 dBA and that at night the upper limit should be 35 dBA. To ensure that this is achieved the British building regulations specify the thickness and weight required in party walls, so that the transmission loss for frequencies around 1 kHz is about 50 dB. Before work starts, the building plans are checked to see if they meet the regulations, but no consideration is taken of other paths of transmission of sound, and no sound level readings are required to be taken in the completed building!

Existing legislation and codes of practice lay guidelines for noise levels, and although these regulations are sometimes rather arbitrary and are open to criticism they do ensure that the potential noise level has to be considered in the design of machinery, aircraft, cars and buildings.

EXERCISES FOR CHAPTER 1

The sound speed and mean density in air and water may be taken as 340 m/s, 1.2 kg/m^3 and 1450 m/s, 10^3 kg/m^3 respectively.

1. A noise is generated by 80 pure tones, of different frequencies but identical power. Each pure tone has a sound pressure level of 60 dB. Determine the sound pressure level of the noise.

2. The sound pressure level due to a press installed in a workshop is 90 dB. If this press operates when the workshop background SPL is 87 dB, what is the resultant SPL?

3. In an experiment to measure the sound pressure level of a source a reading is first taken of the background noise. Show that if the difference between the background level and the level with the source on is greater than 10 dB the background level will not significantly affect the measurement of the source noise level.

If the background level is only 3 dB less than the reading with the source switched on, what correction must be applied to the total noise level to get the noise level due to the source?

4. Show that the transverse displacement, ξ, of a string with tension T and mass per unit length m satisfies the one-dimensional wave equation

$$m \frac{\partial^2 \xi}{\partial t^2} - T \frac{\partial^2 \xi}{\partial x^2} = 0.$$

Deduce that disturbances propagate along the string with speed $\sqrt{(T/m)}$.

5. Determine the maximum particle velocity and displacement associated with a plane sound wave with a pressure level of 100 dB at a frequency at 1 kHz (a) in air and (b) in water.

6. A train travelling at 60 mph enters a long tunnel with the same cross-sectional area as the train. Determine the strength of the pressure wave generated in the tunnel.

7. An underwater sonar beam of diameter 0.5 m carries 100 watts of acoustic power in a plane wave of frequency 20 kHz. Determine (a) the wavelength, (b) the sound pressure level in dB, (c) the maximum particle velocity in the beam of sound and (d) the maximum particle displacement.

8. Compare the intensities of a plane sound wave in air and water for (a) the same acoustic pressure and (b) the same particle velocity.

Chapter 2

Three-Dimensional Sound Waves

2.1 THE BASIC EQUATIONS

The one-dimensional waves discussed in Chapter 1 are a special case of waves in three-dimensional space—special in that they were launched by a source with uniform properties over two of the dimensions. For example, the piston vibrating axially in a cylinder of the same diameter caused an axial (one-dimensional) wave field. The volume inside the cylinder can support much more complicated sound fields and waves that vary with all three space co-ordinates are the usual response to sources that are smaller than the characteristic dimension of the space in which they are confined. In unconfined spaces all fields are three-dimensional. The flow parameters are then functions of three space co-ordinates and time. We can choose to write the space co-ordinates in cartesian, vector or tensor form i.e. as (x_1, x_2, x_3), $\mathbf{x}$ or x_i, where x_1 is the component of the position vector $\mathbf{x}$ on the 1-axis etc., and the suffix i can take any chosen value from 1 to 3 to indicate x_1, x_2 or x_3 as we please.

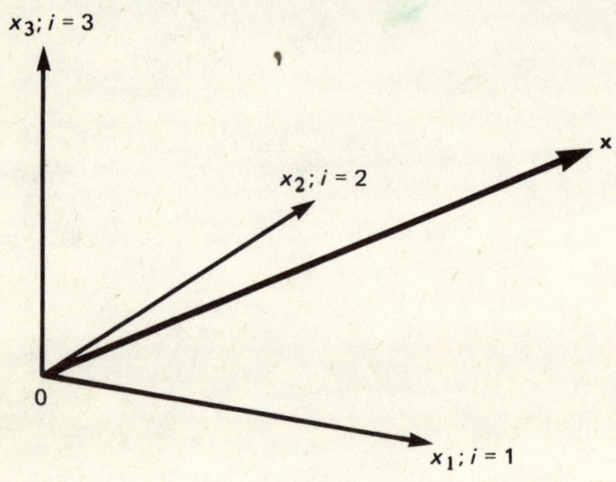

Fig. 2.1 — x_i is the component of the position vector $\mathbf{x}$ on the i-axis.

The material particles displaced by the sound wave will now be free to move in any direction and the velocities at which they move will be the vector field $\mathbf{v}(\mathbf{x}, t)$, or equivalently $(v_1(\mathbf{x}, t), v_2(\mathbf{x}, t), v_3(\mathbf{x}, t))$ or $v_i(\mathbf{x}, t)$. This is the velocity at which the material particle occupying the position $\mathbf{x}$ at time t is moving.

The wave equation is established by considering the conservation of mass and momentum. Consider first mass conservation for a small volume δV enclosed by a surface S. The rate of increase of mass within the volume δV must be equal to the rate at which mass flows into the volume across the bounding surface. The linearised form of this equation is

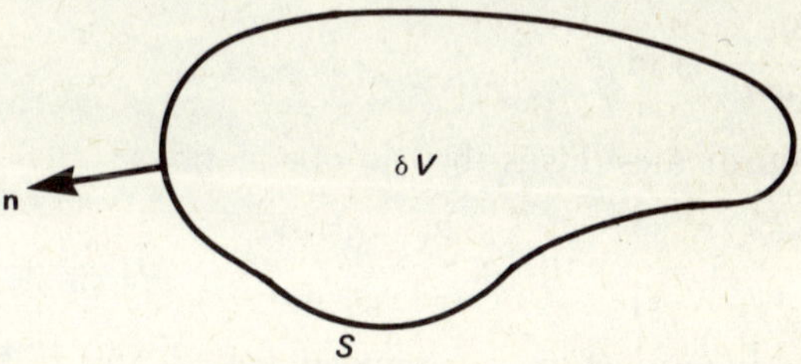

Fig. 2.2 — The control volume δV bounded by a surface S.

$$\delta V \frac{\partial \rho'}{\partial t} = -\int_S \rho_0 \mathbf{v} \cdot \mathbf{n} \, \mathrm{d}S.$$

Gauss's theorem can be used to rewrite the surface integral as $-\rho_0 \operatorname{div} \mathbf{v} \, \delta V$, the linearised equation of mass conservation then follows.

$$\frac{\partial \rho'}{\partial t} + \rho_0 \operatorname{div} \mathbf{v} = 0, \tag{2.1}$$

or, equivalently, in cartesian form,

$$\frac{\partial \rho'}{\partial t} + \rho_0 \frac{\partial v_1}{\partial x_1} + \rho_0 \frac{\partial v_2}{\partial x_2} + \rho_0 \frac{\partial v_3}{\partial x_3} = 0$$

i.e.

$$\frac{\partial \rho'}{\partial t} + \rho_0 \frac{\partial v_i}{\partial x_i} = 0. \tag{2.2}$$

The repetition of the suffix i in $(\partial v_i / \partial x_i)$ must always imply a trio of terms, the sum of the elements with the index i taking successively the values 1, 2 and 3. This is the summation convention of tensor calculus.

$$\frac{\partial v_i}{\partial x_i} = \frac{\partial v_1}{\partial x_1} + \frac{\partial v_2}{\partial x_2} + \frac{\partial v_3}{\partial x_3}.$$

Other examples are

$$v_i v_i = v_1^2 + v_2^2 + v_3^2 = |\mathbf{v}|^2 = \mathbf{v}\cdot\mathbf{v},$$
$$x_j x_j = x_1^2 + x_2^2 + x_3^2 = |\mathbf{x}|^2 = \mathbf{x}\cdot\mathbf{x},$$
$$v_i v_i v_j = |\mathbf{v}|^2 v_j.$$

but $v_i v_i v_i$ is meaningless! Once a suffix has been repeated, it is used up; the summation is compulsory.

The momentum equation states that the rate at which the momentum increases in unit volume of weakly disturbed matter, $\rho_0(\partial\mathbf{v}/\partial t)$, must be equal to the force applied to that unit volume by neighbouring material. That force is $-\text{grad } p'$ and hence

$$\rho_0 \frac{\partial \mathbf{v}}{\partial t} + \text{grad } p' = 0. \tag{2.3}$$

In cartesian co-ordinates this is the trio of equations

$$\rho_0 \frac{\partial v_1}{\partial t} + \frac{\partial p'}{\partial x_1} = 0, \quad \rho_0 \frac{\partial v_2}{\partial t} + \frac{\partial p'}{\partial x_2} = 0, \quad \rho_0 \frac{\partial v_3}{\partial t} + \frac{\partial p'}{\partial x_3} = 0,$$

which can be written much more compactly as

$$\rho_0 \frac{\partial v_i}{\partial t} + \frac{\partial p'}{\partial x_i} = 0 \tag{2.4}$$

with i taking on any desired value from 1 to 3.

When (2.1) is differentiated with respect to t and the divergence of equation (2.3) is subtracted from it we have

$$\frac{\partial^2 \rho'}{\partial t^2} - \nabla^2 p' = 0$$

where

$$\nabla^2 p' = \text{div}(\text{grad } p') = \frac{\partial^2 p'}{\partial x_1^2} + \frac{\partial^2 p'}{\partial x_2^2} + \frac{\partial^2 p'}{\partial x_3^2} = \frac{\partial^2 p'}{\partial x_i\, \partial x_i}.$$

Using $p' = c^2 \rho'$ to eliminate ρ' leads to the wave equation

$$\frac{1}{c^2}\frac{\partial^2 p'}{\partial t^2} - \nabla^2 p' = 0. \tag{2.5}$$

All the small quantities describing the acoustic field satisfy this wave equation. It is certainly true for ρ' which can replace p'/c^2. Differentiation of (2.5) shows that all the components of grad p' satisfy the wave equation, and hence from (2.3) we can deduce that the velocity field also satisfies the same equation;

$$\frac{1}{c^2}\frac{\partial^2 \mathbf{v}}{\partial t^2} - \nabla^2 \mathbf{v} = 0.$$

The velocity potential

We can deduce a very useful property of the velocity field induced by acoustic disturbances by taking the curl of equation (2.3). We find that

$$\frac{\partial}{\partial t}\,\text{curl}\,\mathbf{v} = 0,$$

because curl grad p' is always zero. In other words the vorticity, curl $\mathbf{v}$, remains constant. Since we are considering fluid disturbed from an initial state of rest this constant must be zero;

$$\text{curl}\,\mathbf{v} = 0, \tag{2.6}$$

and the flow is irrotational. The reason for this is essentially because the forces arising from the pressure gradient are conservative and act through the mass centre of a particle. They cannot therefore induce spin or vorticity.

It is a fundamental property of vector fields that whenever curl $\mathbf{v} = 0$ there exists a function $\varphi(\mathbf{x}, t)$ such that

$$\mathbf{v} = \text{grad}\,\varphi. \tag{2.7}$$

Equivalently,

$$v_i = \frac{\partial \varphi}{\partial x_i}$$

is the shorthand form of (2.7), which written out in full means,

$$v_1 = \frac{\partial \varphi}{\partial x_1}, \quad v_2 = \frac{\partial \varphi}{\partial x_2}, \quad v_3 = \frac{\partial \varphi}{\partial x_3}. \tag{2.8}$$

The use of the velocity potential guarantees the irrotationality constraint because curl grad $\varphi \equiv 0$. φ is called the velocity potential and it is often convenient to express all the flow variables in terms of φ. The momentum equation re-written in terms of φ becomes

$$\text{grad}\left(\rho_0 \frac{\partial \varphi}{\partial t} + p'\right) = 0.$$

This equation shows that the sum of the pressure perturbation and $\rho_0(\partial\varphi/\partial t)$ is a constant, (which can be set to zero since all we require of φ is that it be such that $\mathbf{v} = \text{grad}\,\varphi$; φ is essentially undetermined to within an arbitrary function of time) and this fact is often the clue to the shortest way of deducing the value of the pressure field in unsteady weak motion.

$$p' = -\rho_0 \frac{\partial \varphi}{\partial t}. \tag{2.9}$$

Moreover p' satisfies the wave equation, and because of (2.9) this is also true of $\partial\varphi/\partial t$, and therefore of φ;

$$\frac{1}{c^2}\frac{\partial^2\varphi}{\partial t^2} - \nabla\varphi = 0. \tag{2.10}$$

There often occur regions in a sound wave where the motion is indistinguishable from incompressible potential flow, and problems can sometimes be greatly simplified once this fact is recognised. The irrotational motion of incompressible material is one in which the velocity potential φ satisfies Laplace's equation, $\nabla^2\varphi = 0$. It is the limit of the compressible case when the speed of sound c is allowed to become infinite. Then finite acceleration-producing pressure changes can only be admitted if c^2 is large enough that the product $c^2\rho'$, which equals p', remains finite as ρ', the density change resulting from compressibility, vanishes.

The wave equation limits to Laplace's equation as c tends to infinity

$$\lim_{c\to\infty}\left\{\frac{1}{c^2}\frac{\partial^2\varphi}{\partial t^2} - \nabla^2\varphi = 0\right\} = \{-\nabla^2\varphi = 0\}.$$

Sufficiently slow acoustic motions are also a solution of Laplace's equation, steady flow being the extreme case

$$\lim_{(\partial^2/\partial t^2)\to 0}\left\{\frac{1}{c^2}\frac{\partial^2\varphi}{\partial t^2} - \nabla^2\varphi = 0\right\} = \{-\nabla^2\varphi = 0\}.$$

All acoustic motions in the vicinity of a singularity are solutions of Laplace's equation. Near a singularity $\varphi \sim r^{-\eta}$ where η is some positive number. $(\partial^2/\partial t^2)$ will also have the same behaviour. But $\nabla^2\varphi$ is more singular because each differentiation with respect to a space co-ordinate introduces a further factor r in the denominator and this is crucial near the singularity where r vanishes. Therefore if $\varphi \sim r^{-\eta}$

$$\frac{\partial^2\varphi}{\partial t^2} = 0(r^2\nabla^2\varphi) \quad \text{as} \quad r \to 0$$

and

$$\lim_{r\to 0}\left\{\frac{1}{c^2}\frac{\partial^2\varphi}{\partial t^2} - \nabla^2\varphi = 0\right\} = \{-\nabla^2\varphi = 0\}.$$

Slow motion, incompressible motion and the flow in the vicinity of a small source, or singularity, have identical characteristics. The only way in which these potential fields differ from those studied in steady aerodynamics is the way in which pressure is related to the velocity field. In aerodynamics that relation is usually through the non-linear $\frac{1}{2}\rho u^2$ term in Bernoulli's equation. In acoustics of a still medium and in all linear wave motions, the pressure is related to velocity through the linear unsteady term in what is otherwise the same Bernoulli equation. Specifically, $p' = -\rho_0(\partial\varphi/\partial t)$.

The energetics of acoustic motions

We can develop the arguments of section 1.8 to apply to three-dimensional fields with very little added complication.

When the acoustic wave equation for the potential is multiplied by the quantity $(\rho_0/c^2)(\partial\varphi/\partial t)$, and the terms re-arranged, an equation describing an important energy conservation principle is obtained.

$$\frac{\rho_0}{c^2}\frac{\partial\varphi}{\partial t}\left\{\frac{\partial^2\varphi}{\partial t^2} - c^2\nabla^2\varphi = 0\right\}. \tag{2.11}$$

The first term in this equation can be rewritten because

$$\frac{\partial\varphi}{\partial t}\frac{\rho_0}{c^2}\frac{\partial^2\varphi}{\partial t^2} = \frac{\partial}{\partial t}\left\{\frac{1}{2}\frac{\rho_0}{c^2}\left(\frac{\partial\varphi}{\partial t}\right)^2\right\} = \frac{\partial}{\partial t}\left\{\frac{p'^2}{2\rho_0 c^2}\right\}. \tag{2.12}$$

Manipulation of the second term shows that

$$\begin{aligned}\frac{\rho_0}{c^2}\frac{\partial\varphi}{\partial t}c^2\nabla^2\varphi &= \rho_0\frac{\partial\varphi}{\partial t}\frac{\partial^2\varphi}{\partial x_i\,\partial x_i}\\ &= \frac{\partial}{\partial x_i}\left\{\rho_0\frac{\partial\varphi}{\partial t}\frac{\partial\varphi}{\partial x_i}\right\} - \rho_0\frac{\partial^2\varphi}{\partial t\,\partial x_i}\frac{\partial\varphi}{\partial x_i}\\ &= -\frac{\partial}{\partial x_i}(p'v_i) - \frac{\partial}{\partial t}(\tfrac{1}{2}\rho_0 v^2)\end{aligned} \tag{2.13}$$

where v^2 is written for $\nabla\varphi\cdot\nabla\varphi$.

Equation (2.11), which is the difference of (2.12) and (2.13) can therefore be re-expressed exactly as

$$\frac{\partial}{\partial t}\left\{\frac{1}{2}\frac{p'^2}{\rho_0 c^2} + \tfrac{1}{2}\rho_0 v^2\right\} + \frac{\partial}{\partial x_i}\{p'v_i\} = 0. \tag{2.14}$$

Note that equation (1.16), the one-dimensional form of this equation, is hardly any simpler. We have already shown in section 1.8 that the first term in this equation relates to the potential energy that can be stored in the fluid by compressing it.

$$e_p = \frac{1}{2}\frac{p'^2}{\rho_0 c^2}.$$

The second term in equation (2.14) is the rate of increase in the kinetic energy density, e_k, of a fluid that is acoustically energised.

$$e_k = \tfrac{1}{2}\rho_0 v^2.$$

The third term in (2.14) relates to energy flux. The rate at which work is being done by unit area of fluid supporting an externally induced normal stress p' and moving with velocity v_i is $p'v_i$.

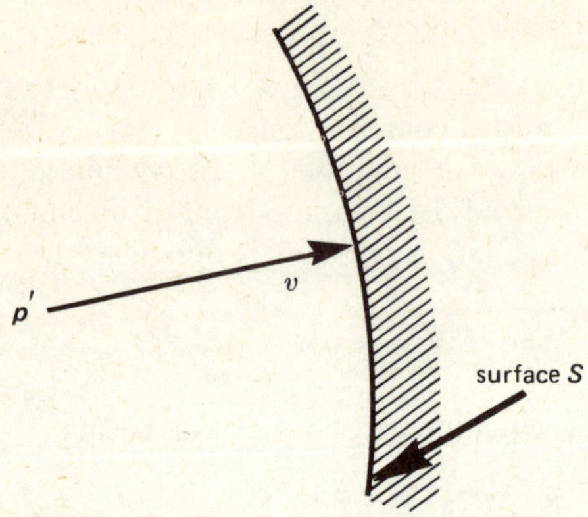

Fig. 2.3 — $p'v$ is the rate and direction at which acoustic energy crosses unit area of space.

This energy flux vector is called the *intensity*. It is the rate (and direction) at which acoustic energy crosses unit area of space.

$$\mathbf{I} = p'\mathbf{v} \quad I_i = p'v_i.$$

Equation (2.14) can thus be seen to be a clear statement of energy conservation,

$$\frac{\partial}{\partial t}(e_p + e_k) = -\frac{\partial I_i}{\partial x_i} = -\operatorname{div}\mathbf{I}. \tag{2.15}$$

When integrated over a volume V bounded by a surface S this is

$$\frac{\partial}{\partial t}\int_V (e_p + e_k)\,\mathrm{d}V = -\int_S I_n\,\mathrm{d}S;$$

energy in volume V decreases because of its transport out across the surface S.

In summary, we have seen that the energy in an acoustically irradiated fluid is composed of two distinct elements, one potential and the other kinetic. That energy increases only as a result of energy fluxing into the fluid element due to the negative divergence (convergence) of the energy flux vector—the acoustic intensity. Acoustic fields are energy conserving. If the mean level of acoustic activity in a region is constant, then $(\partial/\partial t)(e_p + e_k)$ vanishes and the mean intensity vector $\bar{\mathbf{I}}$, is then divergence-free, or solenoidal. Over any closed surface the net energy flow into, or out of, that surface must vanish;

$$\operatorname{div}\bar{\mathbf{I}} = \frac{\partial \bar{I}_i}{\partial x_i} = 0 \quad \text{and} \quad \int_S \bar{I}_n\,\mathrm{d}S = 0. \tag{2.16}$$

2.2 SOME SIMPLE THREE-DIMENSIONAL WAVE FIELDS

The simplest of all the solutions of the three-dimensional wave equation is the relatively trivial case of one-dimensional waves already discussed in Chapter 1. In particular, equation (1.9) gives the general form of waves propagating to the right and left:

$$p'(x_1, t) = f(x_1 - ct) + g(x_1 + ct).$$

This pressure disturbance corresponds to waves propagating along the x_1 co-ordinate axis.

The pressure perturbation in a plane wave of frequency ω propagating in the x_1-direction can be written as

$$p'(\mathbf{x}, t) = A\, \mathrm{e}^{i(\omega t - kx_1)} \quad \text{where } k = \frac{\omega}{c}.$$

A more general plane wave has the form

$$p'(\mathbf{x}, t) = A\, \mathrm{e}^{i(\omega t - \mathbf{k}\cdot\mathbf{x})}.$$

$\mathbf{k} = (k_1, k_2, k_3)$ is called the wave number vector. Direct substitution shows that p' satisfies the wave equation provided $|\mathbf{k}|^2 = \omega^2/c^2$. This is a plane wave propagating in the direction of $\mathbf{k}$ with phase speed c.

The next simplest case is one of a spherical wave centred at the co-ordinate origin; that is, a wave in which all the flow parameters are functions of the radial distance, r, and t only,

$$\text{i.e.} \quad p' = p'(r, t), \quad \mathbf{v} = \mathbf{v}(r, t) \text{ etc.}$$

Since the pressure gradient acts radially, the fluid velocity can only be in the radial direction, and we will denote this radial velocity by $u(r, t)$.

The equation satisfied by $p'(r, t)$ can be obtained by writing ∇^2 in the wave equation in spherical polar co-ordinates;

$$\nabla^2 p' = \frac{1}{r^2}\frac{\partial}{\partial r}\left(r^2 \frac{\partial p'}{\partial r}\right) + \frac{1}{r^2 \sin\theta}\frac{\partial}{\partial \theta}\left(\sin\theta \frac{\partial p'}{\partial \theta}\right) + \frac{1}{r^2 \sin^2\theta}\frac{\partial^2 p'}{\partial \varphi^2}.$$

Since p' is independent of θ and φ the wave equation reduces to

$$\frac{1}{c^2}\frac{\partial^2 p'}{\partial t^2} = \frac{1}{r^2}\frac{\partial}{\partial r}\left(r^2 \frac{\partial p'}{\partial r}\right). \tag{2.17}$$

Alternatively equation (2.17) can be derived directly by considering mass and momentum conservation in a radially symmetric field. The linearised equation of conservation of mass can be derived simply by considering a spherical shell centred on the origin:

$$4\pi r^2\, \delta r \frac{\partial \rho'}{\partial t} = -\rho_0\, \delta(4\pi r^2 u)$$

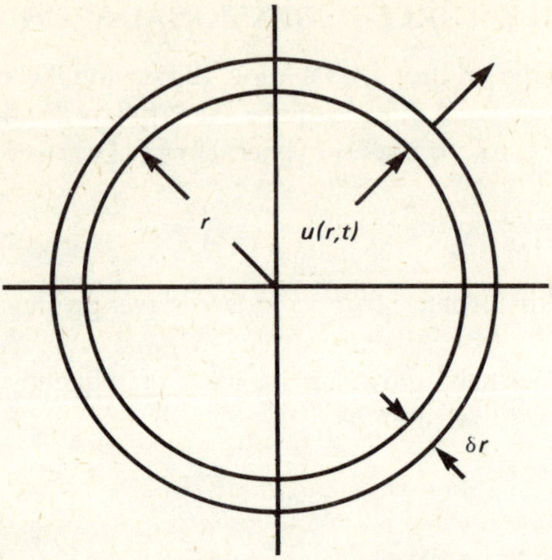

Fig. 2.4 — A spherical shell of thickness δr centred on the origin.

which simplifies to

$$r^2 \frac{\partial \rho'}{\partial t} = -\rho_0 \frac{\partial}{\partial r}(ur^2). \tag{2.18}$$

The radial component of the equation of conservation of momentum (2.4) gives

$$\rho_0 \frac{\partial u}{\partial t} = -\frac{\partial p'}{\partial r}. \tag{2.19}$$

The additional relationship $p' = c^2\rho'$ enables us to eliminate u and ρ' from equations (2.18) and (2.19) and obtain

$$\frac{r^2}{c^2}\frac{\partial^2 p'}{\partial t^2} = \frac{\partial}{\partial r}\left(r^2 \frac{\partial p'}{\partial r}\right),$$

which is a restatement of equation (2.17). A little algebra shows that this equation is precisely the same as

$$\frac{1}{c^2}\frac{\partial^2}{\partial t^2}(rp') - \frac{\partial^2}{\partial r^2}(rp') = 0 \tag{2.20}$$

and we see that in a spherically symmetric sound field, or wave, the product rp' satisfies the one-dimensional wave equation (cf. equations (2.20) and (1.8)) and the general solution for rp' is therefore

$$rp'(r, t) = f(r - ct) + g(r + ct)$$

so that

$$p'(r, t) = \frac{f(r - ct)}{r} + \frac{g(r + ct)}{r} \tag{2.21}$$

where, as in the one-dimensional waves already discussed, f and g can be any functions. f describes a wave radiating outwards to infinity and g corresponds to waves converging on the origin.

In most circumstances where the sound is generated in open space we insist that the sound must not anticipate its cause. The wave field $(1/r)g(r + ct)$ is therefore ruled out because it is a wave system travelling in from infinity. It has existed far away for an indefinitely long time in the past. The requirement that information about the current source activity should not be contained in past waves that anticipate the present is called the *causality* condition. In simple fields this is indistinguishable from the *radiation condition* first formulated by Sommerfeld who insisted that in open space only those solutions to the wave equation

$$\frac{\partial^2 p'}{\partial t^2} - c^2\nabla^2 p' = 0 \tag{2.22}$$

that consisted entirely of outward travelling waves could have any real existence. Specifically, Sommerfeld required that the limit

$$\lim_{r \to \infty} r\left\{\frac{\partial p'}{\partial t} + c\,\frac{\partial p'}{\partial r}\right\} = 0 \tag{2.23}$$

be applied to all solutions of (2.22) as a test for their physical reasonableness. This test admits $(1/r)f(r - ct)$ as a possible three-dimensional sound field whereas $(1/r)g(r + ct)$ is excluded.

Once we have determined the pressure the velocity can be calculated from it by using the linearised momentum equation (2.19). For example, for an outward propagating spherically symmetric wave of frequency ω,

$$p' = \frac{A\,\mathrm{e}^{i\omega(t - r/c)}}{r}$$

where A is a complex constant. Equation (2.19) gives

$$\begin{aligned} u(r, t) &= \frac{A}{i\omega\rho_0}\left\{\frac{i\omega}{cr} + \frac{1}{r^2}\right\}\mathrm{e}^{i\omega(t - r/c)} \\ &= \frac{A}{r}\,\mathrm{e}^{i\omega(t - r/c)}\,\frac{\mathrm{e}^{-i\phi}\sqrt{1 + (c/\omega r)^2}}{\rho_0 c} \end{aligned} \tag{2.24}$$

where φ, the phase angle, is $\tan^{-1}(c/\omega r)$, i.e.

$$u(r, t) = \frac{\sqrt{1 + (c/\omega r)^2}}{\rho_0 c}\,p'\!\left(r, t - \frac{\varphi}{\omega}\right). \tag{2.25}$$

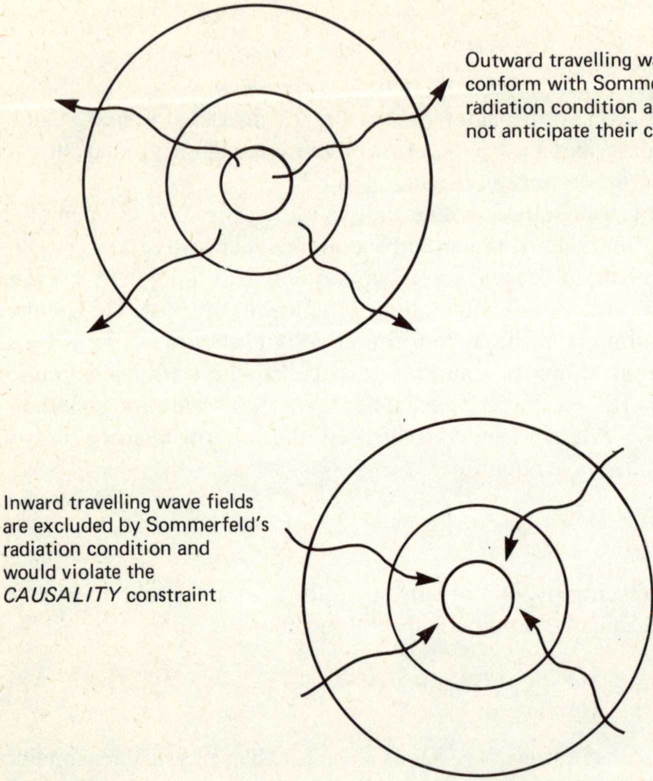

Fig. 2.5 — Spherical wave fields.

We see that in general the pressure perturbation and velocity are not in phase and that the impedance, the ratio of p' to u is generally less than $\rho_0 c$, the plane wave impedance. In fact for very small r the pressure leads the velocity by 90° and $p' = \rho_0 r(\partial u/\partial t)$. The phase lead decreases as r is increased and the impedance approaches $\rho_0 c$. For large $r(\gg c/\omega)$ the pressure and velocity are very nearly in phase and the spherical wave behaves like a plane wave. This is to be expected since at large distances the wave fronts are virtually plane.

Example

The amplitude of the pressure perturbation at a distance 1 m from a source in air is 1 N/m^2 and the frequency of the wave is 100 Hz. Calculate the amplitude of particle velocity at 0.01 m from the source if (a) the waves are plane, and (b) the waves are spherically symmetric.

(a) In the plane wave $p' = A\,\mathrm{e}^{i\omega(t - x/c)}$ and $u = p'/\rho_0 c$.

Since the pressure amplitude is 1 N/m^2, the velocity amplitude = $(\rho_0 c_0)^{-1}$ × 1 N/m^2 = 2.5 10^{-3} m/s everywhere.

(b) In the spherically symmetric wave

$$p' = \frac{A}{r} e^{i\omega(t - r/c)} \quad \text{and} \quad u(r, t) = \frac{\sqrt{1 + (c/\omega r)^2}}{\rho_0 c} p'\left(r, t - \frac{\varphi}{\omega}\right).$$

Since the pressure amplitude is 1 N/m^2, when $r = 1$ m, $|A| = 1$ and the

$$\text{velocity amplitude} = \frac{\sqrt{1 + (c/\omega r)^2}}{\rho_0 c r}$$

$$= 13.3 \text{ m/s for } r = 0.01 \text{ m}.$$

The sound scattered by a small bubble

As our first practical example of a centred three-dimensional field we consider a small spherical inclusion, or void, in an otherwise homogeneous dense medium of high elastic modulus. If the pressure in the vicinity of the hole is made to vary, the void will respond by producing a scattered sound wave centred on the hole. In practice the void might be a gaseous inclusion in a metallic casting arising from a manufacturing fault. When the casting is subjected to ultrasonic testing, sound waves modulate the pressure in the vicinity of the void which responds by scattering a centred wave. Detection of this wave pinpoints the position of the fault. Alternatively the void might be a bubble of gas in a liquid. If the fluid is in unsteady motion the pressure in the vicinity of the bubble will vary, and the bubble will respond to this pressure variation by pulsating and thereby producing a secondary centred sound wave. Water flows with cavitation generate sound that is centred on interior bubbles in this way. Any boiling in a coolant will also produce a centred sound field for precisely the same reason, and monitoring the sound field is a way of detecting coolant malfunction.

The common feature in these two examples is that the spherical void is often unable to support any significant compressive stress and therefore comprises a volume in which the pressure perturbation must always vanish. Of course there will be some small pressure arising from the compression of the gas in the bubble, but this is often quite insignificant in comparison with the pressure that is generated by similar volumetric strain in the surrounding medium.

Nevertheless the bubble volume can sometimes pulsate with large amplitude and at very low frequency the pressure internal to the bubble cannot be neglected. These cases will be discussed in Chapter 6 (section 6.5) where we examine the bubble scattering properties more thoroughly. This simplified treatment will be seen as the limiting high-frequency behaviour of the general case.

The form of the centred sound wave is extremely simple and can be found from the observation that the spherical void supports a spherical sound wave whose pressure variation $p'(r, t)$, say, supplements the unsteady imposed environmental pressure fluctuation $p_i'(t)$. Because the surface of the void is unable to sustain any pressure variation the sum of the incident and scattered pressures must vanish at the boundary surface $\mathbf{r} = a$.

$$p_i'(t) + p'(a, t) = 0. \tag{2.26}$$

$p'(r, t)$ is a centred outgoing sound wave whose form we have already found in (2.21)

$$p'(r, t) = \frac{f(r - ct)}{r}, \quad r \geqslant a.$$

Now from (2.26)

$$\frac{f(a - ct)}{a} = -p_i'(t),$$

and hence

$$p'(r, t) = -\frac{a}{r}\, p_i'\left(t - \frac{r - a}{c}\right). \tag{2.27}$$

The scattered pressure wave travels radially outwards with its amplitude inversely proportional to the distance travelled. The pressure time history at the void, i.e. at $\mathbf{r} = a$, is faithfully mimicked at a time $(r - a)/c$ later by the sound pressure at r.

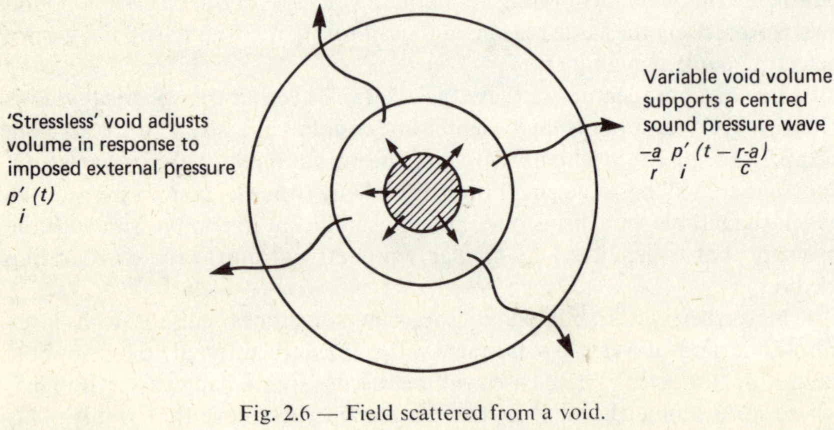

Fig. 2.6 — Field scattered from a void.

The sound generated by a pulsating sphere

The setting of a spherical wave's amplitude is slightly more complicated when the velocity, not the pressure, is specified. A spherical shell of variable radius is a form of loudspeaker that is easy to handle analytically and its study can reveal some of the parameters determining a loudspeaker's efficiency.

Suppose that the radial velocity at radius a on the surface of the spherical radiator is specified to be $u_a(t)$. This surface vibration will generate the centred wave $(a/r)p'(a, t-(r-a)/c)$ and we need to know how pressure is related to the velocity. For this we use the radial component of the momentum balance (2.19)

$$\rho_0 \frac{\partial u}{\partial t} = -\frac{\partial p'}{\partial r}$$

$$= \frac{a}{r^2} p'\left(a, t - \frac{r-a}{c}\right) + \frac{a}{rc}\frac{\partial p'}{\partial t}\left(a, t - \frac{r-a}{c}\right). \tag{2.28}$$

At this stage details of the waveform complicate the formulae and for that reason we choose the simplest case for illustration. We again consider a time harmonic wave with angular frequency ω in which each wave quantity is proportional to $e^{i\omega t}$. Let $\hat{p}$ and $\hat{u}$ now be the complex amplitudes of the perturbation pressure and radial velocity respectively; of course, we imply the 'real part' but do not state this explicitly until the final results. I.e.

$$p'(r,t) = \mathcal{R}\,\hat{p}(r)\,e^{i\omega t},\ u(r,t) = \mathcal{R}\hat{u}(r)\,e^{i\omega t}, \text{ and } u_a(t) = \mathcal{R}\hat{u}_a\,e^{i\omega t}.$$

The differential operator $\partial/\partial t$ is then identical to the algebraic factor $i\omega$ and equation (2.28) is

$$i\omega\rho_0\hat{u} = e^{-i\omega(r-a)/c}\left\{1 + \frac{i\omega r}{c}\right\}\frac{a}{r^2}\hat{p}(a).$$

At $r = a$, $u = u_a$, so that,

$$\hat{p}(a) = \frac{a i\omega\rho_0\hat{u}_a}{1 + \dfrac{i\omega a}{c}}, \tag{2.29}$$

a form that can be derived directly from equation (2.24), and the complete field induced by the pulsating spherical boundary is

$$\hat{p}(r) = \frac{a}{r}\frac{\dfrac{i\omega a}{c}}{1 + \dfrac{i\omega a}{c}}\rho_0 c\hat{u}_a\,e^{-i\omega(r-a)/c}. \tag{2.30}$$

Generally the complex amplitude ratio of the surface pressure perturbation to its radial velocity is called the radiation impedance, Z, and this is often presented as a fraction of a plane acoustic wave's impedance, $\rho_0 c$. That is termed the specific radiation impedance and is found from equation (2.29) to be

$$\frac{Z}{\rho_0 c} = \frac{\dfrac{i\omega a}{c}}{1 + \dfrac{i\omega a}{c}} = \frac{1 + i\left(\dfrac{c}{\omega a}\right)}{1 + \left(\dfrac{c}{\omega a}\right)^2}. \tag{2.31}$$

When $\omega a/c$ is large the specific radiation impedance is purely real and equal to unity, but when $\omega a/c$ is small the specific radiation impedance is purely imaginary and equal to the small quantity $i\omega a/c = i2\pi a/\lambda$.

The size of the sphere's circumference, $2\pi a$, in comparison with the acoustic wavelength λ, determines the ability of the sphere's vibration to radiate sound. That ratio is equal to the Helmholtz number, $\omega a/c$, which is sometimes referred to as the compactness ratio. At very high values of the Helmholtz number, $\omega a/c \gg 1$, the sphere is large enough to be called non-compact. Then the entire disturbance field set up by the pulsations of the sphere is simple and wave-like.

Equation (2.30) then gives

$$p'(r, t) = \frac{a}{r}\rho_0 c u_a\left(t - \frac{r - a}{c}\right), \quad \frac{\omega a}{c} \gg 1 \tag{2.32}$$

and $p' = \rho_0 c u$ everywhere, as in a plane sound wave, u being the radial particle velocity.

On the other hand, if the source scale is much smaller than the wavelength of sound it is said to be compact, i.e. $\omega a/c \ll 1$. The field in the immediate vicinity of the vibrating surface is then not in the least bit wave-like. It is virtually independent of the material's compressibility as expressed in the speed of sound, the appropriate form of (2.30) then being

$$\hat{p}(r) = \frac{i\omega a^2}{r}\rho_0 \hat{u}_a \, e^{-i\omega(r-a)/c} \tag{2.33}$$

i.e.

$$p'(r, t) = \frac{a^2}{r}\rho_0 \frac{\partial u_a}{\partial t}\left(t - \frac{r - a}{c}\right), \quad \frac{\omega a}{c} \ll 1. \tag{2.34}$$

Equations (2.32) and (2.33) show that only a small fraction $\omega a/c$ of the pressure that would be generated by the same amplitude of surface motion in the non-compact case is actually generated by a compact source. Small acoustic radiators, or loudspeakers, are very inefficient and their response increases with frequency. Conversely, large sources are efficient with the entire field induced by a non-compact vibration being simply a superposition of acoustic waves. Those acoustic waves are virtually one-dimensional, their curvature being very small in comparison with their wavelength and there is no preference given to the higher frequencies by a non-compact source.

2.3 A MORE ELABORATE THREE-DIMENSIONAL WAVE FIELD

If φ is a solution of the wave equation then so is $\partial\varphi/\partial x_i$ or $\nabla\varphi$; the wave equation can be differentiated to show that

$$\frac{\partial^2}{\partial t^2}\frac{\partial\varphi}{\partial x_i} - c^2\nabla^2\frac{\partial\varphi}{\partial x_i} = 0 \tag{2.35}$$

provided that

$$\frac{\partial^2\varphi}{\partial t^2} - c^2\nabla^2\varphi = 0.$$

A whole variety of solutions can therefore be generated by elementary differential operations. One such pressure field can be obtained by differentiating (2.21) with respect to x_1:

$$\begin{aligned} p'(\mathbf{x}, t) &= \frac{\partial}{\partial x_1}\left\{\frac{f(r - ct)}{r}\right\} \\ &= \cos\theta\,\frac{\partial}{\partial r}\left(\frac{f(r - ct)}{r}\right) \end{aligned} \tag{2.36}$$

where θ is the angle between the 1-axis and the position vector $\mathbf{x}$, $\cos\theta = x_1/r$.

The sound of an oscillating rigid sphere

Consider a rigid sphere whose centre moves along the x_1 axis in a small amplitude harmonic motion about the origin of co-ordinates. The velocity of the sphere is the real part of $(U\,e^{i\omega t}, 0, 0)$, and so the radial velocity at a point defined by the angle θ on the surface of the sphere is the real part of $u = U\cos\theta\,e^{i\omega t}$, the positive x_1 axis forming the line $\theta = 0$.

The sound field induced by the sphere's vibration is actually the field of the form expressed in equation (2.36) with harmonic time dependence;

$$p'(\mathbf{x}, t) = A\cos\theta\,\frac{\partial}{\partial r}\left\{\frac{e^{i\omega(t - r/c)}}{r}\right\}.$$

The complex amplitude A can be determined by applying the radial momentum equation at the surface of the sphere.

$$\rho_0\frac{\partial u}{\partial t} = i\omega\rho_0 u = -\frac{\partial p'}{\partial r} = -A\cos\theta\,\frac{\partial^2}{\partial r^2}\left\{\frac{e^{i\omega(t - r/c)}}{r}\right\}$$

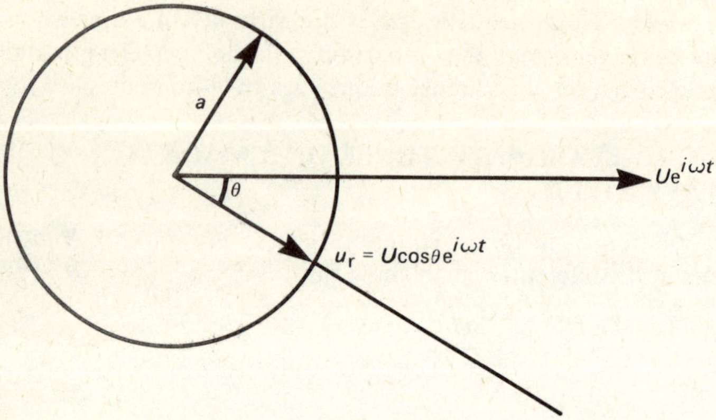

Fig. 2.7 — An oscillating rigid sphere.

so that

$$u = \frac{-\cos\theta}{i\omega\rho_0}\left\{\frac{2}{r^3} + \frac{2i\omega}{cr^2} - \frac{\omega^2}{c^2 r}\right\} A\,\mathrm{e}^{i\omega(t-r/c)}. \tag{2.37}$$

Now on $r = a$, $u = U\cos\theta\,\mathrm{e}^{i\omega t}$ and hence it follows that

$$A = \frac{-i\omega\rho_0 U a^3\,\mathrm{e}^{i\omega a/c}}{2(1 + i\omega a/c) - \omega^2 a^2/c^2} \tag{2.38}$$

and

$$p'(r, \theta, t) = \frac{-i\omega\rho_0 U a^3 \cos\theta}{2(1 + i\omega a/c) - \omega^2 a^2/c^2}\frac{\partial}{\partial r}\left\{\frac{\mathrm{e}^{i\omega(t-(r-a)/c)}}{r}\right\}. \tag{2.39}$$

Again we see in (2.39) the appearance of the parameter $\omega a/c$, the compactness ratio. A compact sphere radiates sound very inefficiently the pressure being proportional to the acceleration of the body and the pressure on the sphere then provides the force needed to overcome the inertia of the basically incompressible material moving around it. With an error of order $\omega^3 a^3/c^3$ smaller than the retained terms,

$$p'(r, \theta, t) = -\tfrac{1}{2} i\omega\rho_0 U a^3 \cos\theta \frac{\partial}{\partial r}\left\{\frac{\mathrm{e}^{i\omega(t-r/c)}}{r}\right\} \tag{2.40}$$

$$= -\frac{\omega^2 a^2}{c^2}\tfrac{1}{2}\rho_0 c U \cos\theta \frac{a}{r}\mathrm{e}^{i\omega(t-r/c)}\left(1 - \frac{ic}{\omega r}\right). \tag{2.41}$$

There is what is termed a near and far field structure to this sound. Much more than a wavelength away, $c/\omega r \ll 1$, the far field form is established. There sound, with the pressure in phase with the velocity, radiates to infinity carrying energy away from the sphere. But in the near field of a compact source

$c/\omega r \gg 1$ and the motion is hardly influenced by the speed of sound, or, therefore, by the material's compressibility. In particular, on the surface of the compact sphere

$$p'(a, \theta, t) = \tfrac{1}{2} i\omega \rho_0 a U \cos\theta\, \mathrm{e}^{i\omega t}\left\{1 + \frac{1}{2}\left(\frac{\omega a}{c}\right)^2 - \frac{i}{2}\left(\frac{\omega a}{c}\right)^3\right\}; \quad \frac{\omega a}{c} \ll 1. \tag{2.42}$$

The reaction of the material surrounding the compact vibrating sphere produces a force to oppose the motion, the value of the axial component of that force being

$$-\int_0^\pi 2\pi a \sin\theta p'(a, \theta, t) \cos\theta a \,\mathrm{d}\theta$$

$$= -\pi i\omega a^3 \rho_0 U\, \mathrm{e}^{i\omega t}\left\{1 + \frac{1}{2}\left(\frac{\omega a}{c}\right)^2 - \frac{i}{2}\left(\frac{\omega a}{c}\right)^3\right\}\int_0^\pi \sin\theta \cos^2\theta \,\mathrm{d}\theta$$

$$= -\rho_0 \tfrac{2}{3}\pi a^3 i\omega U\, \mathrm{e}^{i\omega t}\left\{1 + \frac{1}{2}\left(\frac{\omega a}{c}\right)^2 - \frac{i}{2}\left(\frac{\omega a}{c}\right)^3\right\}. \tag{2.43}$$

The force on the compact sphere is mainly a drag, i.e. a force directed opposite to the direction of motion. Its magnitude is equal to the force required to overcome the inertia of half the mass of fluid displaced by the sphere.

But there is also a very much smaller in-phase component of the surface pressure that will give rise to a drag force proportional to the velocity. That in-phase drag is

$$\tfrac{1}{3}\pi a^2 \rho_0 c U\, \mathrm{e}^{i\omega t}\left(\frac{\omega a}{c}\right)^4 \tag{2.44}$$

and the rate at which work must be done to move the sphere against the drag is

$$\tfrac{1}{3}\pi a^2 \rho_0 c \overline{U^2}\left(\frac{\omega a}{c}\right)^4, \tag{2.45}$$

where $\overline{U^2}$ is the mean square value of the sphere's velocity. This rate of work contains the small factor $(\omega a/c)^4$ and is a measure of the small amount of power that is radiated into the sound field by the rigid motion of a very compact sphere.

In contrast, the motion of a grossly non-compact sphere radiates sound very effectively and the surface pressure is then in phase with surface velocity. In the limit $\omega a/c \gg 1$ equation (2.39) gives

$$p'(r, \theta, t) = \rho_0 c U \cos\theta \frac{a}{r} \mathrm{e}^{i\omega(t - (r-a)/c)}$$

or

$$p'(r, \theta, t) = \frac{a}{r} p'\left(a, \theta, t - \frac{r-a}{c}\right). \tag{2.46}$$

The pressure on the surface of the sphere is communicated outwards as acoustic rays, the strength falling off in the manner necessary to conserve energy, i.e. as r^{-1}.

The drag on the non-compact sphere is

$$\int_0^\pi 2\pi a \sin \theta p'(a, \theta, t) \cos \theta a \, d\theta = 2\pi a^2 \rho_0 c U \, e^{i\omega t} \int_0^\pi \sin \theta \cos^2 \theta \, d\theta$$

$$= \tfrac{4}{3}\pi a^2 \rho_0 c U \, e^{i\omega t} \tag{2.47}$$

and the rate at which energy has to be supplied to move the sphere against this drag force is the mean value of this times the velocity, i.e.

$$\tfrac{4}{3}\pi a^2 \rho_0 c \overline{U^2} \tag{2.48}$$

where $\overline{U^2}$ is again the mean square value of the translational velocity of the non-compact sphere.

2.4 TWO-DIMENSIONAL SOUND WAVES

Just as the one-dimensional wave-fields $f(x_1 - ct)$ and $g(x_1 + ct)$ are trivial solutions of the three-dimensional wave equation, so is the set of solutions involving variation with only two of the three space co-ordinates. They constitute two-dimensional wave fields such as the ripples that form when a layer of shallow water is disturbed from rest.

The fact that shallow water waves conform exactly with two-dimensional sound waves can be used in an analogy where sound can be simulated and waves visualised; sound waves can very rarely be seen even with the most modern and sensitive of optical equipment.

The principles underlying the incompressible motion of long waves on water are exactly those we have considered already for sound, but the wave speed is set by different constraints. Consider the situation illustrated in the figure below, where the shallow layer of water is slightly disturbed from rest, the disturbance varying only slowly with position $\mathbf{x}$ along the layer. $\mathbf{x}$ is a horizontal two-dimensional vector.

The pressure varies in the water due only to hydrostatic effects, the vertical acceleration being negligible compared to gravity:

$$p = p_0 + \rho g(\xi + h - z). \tag{2.49}$$

is the pressure at height z above the bottom of the layer, so that the horizontal pressure gradient, which is the negative of the force per unit volume acting to accelerate the water, is independent of height.

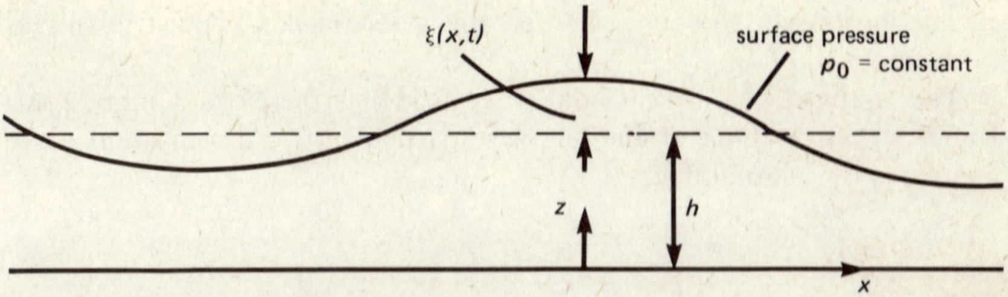

Fig. 2.8 — Long wavelength waves on water.

$$\rho g \frac{\partial \xi}{\partial x_\alpha} = \frac{\partial p}{\partial x_\alpha} = -\rho \frac{\partial v_\alpha}{\partial t} \quad \text{for } \alpha = 1 \text{ or } 2 \tag{2.50}$$

All particles of water in the vertical column at ($\mathbf{x}$, t) move with the same horizontal velocity $\mathbf{v}$.

Conservation of mass requires that any divergence in the mass flux flowing in the layer is accompanied by a depletion in the mass per unit area i.e. the layer must get thinner at the required rate:

$$\rho h \frac{\partial v_\alpha}{\partial x_\alpha} = -\rho \frac{\partial \xi}{\partial t}, \quad \text{summed over } \alpha = 1 \text{ and } 2. \tag{2.51}$$

This and equation (2.50) combine on cross differentiation into the two-dimensional wave equation

$$\frac{\partial^2 \xi}{\partial t^2} - c^2 \frac{\partial^2 \xi}{\partial x_\alpha^2} = 0, \quad \text{where } c^2 = gh. \tag{2.52}$$

So long as the waves are longer than the mean layer depth, this equation describes their behaviour well, and they are a close analogue of two-dimensional sound waves. But if the layer is deeper than the wavelength is long, then the modelling fails and the surface waves that do form travel with a frequency-dependent wave speed, and groups of waves of different frequency therefore disperse during propagation and the wave system is said to be dispersive. Other differences are due to surface tension effects at the free surface effects which become extreme for very short wavelength ripples.

The two-dimensional case is exceptional though in that the structure of the wave-field is somewhat more complicated. In both the one-dimensional wave, and in the three-dimensional case, a knowledge of the time history of the wave at any one point gives a knowledge of the time variations that will occur further along the wave as it travels out, the fields being $f(x_1 - ct)$ and $(1/r)f(r - ct)$ respectively in the simplest cases. But that is not true in a two-dimensional wave field. In fact Huygen's principle, that the future behaviour of waves can be determined entirely by the propagation of wave fronts, a principle most useful in one- and three-dimensional fields, is not true in two dimensions.

The simplest two-dimensional wave field has therefore a more complicated structure. One of the simplest particular two-dimensional wave solutions is the centred field

$$p'(\mathbf{x}, t) = \frac{1}{\sqrt{c^2t^2 - r^2}}, \quad r < ct$$
$$= 0 \qquad\qquad r > ct \tag{2.53}$$

where now $r = (x_1^2 + x_2^2)^{\frac{1}{2}}$.

It can now be checked that this is a solution by direct differentiation, a step that is made easier by the use of

$$\nabla^2 p' = \frac{1}{r}\frac{\partial}{\partial r}\left(r\frac{\partial p'}{\partial r}\right)$$

for a two-dimensional field with no angular variation.

The simplest solution involving general time dependence has the form

$$p'(\mathbf{x}, t) = \int_{-\infty}^{t-r/c} \frac{f(\tau)\,\mathrm{d}\tau}{\sqrt{c^2(t-\tau)^2 - r^2}} \tag{2.54}$$

which is considerably more complicated than the simplest members of one- and three-dimensional wave fields. Of course as the waves propagate very far from their origin, they become less and less curved and tend to behave locally like one-dimensional fields. That must be true whether the field is one-, two- or three-dimensional. The three-dimensional case is easily treated. We can write $r = r_0 - \Delta$ so that over a range of Δ,

$$\frac{1}{r} \sim \frac{1}{r_0} + \frac{\Delta}{r_0^2} + \cdots. \tag{2.55}$$

r_0 is the value of r at the centre of the region of interest.

$1/r$ in $f(r - ct)/r$ can then be approximated to by the constant $1/r_0$ whenever r_0 is big enough. The field then behaves locally as $(1/r_0)f(r - ct)$, i.e. it is to a good approximation indistinguishable from a one-dimensional field propagating radially outwards from the origin its amplitude changing *very* slowly as r_0 tends to infinity.

The same must be true of the two-dimensional case but for more subtle reasons. The integrand in (2.54) can be factorised

$$c^2(t-\tau)^2 - r^2 = \{c(t-\tau) + r\}\{c(t-\tau) - r\}. \tag{2.56}$$

The integrand is singular when $c(t - \tau) = r$, from which it is suggested that nearly all the important contributions to the integral come from values of τ in this range. Therefore if we approximate the values of $c(t - \tau)$ in the first bracket in (2.56), which is smoothly varying and not singular, by its value at the centre of the most active region, i.e. r, we have

$$c^2(t-\tau)^2 - r^2 = 2r\{c(t-\tau) - r\} \tag{2.57}$$

and the 'far field' form of the integral in (2.54) is then

$$p'(\mathbf{x}, t) \underset{r\to\infty}{\sim} \frac{F(r-ct)}{\sqrt{r}}$$

where

$$F(r-ct) = \frac{1}{\sqrt{2c}} \int_{-\infty}^{t-r/c} \frac{f(\tau)\,\mathrm{d}\tau}{\sqrt{t-r/c-\tau}}. \tag{2.58}$$

Again as r tends to infinity the variation of $1/\sqrt{r}$ is locally negligible and the wave behaves as if it were one-dimensional in the far distant field.

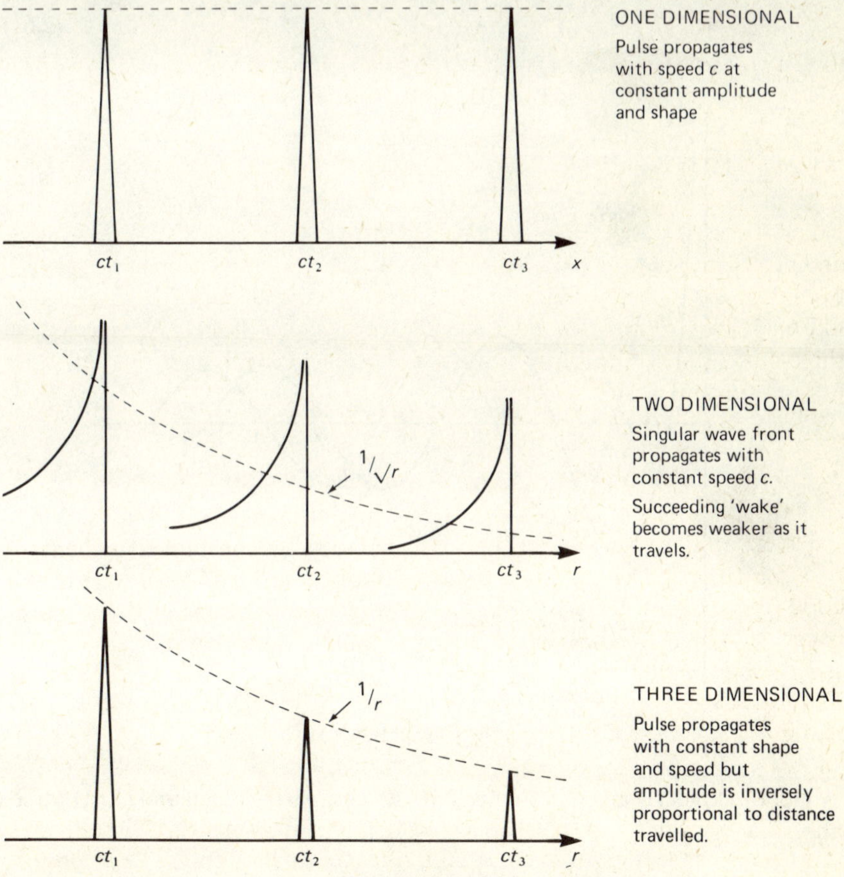

Fig. 2.9 — Illustration of simple wave fields.

When the sound field has frequency ω, the general cylindrically symmetric pressure field given in (2.54) becomes

$$p'(r, t) = A \int_{-\infty}^{t - r/c} \frac{e^{i\omega\tau}\, d\tau}{\sqrt{c^2(t-\tau)^2 - r^2}} \tag{2.59}$$

where A is a constant. By changing variables and writing the integral in terms of $s = c(t - \tau)/r$, we can express this as

$$p'(r, t) = \frac{A\, e^{i\omega t}}{c} \int_1^{\infty} \frac{e^{-i\omega s r/c}}{(s^2-1)^{\frac{1}{2}}}\, ds. \tag{2.60}$$

This integral cannot be evaluated explicitly. Instead it is given a name and tabulated in Mathematical Tables. A zeroth order Hankel function, $H_0^{(2)}(X)$, is defined to be

$$H_0^{(2)}(X) = \frac{2i}{\pi} \int_1^{\infty} \frac{e^{-iXs}}{(s^2-1)^{\frac{1}{2}}}\, ds. \tag{2.61}$$

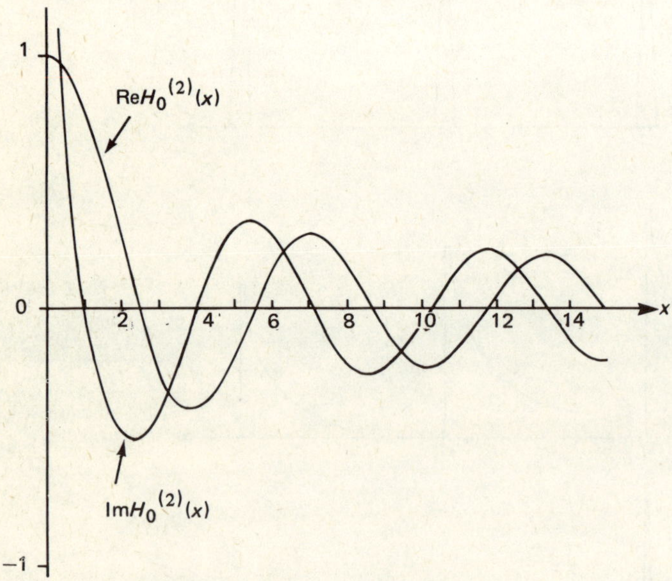

Fig. 2.10 — The Hankel function $H_0^{(2)}(x)$.

A plot of $H_0^{(2)}(X)$ is given in Figure 2.10. By comparing (2.60) and (2.61) we see that

$$p'(r, t) = \frac{A\pi}{2ic} e^{i\omega t} H_0^{(2)}\left(\frac{\omega r}{c}\right). \tag{2.62}$$

The far-field form of this pressure perturbation is, from (2.58),

$$p'(r, t) \sim \frac{A}{\sqrt{2rc}} \int_{-\infty}^{t-r/c} \frac{e^{i\omega\tau}\, d\tau}{\sqrt{t-\tau-r/c}}, \tag{2.63}$$

$$= \left(\frac{2}{rc}\right)^{\frac{1}{2}} A\, e^{i\omega(t-r/c)} \int_0^{\infty} e^{-i\omega s^2}\, ds$$

after a simple variable change to $s = (t - \tau - r/c)^{\frac{1}{2}}$. The s-integral can then be evaluated to show

$$p'(r, t) \underset{r\to\infty}{\sim} \sqrt{\frac{\pi}{2rc\omega}}\, A\, e^{i\omega(t-r/c)-i\pi/4}. \tag{2.64}$$

Again we see that the distant sound field has the form of outward propagating waves, whose amplitude decays with the inverse square-root of r.

We have seen that the cylindrically symmetric waves described by equation (2.62) have a simple form for large values of r. In fact their structure also simplifies near the source, where r is small. The behaviour of the Hankel function for small X can be demonstrated by changing the integration variable in (2.61) to $p = Xs$. Then

$$H_0^{(2)}(X) = \frac{2i}{\pi} \int_X^{\infty} \frac{e^{-ip}}{(p^2 - X^2)^{\frac{1}{2}}}\, dp,$$

and for small X

$$H_0^{(2)}(X) \sim \frac{2i}{\pi} \int_X^{\infty} \frac{e^{-ip}}{p}\, dp.$$

The integrand is of order p^{-1} near the origin and so as X tends to zero the integral diverges like $\log_e X$.

$$\lim_{X\to 0} H_0^{(2)}(X) = -\frac{2i}{\pi} \log_e X. \tag{2.65}$$

Hence from (2.62)

$$p'(r, t) \sim -\frac{A}{c}\, e^{i\omega t} \log_e\!\left(\frac{\omega r}{c}\right) \quad \text{for small } r. \tag{2.66}$$

The pressure perturbation and hence the velocity potential have a logarithmic singularity at the origin. This is the characteristic form for the velocity potential due to two-dimensional sources in *steady* aerodynamics. We noted in section 2.1 that near singularities and sources the velocity potential in acoustic problems should reduce to the form for steady potential flows. The expansion in (2.66) verifies that this criterion is satisfied by the two-dimensional sound field in (2.62).

Equation (2.62) gives the general form for cylindrically symmetric waves with frequency ω, other harmonic solutions of the wave equation can be determined by differentiation.

Example

An infinity long rigid cylinder vibrates transversely with frequency ω so that at time t its centre-line lies along the line $x_1 = \varepsilon\, e^{i\omega t}$, $x_2 = 0$. Show that the pressure perturbation at $\mathbf{x}$ is

$$\frac{\rho_0 \varepsilon \omega c \cos\theta}{H_0^{(2)\prime\prime}\left(\dfrac{\omega a}{c}\right)} H_0^{(2)\prime}\left(\frac{\omega r}{c}\right) e^{i\omega t},$$

where $r = (x_1^2 + x_2^2)^{\frac{1}{2}}$, $\cos\theta = x_1/r$, $H_0^{(2)}(z)$ is a zeroth order Hankel function and the dash denotes its derivative. On the cylinder continuity requires that the normal velocities of the fluid and the cylinder be equal, and so the radial component of the fluid velocity, u, satisfies

$$u|_{r=a} = i\omega\varepsilon\, e^{i\omega t} \cos\theta. \tag{2.67}$$

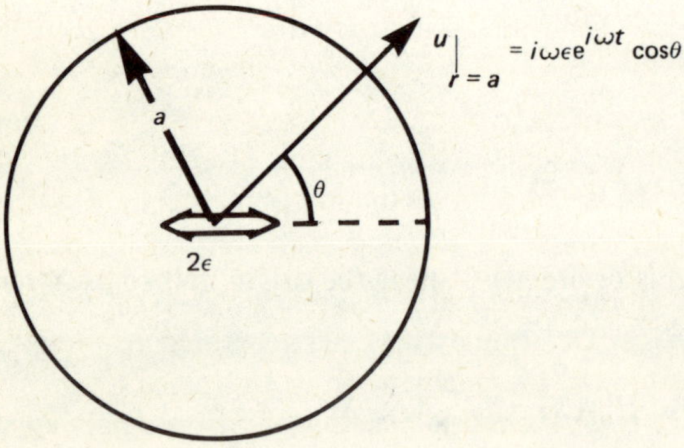

Fig. 2.11 — An oscillating cylinder.

Hence we can deduce from the linear momentum equation in the fluid that

$$\left.\frac{\partial p'}{\partial r}\right|_{r=a} = \rho_0 \omega^2 \varepsilon \cos\theta\, e^{i\omega t}. \tag{2.68}$$

The pressure perturbation must have a cos θ dependence. A solution of the wave equation proportional to cos θ can be obtained by differentiating (2.62) with respect to x_1.

$$p'(\mathbf{x}, t) = \frac{\partial}{\partial x_1}\left\{A\, e^{i\omega t} H_0^{(2)}\left(\frac{\omega r}{c}\right)\right\}$$

$$= \frac{A\omega}{c} \cos\theta\, e^{i\omega t} H_0^{(2)\prime}\left(\frac{\omega r}{c}\right).$$

Equation (2.68) determines the value of the constant A and leads to

$$p'(\mathbf{x}, t) = \frac{\rho_0 \omega \varepsilon c \cos\theta}{H_0^{(2)\prime\prime}\left(\dfrac{\omega a}{c}\right)} e^{i\omega t} H_0^{(2)\prime}\left(\frac{\omega r}{c}\right). \tag{2.69}$$

EXERCISES FOR CHAPTER 2

1. How much power is radiated in the various harmonic sound fields whose pressures are

$$p'(\mathbf{x}, t) = p_0 l^{n+1} \frac{\partial^n}{\partial x_1^n}\left\{\frac{\cos\omega(t - r/c)}{r}\right\}$$

for different integer values of n? $p_0 l^{n+1}$ represents the source strength, p_0 being a reference pressure and l an appropriate length scale.

2. A spherical loudspeaker of radius 30 cm vibrates with simple harmonic motion of amplitude 1 mm. Determine an expression for the acoustic power radiated by the loudspeaker as a function of frequency. How much power is radiated at frequencies of 10, 100 and 1000 Hz?

3. A boundary vibrates in air at a frequency of 100 Hz with a displacement amplitude 10^{-3} m. Determine the SPL generated by that vibration at a distance 10 m from the boundary when
a) the boundary is a plane surface with uniform normal displacement
b) the boundary is a circular cylinder of radius 5 cm with axially uniform displacement and
c) the boundary is a sphere of radius 5 cm with symmetric radial displacement.
What are the corresponding SPL if the vibration occurs at 10 kHz?

4. What is the maximum pressure in the sound wave generated by a rigid door slammed to stop impulsively from a 'tip' speed of 10 m/s?

5. A spherical air filled soap bubble fractures impulsively and radiates a 'pop-like' sound. If the surface tension is T and the bubble radius a determine the subsequent pressure field outside the bubble. Show that the total energy radiated is $8\pi a T^2/3\rho_0 c^2$.

6. The radius of an isolated spherical bubble formed by boiling water expands uniformly from nothing to 10^{-3} m in 10^{-3} s. Determine the shape and amplitude of the sound wave generated by this uniform expansion that is heard in water 1 m away.

Chapter 3

Waves in Pipes

3.1 PLANE WAVES

Many practical noises concern sound waves propagating in pipes. If the wavelength of the sound is large in comparison with the diameter of the pipe the sound propagates as a one-dimensional wave, and has the form given in equation (1.9). For definiteness consider a harmonic pressure wave of frequency ω, and strength I travelling along a duct to the right. I.e.

$$p' = I\,e^{i\omega(t-x/c)}.$$

I is a complex amplitude and the 'real part of' is again implied, i.e.

$$p' = \mathscr{R} I\,e^{i\omega(t-x/c)}.$$

We will investigate what happens to this wave if the cross-sectional area of the pipe suddenly changes. A transmitted wave travels on down the pipe and another wave will be reflected back from the abrupt change.

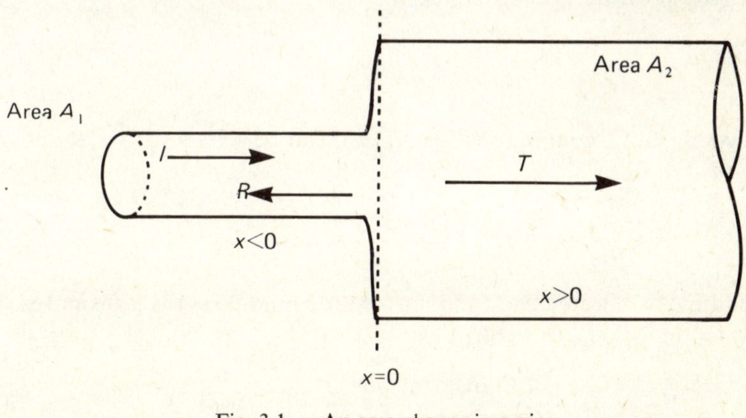

Fig. 3.1 — An area change in a pipe.

We choose a co-ordinate system so that the area variation occurs at $x = 0$, and suppose the area changes form A_1 to A_2. Denote the strength of the pressure perturbation in the reflected wave by R, and let T be the transmitted wave that travels into the region $x > 0$. Then on the left of the junction the pressure is the sum of that in the incident and reflected wave, while on the right it is that in the transmitted wave. I.e.

$$\begin{aligned} p' &= I\,e^{i\omega(t-x/c)} + R\,e^{i\omega(t+x/c)} \quad &\text{in } x < 0 \\ &= T\,e^{i\omega(t-x/c)} \quad &\text{in } x > 0. \end{aligned}$$

R and T can be calculated from conservation relations at $x = 0$. Firstly the mass flux into the junction must equal the mass flux out, and hence

$$\rho_0 A_1 u_1 = \rho_0 A_2 u_2 \tag{3.1}$$

where u_1 is the (uniform) velocity just to the left of the junction and u_2 the velocity to the right. On the left the velocity is the sum of elements, one the incident and the other the reflected wave component, while on the right the velocity is due solely to the transmitted wave. The relations (1.11) and (1.12) can be used to express these velocity components in terms of the pressure and equation (3.1) then becomes

$$\frac{A_1}{\rho_0 c}(I - R) = \frac{A_2}{\rho_0 c}\,T. \tag{3.2}$$

A second condition that must hold at the junction of negligible volume and therefore of negligible energy storing capacity is that the energy flux into the junction must be equal to the flux out of it, i.e.

$$A_1 p'_1 u_1 = A_2 p'_2 u_2,$$

but from (3.1) $A_1 u_1 = A_2 u_2$ and so this energy condition simplifies to one which says pressure is continuous;

$$\begin{aligned} p'_1 &= p'_2 \\ I + R &\doteq T. \end{aligned} \tag{3.3}$$

The two algebraic equations (3.2) and (3.3) can be solved to give R and T in terms of I.

$$R = \frac{A_1 - A_2}{A_1 + A_2}\,I, \quad T = \frac{2A_1}{A_1 + A_2}\,I.$$

Some sound is reflected back and not all the sound power is transmitted. The transmission loss, L_T, is defined as

$$L_T = 10\log_{10}\left(\frac{\text{incident power}}{\text{transmitted power}}\right). \tag{3.4}$$

In this example

$$L_T = 10 \log_{10}\left(\frac{A_1 I^2}{A_2 T^2}\right)$$

$$= 10 \log_{10}\left(\frac{(A_1 + A_2)^2}{4A_1 A_2}\right) \tag{3.5}$$

It is interesting to note that the transmission loss is symmetric in A_1 and A_2; the transmission loss is the same whether the sound is incident from the left where the duct area is A_1, or from the right, where the area is A_2. Such symmetry often occurs in acoustics.

We have seen that a simple change in area can reduce the sound transmission down a duct, but most practical sound mufflers make use of more than one change in area.

A single expansion-chamber 'silencer'

This simple muffler is a model of those in use in car 'silencers' and industrial detuners. It consists of inlet and outlet pipes with cross-sectional area A_1, and an expansion chamber between them of cross-sectional area A_2 and length l.

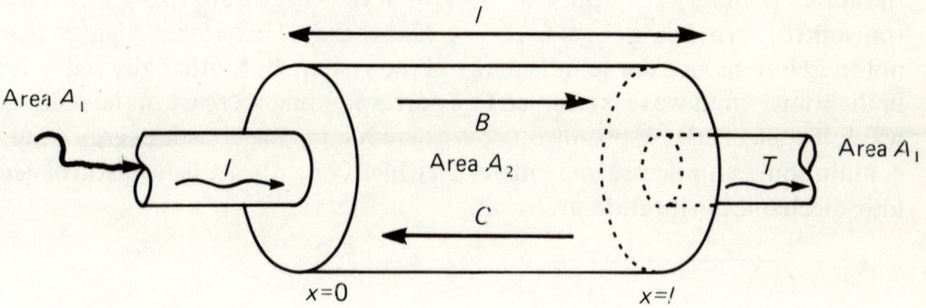

Fig. 3.2 — A single expansion chamber 'silencer'.

We will choose our co-ordinate system so that the first area change occurs at $x = 0$. The second occurs at $x = l$. We can write

$$p' = I\,e^{i\omega(t - x/c)} + R\,e^{i\omega(t + x/c)} \qquad \text{in } x < 0$$

$$p' = B\,e^{i\omega(t - x/c)} + C\,e^{i\omega(t + x/c)} \qquad \text{in } 0 < x < l$$

$$p' = T e^{i\omega(t - x/c)} \qquad \text{in } l < x.$$

Then the conditions of continuity of mass flux and pressure (or energy flux) at $x = 0$ give

$$A_1(I - R) = A_2(B - C)$$

$$I + R = B + C.$$

Conditions at $x = l$ lead to

$$A_1 T\, \mathrm{e}^{-i\omega l/c} = A_2(B\, \mathrm{e}^{-i\omega l/c} - C\, \mathrm{e}^{i\omega l/c})$$

$$T\, \mathrm{e}^{-i\omega l/c} = B\, \mathrm{e}^{-i\omega l/c} + C\, \mathrm{e}^{i\omega l/c}.$$

These algebraic equations when solved for R and T give

$$R = \frac{\left(\dfrac{A_1}{A_2} - \dfrac{A_2}{A_1}\right) Ii \sin \dfrac{\omega l}{c}}{2 \cos \omega l/c + i\left(\dfrac{A_1}{A_2} + \dfrac{A_2}{A_1}\right) \sin \dfrac{\omega l}{c}}$$

and

$$T = \frac{2I\, \mathrm{e}^{i\omega l/c}}{2 \cos \omega l/c + i\left(\dfrac{A_1}{A_2} + \dfrac{A_2}{A_1}\right) \sin \omega l/c}. \tag{3.6}$$

It is immediately apparent that

$$|R|^2 + |T|^2 = |I|^2;$$

the acoustic energy is conserved, the sum of the energies in the reflected and transmitted wave being equal to the incident energy. This simple 'silencer' does not therefore reduce the sound energy in the system, but rather any reduction in the transmitted wave is coupled to a corresponding increase in the reflected wave. If a 'silencer' is to produce an overall reduction in acoustic energy it must contain some sound absorbing material, which converts acoustic disturbances into mechanical vibration and heat.

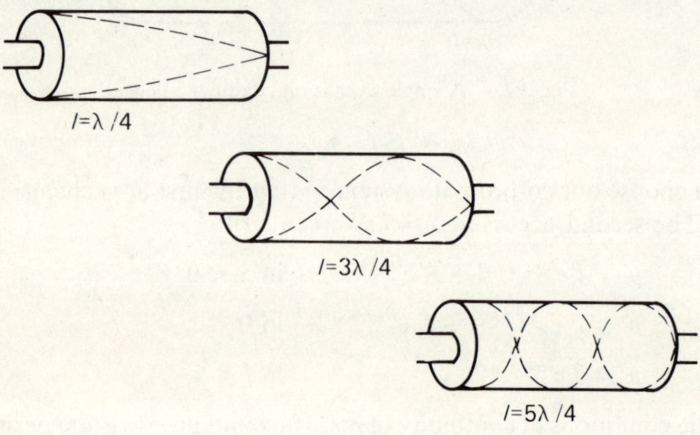

Fig. 3.3 — Duct lengths for maximum transmission loss.

The transmission loss is

$$L_T = 10\log_{10}\left(\frac{|I|^2}{|T|^2}\right);$$

after some algebra it can be shown that

$$L_T = 10\log_{10}\left[1 + \frac{1}{4}\left(\frac{A_1}{A_2} - \frac{A_2}{A_1}\right)^2 \sin^2\left(\frac{\omega l}{c}\right)\right]. \tag{3.7}$$

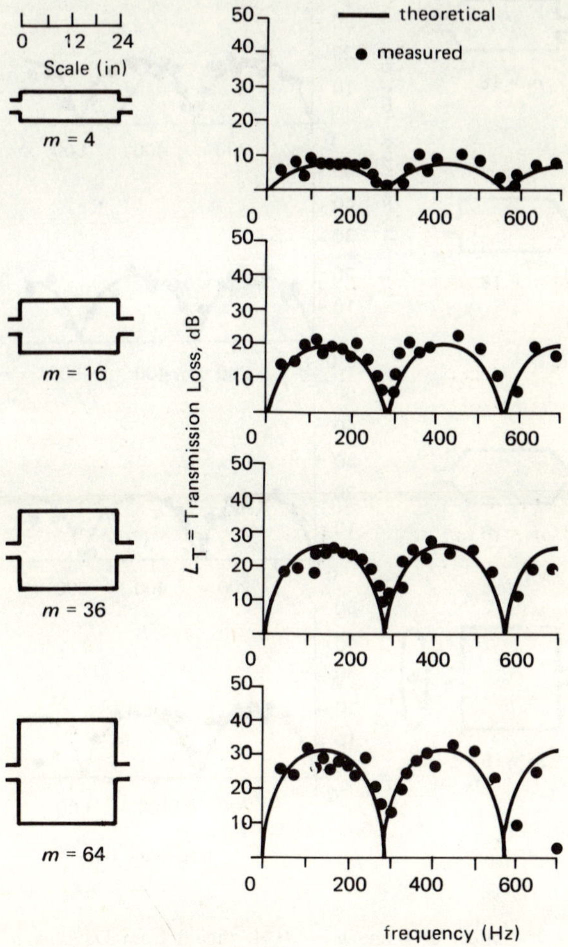

Fig. 3.4 — The effect of expansion ratio $m = A_2/A_1$. Comparison of theoretical and experimental attenuation characteristics for single expansion chamber mufflers. The theoretical curve is given by equation (3.7) (from Davis *et al.* NACA Report 1192 (1954)).

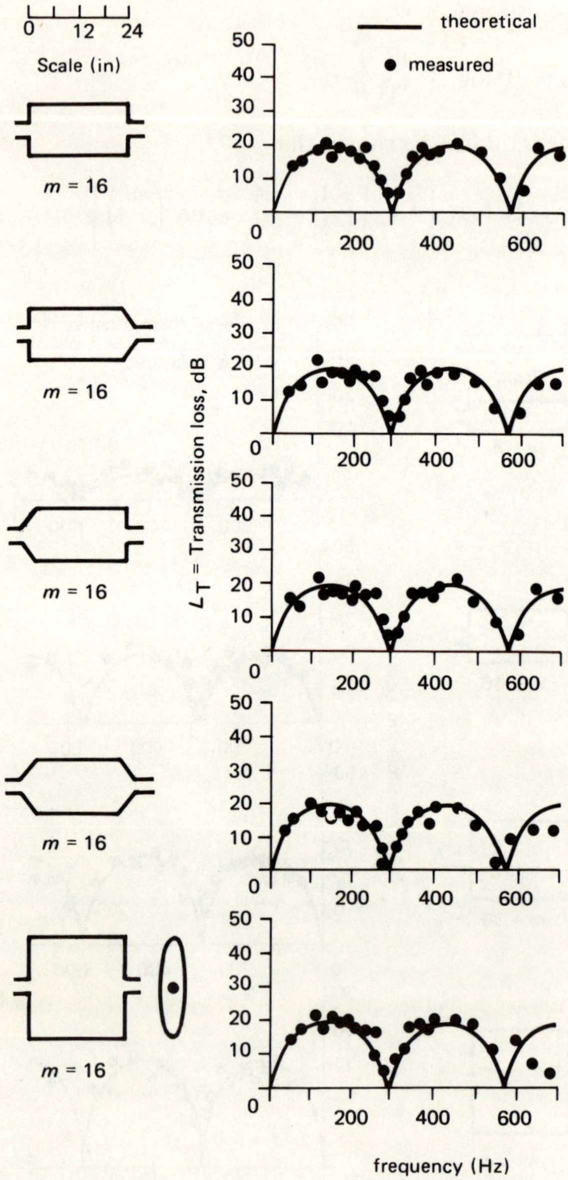

Fig. 3.5 — The effect of shape $m = A_2/A_1$ (again from Davis *et al.* NACA Report 1192 (1954)).

The transmission loss is maximum at frequencies for which

$$\sin \omega l/c = \pm 1 \quad \text{i.e.} \quad \omega = \frac{\pi c}{2l}, \frac{3\pi c}{2l}, \frac{5\pi c}{2l} \ldots \text{etc.}$$

but zero when $\sin \omega l/c = 0$ i.e. when $\omega = \pi c/l$, $2\pi c/l$, $3\pi c/l \ldots$ etc. These frequencies respectively correspond to the $\frac{1}{4}$ and $\frac{1}{2}$ wavelength resonances for a duct of length l, because their wavelengths are $\lambda = 4l$, $4l/3$, $4l/5 \ldots$ for maximum transmission loss and $\lambda = 2l$, l, $2l/3$ etc. for zero transmission loss. The muffler should be 'tuned' so that the maximum attenuation occurs at the dominant frequencies of the oncoming sound. Figures 3.4 and 3.5 show that this simple theory agrees well with experimental results. The discrepancy between the experiment and the theory at high frequencies for the large diameter mufflers is because the sound wave becomes 'three-dimensional' when the wavelength of the sound is comparable with the diameter and the sound field is not restricted to the simple plane axial modes of propagation.

Experiments also show that the geometrical shape of the duct is not important (provided the area change occurs in a distance short in comparison with the wavelength).

3.2 HIGHER ORDER MODES

It was stated in section 3.1 that only one-dimensional waves propagate in a pipe whose width is small in comparison with the wavelength of the sound. This may be verified by solving the wave equation in the pipe.

$$\frac{1}{c^2}\frac{\partial^2 p'}{\partial t^2} - \nabla^2 p' = 0. \tag{3.8}$$

As an illustration we will consider sound of frequency ω in a rigid walled duct of square cross-section with sides of length a. Then solutions to equation (3.8) can be found by considering separable functions of the form

$$p'(\mathbf{x}, t) = f(x_1)g(x_2)h(x_3)\,\mathrm{e}^{i\omega t}, \tag{3.9}$$

where f is a function of x_1 only, g a function of x_2, and h a function of x_3, the axial direction. Substitution for p' into the wave equation gives

$$\frac{f''}{f} = -\frac{g''}{g} - \frac{h''}{h} - \frac{\omega^2}{c^2}. \tag{3.10}$$

f'' denotes the second derivative of f with respect to its argument x_1. The usual procedure for determining the form of a separable solution can now be

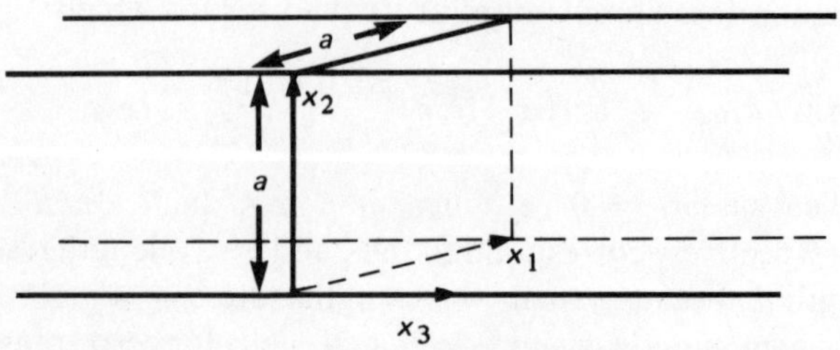

Fig. 3.6 — A duct of square cross-section with sides of length a.

applied. We note that the left-hand side of equation (3.10) is a function of x_1 only, while the right-hand side only depends on x_2 and x_3 and so we must conclude that, since they are equal, both sides must be constant. We will write this constant as $-\alpha_1^2$ and hence

$$\frac{f''}{f} = -\alpha_1^2,$$

so that $f = A_1 \cos \alpha_1 x_1 + B_1 \sin \alpha_1 x_1$. Since the walls are rigid the normal particle velocity must vanish. This condition of no flow into the walls gives

$$\frac{\mathrm{d}f}{\mathrm{d}x_1} = 0 \quad \text{on } x_1 = 0, a.$$

Hence we can deduce that B_1 is zero and $\alpha_1 = m\pi/a$ for some integer m. Therefore

$$f(x_1) = A_1 \cos\left(\frac{m\pi x_1}{a}\right).$$

Similar reasoning gives $g(x_2) = A_2 \cos(n\pi x_2/a)$ for some integer n. Then (3.10) simplifies to

$$\frac{\mathrm{d}^2 h}{\mathrm{d}x_3^2} + \left(\frac{\omega^2}{c^2} - \frac{\pi^2}{a^2}(m^2 + n^2)\right)h = 0, \tag{3.11}$$

and hence

$$h(x_3) = A_{mn}\,\mathrm{e}^{-ik_{mn}x_3} + B_{mn}\,\mathrm{e}^{ik_{mn}x_3}, \tag{3.12}$$

where A_{mn}, B_{mn} are constants and

$$k_{mn} = \sqrt{\frac{\omega^2}{c^2} - \frac{\pi^2}{a^2}(m^2 + n^2)}. \tag{3.13}$$

The axial phase speed $c_p = \omega/k_{mn}$ is now a function of the mode number and therefore the propagation of a group of waves will cause them to disperse.

Equation (3.13), the relation between frequency and wave number is called the dispersion equation. The pressure perturbation in the (m, n) mode therefore has the form

$$p'(\mathbf{x}, t) = \cos\left(\frac{m\pi x_1}{a}\right)\cos\left(\frac{n\pi x_2}{a}\right)[A_{mn}\,\mathrm{e}^{i(\omega t - k_{mn}x_3)} + B_{mn}\,\mathrm{e}^{i(\omega t + k_{mn}x_3)}]. \quad (3.14)$$

A general solution to the wave equation in the duct is given by a superposition of modes of this type. When k_{mn} is real, (3.14) represents travelling waves propagating down the x_3 axis, but not usually at the speed of sound c as the dispersion equation (3.13) makes clear. If however $\omega a < c\pi(m^2 + n^2)^{\frac{1}{2}}$, k_{mn} is purely imaginary, and the strength of the mode varies exponentially with distance along the pipe. Such disturbances are evanescent. The momentum equation shows that the pressure perturbation and particle velocity, $\mathbf{v}$, are 90° out of phase in an evanescent mode. The mean value of the intensity, $\overline{p\mathbf{v}}$, is therefore zero and there is no net transport of energy. Equation (3.14) demonstrates that the plane wave mode ($m = n = 0$) can always propagate, but that higher order modes only propagate if $\omega > c\pi/a$. Hence whenever $2a$ is less than the wavelength only plane waves propagate, and all the other modes are non-propagating or 'cut-off'.

3.3 PIPES OF VARYING CROSS-SECTION

Consider a rigid pipe whose cross-sectional area A is a function of axial distance, x. If the pipe diameter is small in comparison with both the acoustic wavelength and the length scale over which the cross-sectional area changes,

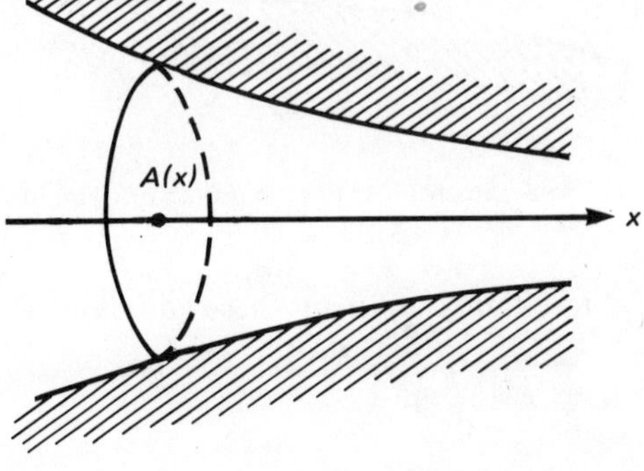

Fig. 3.7 — A pipe of cross-sectional area $A(x)$.

most particle motions are longitudinal. Then the particle velocities in directions perpendicular to the axis are negligible in comparison with those along the axis, and to a good approximation the pressure, density and axial particle velocity, u, are functions of x and t only. Conservation of mass for the shaded control volume gives

$$A\frac{\partial\rho}{\partial t} = -\rho_0\frac{\partial}{\partial x}(uA), \tag{3.15}$$

and the linearised momentum equation is

$$\rho_0\frac{\partial u}{\partial t} = -\frac{\partial p'}{\partial x}. \tag{3.16}$$

Exercise: Derive this momentum equation from a combination of the energy equation

$$A\frac{\partial e}{\partial t} + \frac{\partial}{\partial x}(IA) = 0$$

and equation (3.15).

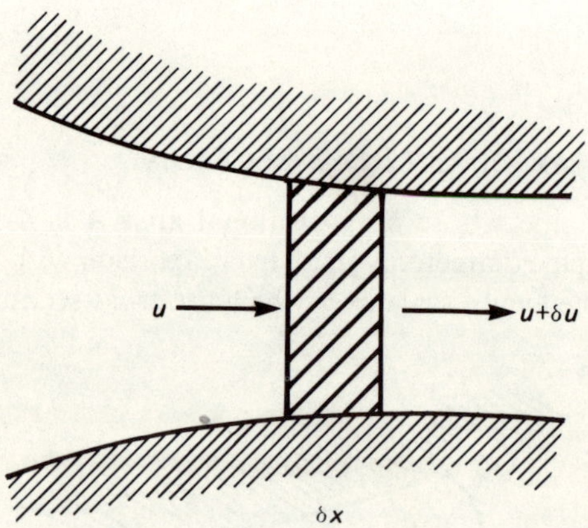

Fig. 3.8 — One dimensional flow in a duct of changing area.

Using $p' = c^2\rho'$ to eliminate ρ from these equations leads to

$$\frac{A}{c^2}\frac{\partial^2 p'}{\partial t^2} = \frac{\partial}{\partial x}\left(A\frac{\partial p'}{\partial x}\right) \tag{3.17}$$

a modified wave equation. For some geometries this equation has simple solutions. For example, for an 'exponential horn' $A(x) = A_0\,e^{\alpha x}$; A_0 is the cross-sectional area of the horn at $x = 0$ and α is called the flare constant. For such an area variation equation (3.17) simplifies to

$$\frac{1}{c^2}\frac{\partial^2 p'}{\partial t^2} = \frac{\partial^2 p'}{\partial x^2} + \alpha\frac{\partial p'}{\partial x}.$$

The pressure perturbation in sound waves of frequency ω then has the form

$$p'(x, t) = \mathrm{e}^{-\alpha x/2}\{A\,\mathrm{e}^{i(\omega t - kx)} + B\,\mathrm{e}^{i(\omega t + kx)}\}, \tag{3.18}$$

where $k = \sqrt{(\omega^2/c^2) - (\alpha^2/4)}$. Disturbances with $\omega > \alpha c/2$ propagate and the pressure but not the energy flux attenuates during propagation, while lower frequency modes are 'cut-off'. Of course practical devices do not have such a simple shape but even so real horns and other wind instruments do show the same characteristics of having a cut-off frequency which depends on the horn flare.

EXERCISES FOR CHAPTER 3

The sound speed and mean density of air and water may be taken as 340 m/s, 1.2 kg/m^3 and 1450 m/s, 10^3 kg/m^3 respectively.

1. A simple expansion chamber 'silencer' is to be used to reduce the sound from a 4-stroke, 4-cylinder petrol engine operating at 4000 rpm. Calculate the minimum length of expansion chamber that will effectively silence the fundamental frequency.

2. A straight-walled vessel has a neck of length l and cross-sectional area A_1 and a bulb of length d, cross-sectional area A_2. Show that for disturbances of frequency ω the pressure perturbation at the neck, p', is related to Q the mass flux into the vessel by

$$p' = \frac{-icQ}{A_1}\frac{A_1 - A_2\tan\alpha\tan\beta}{A_1\tan\alpha + A_2\tan\beta},$$

where $\alpha = \omega l/c$, $\beta = \omega d/c$. When A_1 is much smaller than A_2 and the length of the vessel is much less than the wavelength, show that this relationship simplifies to

$$p' = \frac{-ic^2Q}{\omega V}\left(1 - \frac{\omega^2 Vl}{c^2 A_1}\right), \quad \text{with } V = A_2 d.$$

[Note: For frequencies $\omega \sim c\sqrt{A_1/Vl}$ small pressure changes produce a large mass flux, such a device is called a Helmholtz resonator; see Chapter 6.]

3. A pump that employs the water hammer principle to raise water from a stream to a hillside residence is illustrated below.

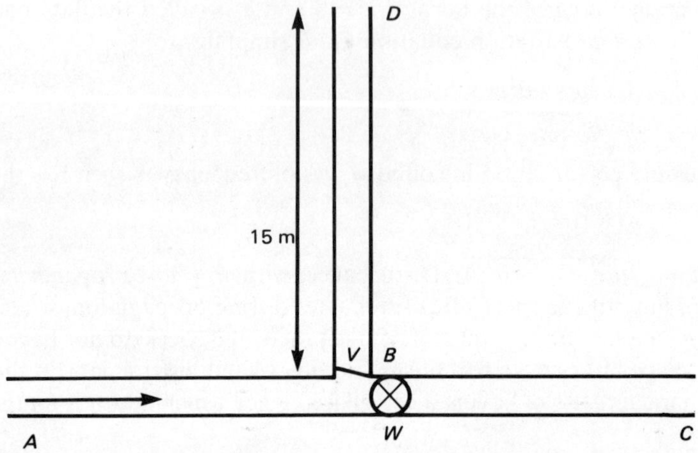

Fig. 3.9 — The water hammer pump.

Initially the valve W is open and water flows through the main pipe ABC. The pipe AB is long and has a cross-sectional area 10^{-2} m^2. The supply pipe BD is of length 15 m and cross-sectional area 10^{-3} m^2. BD is initially full of stationary water which is supported by a non-return valve V of negligible inertia. At all times the water pressure at D is equal to atmospheric pressure.

When the flow in the main pipe reaches a speed of 1 m/s the valve W closes abruptly to block the flow. In response to the closure of the valve W, pressure builds up to a value large compared with the hydrostatic pressure in the pipes, opening the valve V and generating pressure waves which retard the flow in the main pipe and force water through into the supply pipe. When V is open it offers no resistance to the flow and the junction between AB and BD may be modelled as a region of uniform pressure. Describe qualitatively the subsequent fluid motion and pressure waves. Determine how much water is delivered to the residence during the first short period of steady flow in the supply pipe.

4. Show that pressure disturbances in a conical horn have the form

$$p'(r, t) = \frac{f(t - r/c)}{r} + \frac{g(t + r/c)}{r}.$$

A curved diaphragm vibrates at a frequency of 1 kHz in a conical horn and produces an outward propagating spherical wave. If the SPL 5 cm from the apex of the cone is 100 dB, what is the peak particle velocity there?

Chapter 4

Sound Waves Incident on a Flat Surface of Discontinuity

4.1 NORMAL TRANSMISSION FROM ONE MEDIUM TO ANOTHER

When a sound wave crosses an interface between two different fluids some of the acoustic energy is usually reflected. Consider a wave of pressure I and frequency ω incident from the left onto a plane discontinuity that lies normal to the direction of wave travel. On the left the fluid has mean density ρ_0 and sound speed c_0, while on the right the fluid has mean density ρ_1 and sound speed c_1. Only a transmitted wave of pressure T escapes through the interface and a wave may also be reflected from the interface to propagate in the opposite direction to the incident wave. We let the value of the pressure in this reflected wave be R, and choose the co-ordinate system so that the mean position of the interface between the fluids is the plane $x = 0$.

There are two boundary conditions that must be satisfied at all times and at all points on the plane surface separating the two media (i) the pressures on

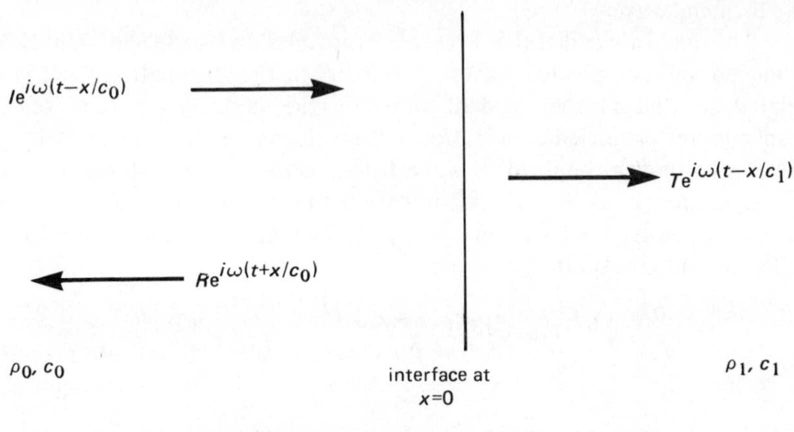

Fig. 4.1 — Normal transmission from one medium to another.

the two sides of the boundary must be equal and (ii) the particle velocities normal to the interface must be equal. These conditions immediately imply that the transmitted wave has the same frequency as the incident wave. Since the sound speeds in the two media are different this means that the two wavelengths $2\pi c_0/\omega$ and $2\pi c_1/\omega$ will not be equal.

In $x < 0$ the pressure is the sum of that in the incident and reflected wave, while in $x > 0$ it is that in the transmitted wave. Since these pressures must be equal at $x = 0$

$$I + R = T. \tag{4.1}$$

Similarly on the left the velocity is again the sum of elements; one the incident and the other the reflected wave component, while on the right the velocity is due solely to the transmitted wave:

$$\frac{I}{\rho_0 c_0} - \frac{R}{\rho_0 c_0} = \frac{T}{\rho_1 c_1}. \tag{4.2}$$

Elimination of T from equations (4.1) and (4.2) yields the result

$$R = \left\{\frac{\rho_1 c_1 - \rho_0 c_0}{\rho_1 c_1 + \rho_0 c_0}\right\} I. \tag{4.3}$$

Waves are evidently reflected from a surface of discontinuity, the amplitude of the reflected wave being $(\rho_1 c_1 - \rho_0 c_0)/(\rho_1 c_1 + \rho_0 c_0)$ times that of the incident wave.

This factor is called the reflection coefficient. The acoustic properties that determine the reflection coefficient are the acoustic impedances $\rho_1 c_1$ and $\rho_0 c_0$ of the two media and not either the density or the speed of sound separately.

The transmitted wave has pressure amplitudes $I + R$ so that the transmitted wave has a pressure equal to $(2\rho_1 c_1)/(\rho_1 c_1 + \rho_0 c_0)$ times that in the incident wave.

This factor is called the 'Pressure Transmission Coefficient'. Unlike the situation for the reflected wave, the velocity in the transmitted wave is *not* related to that in the incident wave by the same factor. The velocity transmission coefficient, the ratio of the velocity in the transmitted wave $T/\rho_1 c_1$, to that in the incident wave $I/\rho_0 c_0$, is $(2\rho_0 c_0)/(\rho_1 c_1 + \rho_0 c_0)$.

The energy flux in the incident wave per unit cross sectional area, i.e. the acoustic intensity, is (cf. section 1.8) $I^2/\rho_0 c_0$. The sum of the energy fluxes in the reflected and transmitted wave is

$$\frac{R^2}{\rho_0 c_0} + \frac{T^2}{\rho_1 c_1} = \frac{(\rho_1 c_1 - \rho_0 c_0)^2 I^2}{(\rho_1 c_1 + \rho_0 c_0)^2 \rho_0 c_0} + \frac{4\rho_1^2 c_1^2 I^2}{(\rho_1 c_1 + \rho_0 c_0)^2 \rho_1 c_1}$$

$$= \frac{I^2}{\rho_0 c_0}.$$

Hence the sum of the energy flux in the reflected and transmitted waves is equal to the incident energy flux. These simple reflection/transmission processes are ones in which the energy flow rates are conserved.

Reflection from a high and low impedance fluid

A typical example of this case is aerial sound waves incident onto a water surface. $\rho_0 c_0$ is then negligible compared with $\rho_1 c_1$ and $R = I$, $T = 2I$. The reflected wave has virtually the same strength as the incident wave, with, in addition, a pressure wave of *twice* the amplitude of the incident wave being transmitted. But the transmitted wave carries negligible energy since the velocity transmission coefficient is effectively zero.

In the opposite case, for sound in water incident onto a free surface with air (a surface that acts as a pressure release surface), the pressure reflection coefficient is then -1 and $T = 0$. Again the acoustic energy is totally reflected. The transmitted pressure field is negligible, though the perturbation velocity is high in this field transmitted across the surface into air.

4.2 SOUND PROPAGATION THROUGH WALLS

Consider now a sound wave normally incident on a plane material layer partitioning a fluid which has uniform acoustic properties, ρ_0 and c_0. Some sound will be reflected from the layer and some will be transmitted through the wall. By continuity the velocity of the wall, u, must be equal to the wave

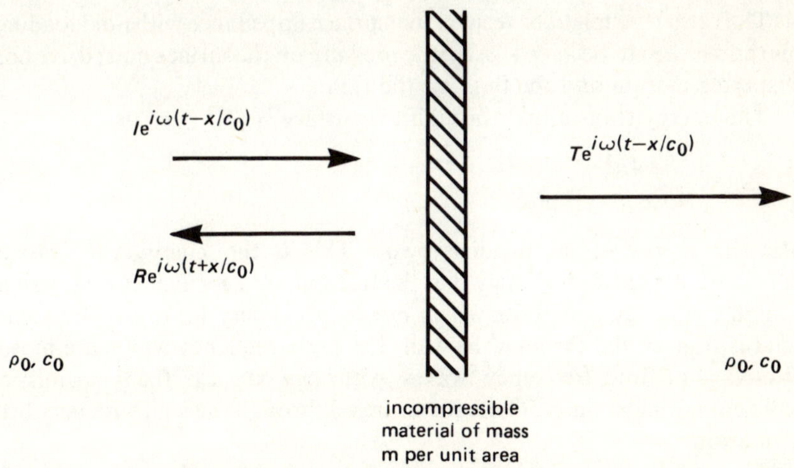

Fig. 4.2 — Normal sound transmission through a wall.

induced particle velocities on either side of the wall. On the left the particle velocity is $(I - R)\,e^{i\omega t}/\rho_0 c_0$ and on the right it is $T\,e^{i\omega t}/\rho_0 c_0$. Hence

$$u = (I - R)\frac{e^{i\omega t}}{\rho_0 c_0} = \frac{T}{\rho_0 c_0} e^{i\omega t}. \tag{4.4}$$

There will be a pressure difference across the wall in order to provide the force necessary to accelerate unit area of the surface material. The pressure on the left is

$$p_1' = (I + R)\,e^{i\omega t}$$

and on the right

$$p_2' = T\,e^{i\omega t},$$

giving a net force of $(I + R - T)\,e^{i\omega t}$ to balance the inertia of unit area of the surface material of mass m per unit area,

$$(I + R - T)\,e^{i\omega t} = m\frac{\partial u}{\partial t}. \tag{4.5}$$

Substitution for u from (4.4) shows

$$I + R - T = m\frac{i\omega T}{\rho_0 c_0}. \tag{4.6}$$

The equations (4.4) and (4.6) can be solved to find R and T in terms of I.

$$R = \frac{i\omega m}{2\rho_0 c_0 + i\omega m} I, \quad T = \frac{2\rho_0 c_0}{2\rho_0 c_0 + i\omega m} I. \tag{4.7}$$

The ratio p_1'/u might be termed the 'surface impedance with fluid loading'. This ratio is seen to be $i\omega m + \rho_0 c_0$; the pressure on the surface must drive both the surface motion and the fluid on the right.

The energy transmitted through the surface is $|T|^2/\rho_0 c_0$ or

$$\frac{4\rho_0^2 c_0^2}{4\rho_0^2 c_0^2 + \omega^2 m^2}$$

times the energy of the incident beam. This is the 'energy transmission coefficient'. We see immediately that the transmission coefficient is very small for high frequency waves for which $\omega m \gg \rho_0 c_0$; very little high frequency sound is transmitted through the wall. The high frequency waves are mostly reflected. For low frequency waves with $\omega m \ll \rho_0 c_0$, the transmission coefficient is almost unity. These waves travel through the wall with very little attenuation.

This result suggests why low frequency sounds get through all but the most massive walls, while high frequency noise is effectively stopped. For

$m = 50\,\text{kg m}^{-2}$, $\rho_0 c_0 \doteq 410\,\text{kg/m}^2\text{s}$ (air) at a frequency of 10 Hz

$$10 \log_{10}\left(\frac{4\rho_0^2 c_0^2}{4\rho_0^2 c_0^2 + \omega^2 m^2}\right) \sim -12\,\text{dB};$$

an attenuation of only 12 dB, while for a frequency of 1 kHz the reduction is about 50 dB.

It can be easily checked that again the sum of the energies in the transmitted and reflected wave is equal to that in the incident wave.

4.3 OBLIQUE WAVES INCIDENT ON A FLEXIBLE SURFACE

In this section we consider the case where an incident sound wave is inclined at an angle θ to the normal of a surface.

A plane harmonic wave propagating in the direction of increasing X in a medium with sound speed c has a pressure perturbation of the form

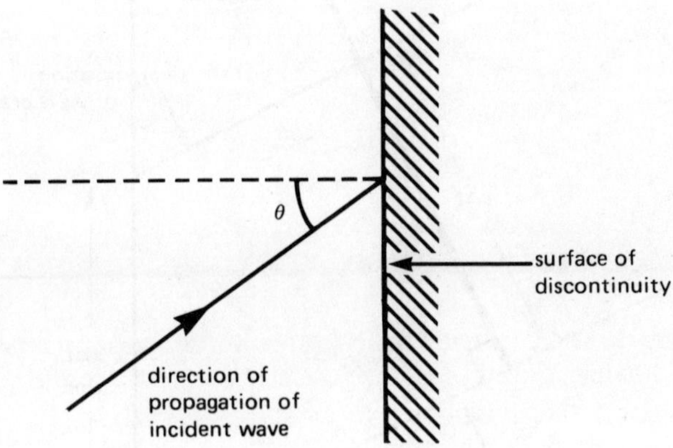

Fig. 4.3 — An obliquely incident sound wave.

$$p' = I\,\mathrm{e}^{i\omega(t - X/c)}. \tag{4.8}$$

If we choose our co-ordinate system so that the surface lies in the plane $x = 0$, then X, the co-ordinate in the direction of propagation, is given by $X = x\cos\theta + y\sin\theta$. By replacing X in (4.8) the pressure perturbation can be written as

$$p' = I\,\mathrm{e}^{i\omega(t - x\cos\theta/c - y\sin\theta/c)}. \tag{4.9}$$

The pressure on the surface $x = 0$ is

$$p' = I\,\mathrm{e}^{i\omega(t - y\sin\theta/c)}.$$

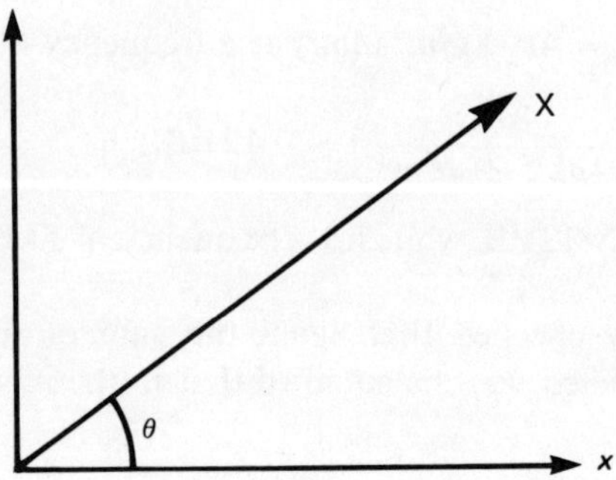

Fig. 4.4 — The co-ordinate system.

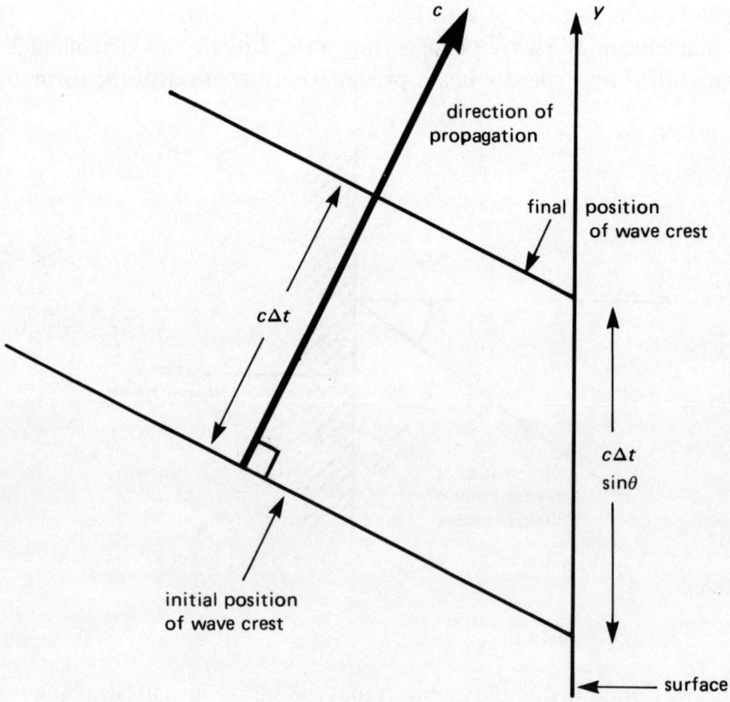

Fig. 4.5—Surface wavelength and wave speed of oblique waves.

There is a wave propagating along the surface in the y-direction with a speed $c/\sin\theta$. [Note: This is always faster than c, the speed of sound in the medium. The surface wave speed is *not* $c \sin\theta$ the component of c parallel to the surface. This sometimes causes confusion but a diagram makes it obvious.] Consider one crest (or trough) of the sound wave. It is travelling with speed c in the direction marked by an arrow, and in a time Δt it has moved a distance $c\Delta t$. Any wave travelling along the surface but wishing to keep abreast with the crest has to travel the larger distance, $c\Delta t/\sin\theta$, in time Δt i.e. has a faster speed $c/\sin\theta$.

A simple form of flexible surface is one for which the surface motion is linearly related to the surface pressure, p', and for a harmonic pressure disturbance we can write

$$p' = Zu, \tag{4.10}$$

where u is the normal surface velocity measured in the direction leading *into* the surface from the half space containing the incident wave. Z is called the surface impedance and is determined by the mechanical properties of the surface. A large variety of surfaces fall into this category. For example on a limp, light surface $p' = 0$, so that such a surface has $Z = 0$. On a rigid or very heavy surface $u = 0$, and $1/Z = 0$; Z is infinitely large. The surface response of tensioned membranes, bending plates and sound absorbent materials can also be written in this form.

Sound incident on the surface will produce a reflected wave. This wave propagates in the direction of decreasing x and it makes an angle θ', say, with the surface normal. The reflected wave can therefore be written

$$R\,\mathrm{e}^{i\omega(t + x\cos\theta'/c - y\sin\theta'/c)}.$$

The surface pressure is the sum of those in the incident and reflected wave:

$$p' = I\,\mathrm{e}^{i\omega(t - y\sin\theta/c)} + R\,\mathrm{e}^{i\omega(t - y\sin\theta'/c)}.$$

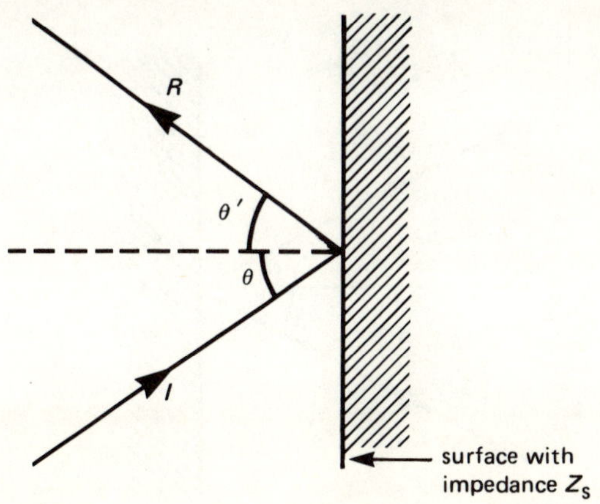

Fig. 4.6 — Reflection from a surface with impedance Z_s.

The surface velocity, which must be equal to the normal fluid particle velocity at the surface (because the fluid is always in contact with the surface) is also the sum of the elements in the two waves. The normal component in the incident wave is $I \cos\theta\, e^{i\omega(t - y\sin\theta/c)}/\rho_0 c$, while that in the reflected wave, which is propagating away from the interface, is $-R\cos\theta'\, e^{i\omega(t - y\sin\theta'/c)}/\rho_0 c$.

$$u = \frac{1}{\rho_0 c}(\cos\theta I\, e^{i\omega(t - y\sin\theta/c)} - \cos\theta' R\, e^{i\omega(t - y\sin\theta/c)}).$$

The surface response equation (4.10) therefore requires that

$$I\, e^{i\omega(t - y\sin\theta/c)} + R\, e^{i\omega(t - y\sin\theta'/c)}$$
$$= \frac{Z}{\rho_0 c}\{\cos\theta I\, e^{i\omega(t - y\sin\theta/c)} - \cos\theta' R\, e^{i\omega(t - y\sin\theta'/c)}\}. \quad (4.11)$$

We see immediately that if this is to be true for all values of y the surface wave speeds of the incident and reflected waves must match, i.e. $\theta' = \theta$; the reflected wave makes the same angle with the normal as the incident wave.

The solution of (4.11) shows that the reflected wave pressure is

$$R = \frac{Z - \rho_0 c/\cos\theta}{Z + \rho_0 c/\cos\theta} I. \quad (4.12)$$

The parameter that determines the strength of the reflected wave is the ratio between the surface impedance Z and $\rho_0 c/\cos\theta$. The quantity $\rho_0 c/\cos\theta$ can be interpreted as an 'oblique incidence' wave impedance. That interpretation is easily justified by considering a plane wave propagating through uniform fluid. Then the particle velocity in the direction of propagation is $p'/\rho_0 c$ and the velocity u in a direction inclined at an angle θ to the wave path is $\cos\theta p'/\rho_0 c$.

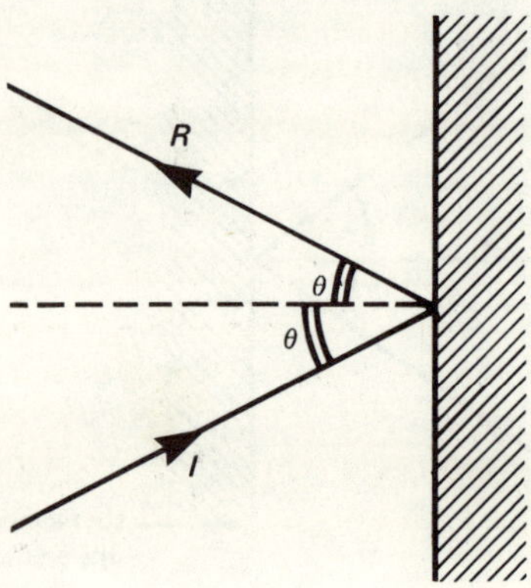

Fig. 4.7 — Angles of incidence and reflection are equal.

Therefore the ratio p'/u which we defined as the surface impedance has in the homogeneous case of no surface, only fluid, a value $\rho_0 c/\cos\theta$. The fluid offers very high impedance indeed to any wave passing tangential to the surface under study because $\cos\theta$ vanishes there.

For a limp, light surface $Z = 0$ and $R = -I$. For a heavy or rigid surface Z is very large and $R \sim I$, away from the region where $\cos\theta = 0$. Perhaps the most striking feature of equation (4.12) is that no matter how strong or massive a mechanical surface might be it is limp to waves at grazing incidence where $\cos\theta = 0$ and the reflection coefficient must be -1. The direct wave and the reflected wave then cancel exactly and there is no sound. Sound waves cannot travel parallel to a flat surface without suffering severe attenuation.

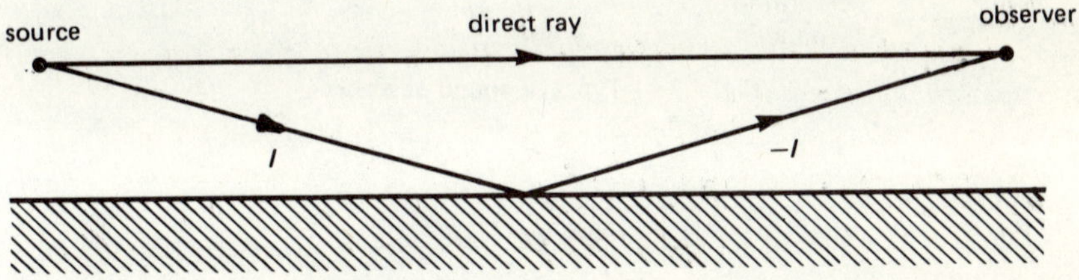

Fig. 4.8 — At grazing incidence the direct and reflected waves cancel.

Sound absorbers

Sound absorbent material is often used to reduce the acoustic energy. Practical sound absorbers are usually characterised by a surface impedance, Z, and are made of porous or perforated material. A sound wave incident on a porous material like foam or expanded polystyrene causes the air inside the pores to vibrate. The relative motion between this air and the 'solid' skeleton dissipates sound energy into heat by friction. There is also an additional energy loss due to the heat exchange between the heated compressed air (or cooled rarefied air) and the solid skeleton. A perforated or honeycomb panel exploits a slightly different method of dissipation. The holes in a perforated panel act like Helmholtz resonators (see Chapter 6). The main property of a Helmholtz resonator is that near the resonant frequency pressure perturbations at the neck produce large velocities into the resonator. These large velocities in a narrow passage result in considerable viscous dissipation. The best attenuation occurs near the resonance frequency and that can be controlled by altering the depth of the perforations and the volume of the backing cavities.

The absorption coefficient, α, is defined to be the proportion of the incident energy dissipated by an absorber. For a plane harmonic wave

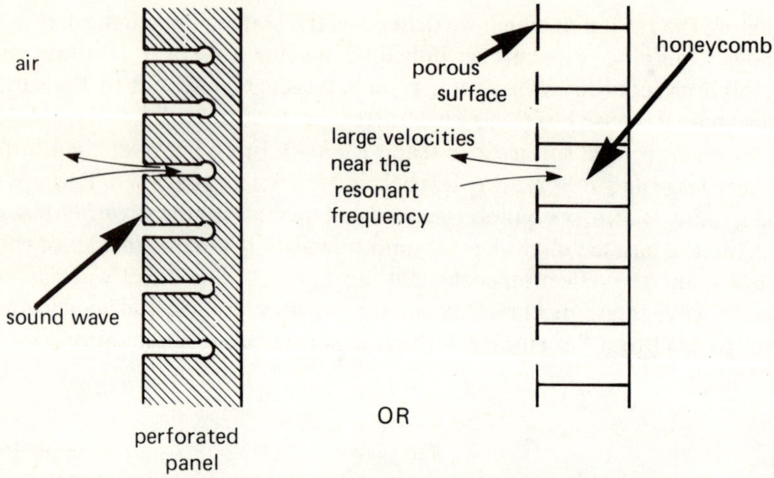

Fig. 4.9 — Types of sound absorbers.

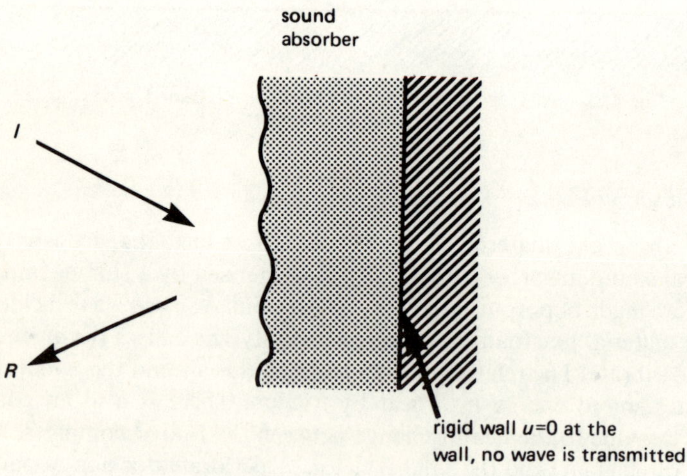

Fig. 4.10 — A sound absorber on a rigid wall.

incident at an angle θ onto a surface and producing a reflected wave R and a transmitted wave, T, α is given by

$$\alpha = 1 - \left|\frac{R}{I}\right|^2 - \left|\frac{T}{I}\right|^2. \tag{4.13}$$

If the sound absorber is backed by a rigid wall, $u = 0$ at the wall, and no acoustic energy is transmitted. Then α can be expressed directly in terms of the surface impedance

$$\alpha = 1 - \left|\frac{R}{I}\right|^2$$

$$= 1 - \left|\frac{Z - \rho_0 c/\cos\theta}{Z + \rho_0 c/\cos\theta}\right|^2 \quad \text{from (4.12).}$$

When Z is purely imaginary no acoustic energy is absorbed. The real part of Z contains the elements that cause sound absorption. The absorption coefficient depends on both the frequency of the sound waves and on the angle at which they strike the wall.

4.4 REFRACTION OF SOUND CROSSING FROM ONE FLUID INTO ANOTHER

Sound obliquely incident on a plane interface separating one homogeneous fluid from another is scattered and partitioned between transmitted and reflected elements. This is the oblique incidence form of the situation examined in section 4.1. To the left of the interface the fluid has mean density ρ_0 and sound speed c_0 while to the right the values are ρ_1 and c_1 respectively. As we have already seen the incident and reflected rays make a common angle, θ, to the surface normal because they must both have the same wave speed along the surface.

$$p' = I\,\mathrm{e}^{i\omega(t - x\cos\theta/c_0 - y\sin\theta/c_0)} + R\,\mathrm{e}^{i\omega(t + x\cos\theta/c_0 - y\sin\theta/c_0)}.$$

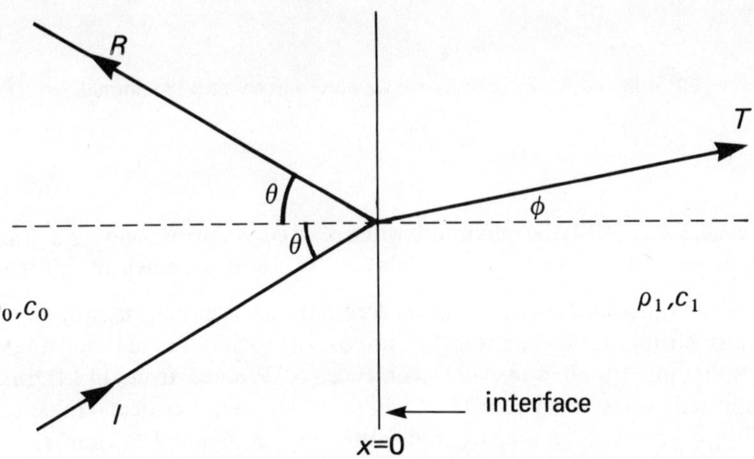

Fig. 4.11 — Sound crossing from one fluid to another.

If we denote the angle the transmitted wave makes with the normal by φ then the pressure perturbation in the transmitted wave is

$$p' = T\,e^{i\omega(t - x\cos\varphi/c_1 - y\sin\varphi/c_1)}.$$

If these waves are to match at the surface $x = 0$ the surface wave speeds must be the same. That requirement immediately yields *Snell's law.*

$$\frac{\sin\theta}{c_0} = \frac{\sin\varphi}{c_1}. \tag{4.14}$$

If $c_0 > c_1$, then $\theta > \varphi$ and transmitted rays are bent towards the normal. There are some directions into which no waves are transmitted for any angle of

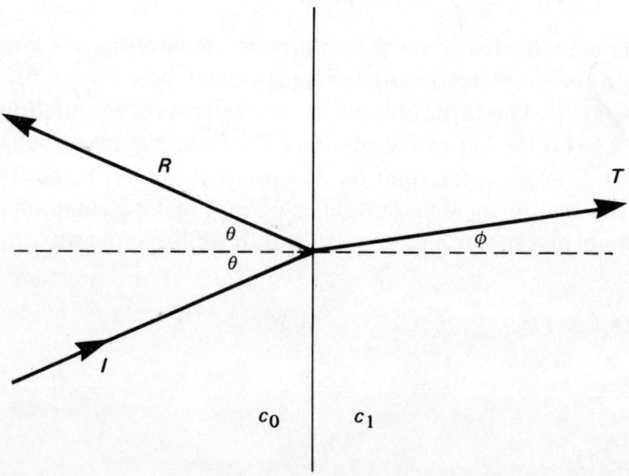

Fig. 4.12 — $c_0 > c_1$, transmitted rays are bent towards the normal.

incidence. From (4.14) the maximum value of φ, φ_{MAX} corresponds to a grazing incident ray and is given by $\varphi_{MAX} = \sin^{-1}(c_1/c_0)$ and is shown in figure 4.13

When the sound speed increases across the surface, $c_1 > c_0$, and rays are bent away from the normal. It is then impossible to find transmission angles φ corresponding to all angles of incidence, θ. We see from (4.14) that a transmitted wave only exists if $\sin\theta c_1 \leqslant c_0$. The critical angle $\theta_c = \sin^{-1}(c_0/c_1)$ produces a grazing transmitted ray. cf. figure 4.15. Larger angles of incidence produce no transmitted wave and the incident waves are then *totally internally reflected.* The resultant field due to waves at such angles of

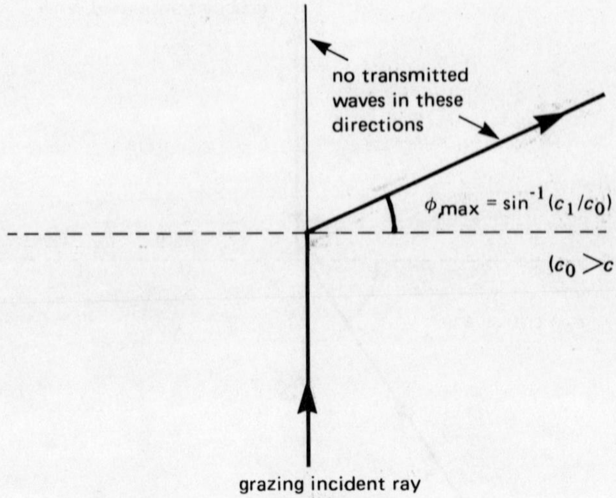

Fig. 4.13 — The limiting transmission angle.

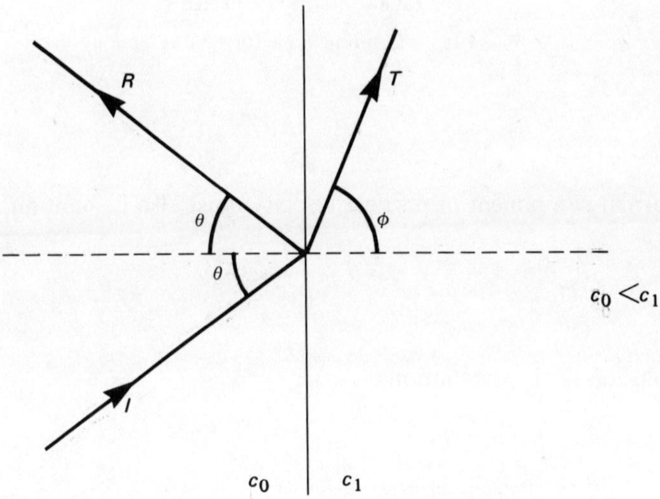

Fig. 4.14 — $c_1 > c_0$ transmitted waves are bent away from the normal.

incidence will be discussed in the next section. Here we will restrict our attention to the case where $\sin\theta c_1 < c_0$ and there is always one transmitted wave.

The pressure must be continuous across the surface of material discontinuity

$$I + R = T.$$

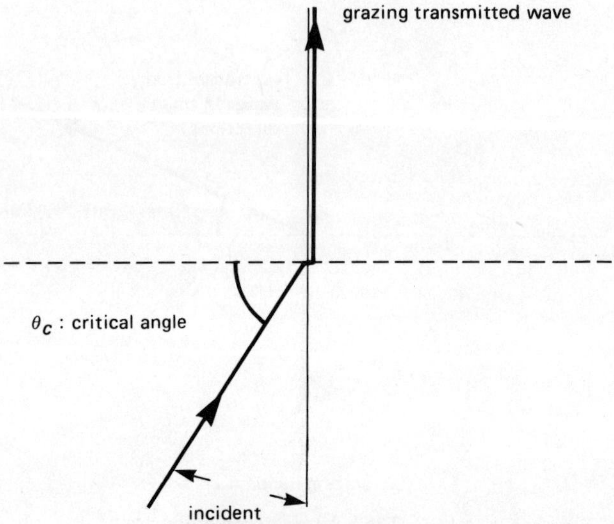

Fig. 4.15 — Limiting angle for total reflection.

The normal component of particle velocity must also be continuous

$$\frac{I}{\rho_0 c_0}\cos\theta - \frac{R}{\rho_0 c_0}\cos\theta = \frac{T}{\rho_1 c_1}\cos\varphi.$$

The solution of these equations is

$$R = \frac{\dfrac{\rho_1 c_1}{\cos\varphi} - \dfrac{\rho_0 c_0}{\cos\theta}}{\dfrac{\rho_1 c_1}{\cos\varphi} + \dfrac{\rho_0 c_0}{\cos\theta}} I \tag{4.15}$$

$$T = \frac{\dfrac{2\rho_1 c_1}{\cos\varphi}}{\dfrac{\rho_1 c_1}{\cos\varphi} + \dfrac{\rho_0 c_0}{\cos\theta}} I. \tag{4.16}$$

Again we see that the reflection and transmission coefficients are determined by the ratio between the two 'oblique incidence wave impedances'.

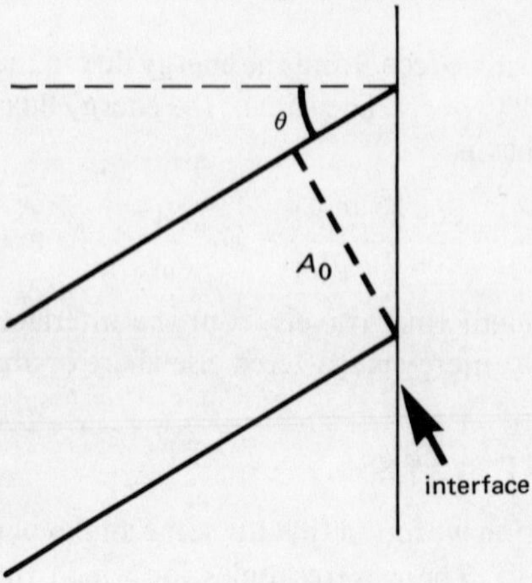

Fig. 4.16 — A 'ray tube' of cross-sectional area A_0.

We can check that the energy flux in the reflected and transmitted wave is equal to the energy flux in the incident wave. We will consider the energy conservation in a 'ray tube' of cross-sectional area A_0. The energy flux is (see section 2.1) the product of the acoustic intensity and the area normal to the direction of propagation. The incident energy is therefore $A_0|I|^2/\rho_0 c_0$.

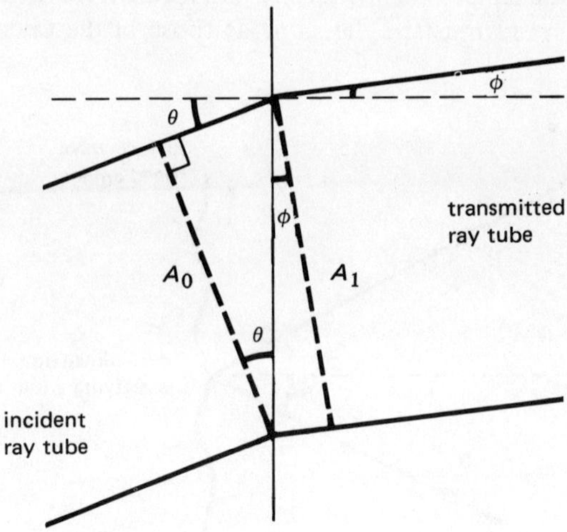

Fig. 4.17 — Area of the transmitted ray.

Similarly the reflected energy is $A_0|R|^2/\rho_0 c_0$. The rays bend and so the cross-sectional area, A_1, of the transmitted ray tube differs from A_0. From the geometry we see that

$$\frac{A_0}{\cos\theta} = \frac{A_1}{\cos\varphi},$$

and hence $A_1 = A_0 \cos\varphi/\cos\theta$ and the energy flux in the transmitted ray tube is $A_0(\cos\varphi/\cos\theta)(|T|^2/\rho_1 c_1)$. The sum of the energy fluxes in the reflected and transmitted ray tubes is

$$A_0\left\{\frac{|R|^2}{\rho_0 c_0} + \frac{\cos\varphi}{\cos\theta}\frac{|T|^2}{\rho_1 c_1}\right\} = A_0\frac{|I|^2}{\rho_0 c_0}, \quad \text{from (4.15) and (4.16).}$$

All the energy incident thus travels from the interface in the reflected and transmitted rays. No energy is scattered elsewhere or absorbed by the surface.

4.5 EVANESCENT WAVES

In the previous section we found that for some angles of incidence there can be no transmitted wave. These were angles for which the surface wave speed $c_0/\sin\theta$ was subsonic relative to the sound speed c_1, i.e. $c_1 > c_0/\sin\theta$. Disturbances to the flow in $x > 0$ must satisfy the two-dimensional wave equation:

$$\frac{1}{c_1^2}\frac{\partial^2 p'}{\partial t^2} = \frac{\partial^2 p'}{\partial x^2} + \frac{\partial^2 p'}{\partial y^2}. \tag{4.17}$$

Since these disturbances are produced by an incident wave their frequency and surface wave speed must be the same as those of the oncoming wave. The

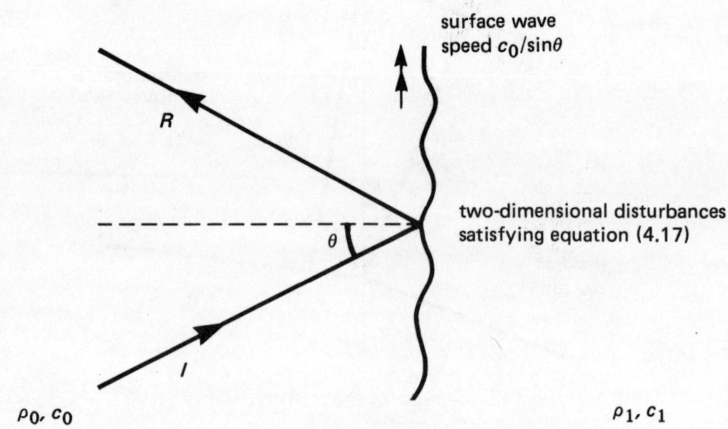

Fig. 4.18 — Reflection from an interface.

pressure on $x = 0$ due to the incident and reflected waves is

$$p' = (I + R)\, e^{i\omega(t - y\sin\theta/c_0)},$$

and so we will now look for a solution of equation (4.17) of the form

$$p' = f(x)\,e^{i\omega(t - y/U)} \quad \text{in } x > 0,$$

where U is the surface wave speed $c_0/\sin\theta$. Substitution of this form for the pressure into (4.17) leads to an ordinary differential equation for f:

$$\frac{\partial^2 f}{\partial x^2} + \frac{\omega^2}{c_1^2}\left(1 - \frac{c_1^2}{U^2}\right)f = 0. \tag{4.18}$$

There are two quite distinct types of solution to this equation.

If $U > c_1$, i.e. the surface wave speed is supersonic,

$$f = S\,e^{i\beta x} + T\,e^{-i\beta x},$$

where

$$\beta = \frac{\omega}{c_1}\left(1 - \frac{c_1^2}{U^2}\right)^{\frac{1}{2}}.$$

This is a propagating wave-type of solution. The wave travelling towards the right has the form

$$p' = T\,e^{i\omega\{t - y/U - (1 - c_1^2/U^2)^{1/2} x/c_1\}}.$$

By comparing this with the form given in (4.9) we see that this is a plane wave travelling in a direction which makes an angle φ with the normal, where

$$\frac{\sin\varphi}{c_1} = \frac{1}{U} = \frac{\sin\theta}{c_0}.$$

Snell's law again! This case was discussed in section 4.4.

If $U < c_1$, then the surface wave speed is subsonic and the solutions of (4.18) have the form

$$f = S\,e^{\gamma x} + T\,e^{-\gamma x}, \tag{4.19}$$

where

$$\gamma = \frac{\omega}{c_1}\sqrt{\frac{c_1^2}{U^2} - 1}.$$

These modes grow or decay exponentially. The solution valid in the region $x > 0$ must be finite for large positive x and hence

$$p' = T\,e^{i\omega(t - y/U) - \gamma x}.$$

This is a wave that propagates along the surface but decays exponentially away from the surface. Such waves are evanescent. Water waves have this form. They have a wave speed which is much slower than the speed of sound and the intense motion seen on the surface of the sea, $x = 0$, rapidly decays with depth.

Let us now go on to consider the energy transported by evanescent waves. The energy flux in the x-direction, through unit area, is given by the product $p'u$, the rate of working on the fluid. u can be calculated from the momentum equation

$$u = \frac{\gamma}{i\omega\rho_1} T\, e^{i\omega(t - y/U) - \gamma x}.$$

The pressure and velocity are therefore always out of phase, and the average of $p'u$ over a cycle is zero. No net energy is transported away from the surface by these evanescent waves.

We can calculate T and find the complete sound field when $c_0/\sin\theta < c_1$ by matching this evanescent wave onto the incident and reflected waves. In the region $x < 0$

$$p' = I\, e^{i\omega(t - x\cos\theta/c_0 - y/U)} + R\, e^{i\omega(t + x\cos\theta/c_0 - y/U)}$$

and in $x > 0$

$$p' = T\, e^{i\omega(t - y/U) - \gamma x}.$$

Therefore continuity of pressure leads to

$$I + R = T.$$

The normal velocity is continuous and hence

$$\frac{I\cos\theta}{\rho_0 c_0} - \frac{R\cos\theta}{\rho_0 c_0} = -i\frac{T\gamma}{\rho_1\omega}.$$

The solution of these two equations is

$$\frac{R}{I} = \frac{\dfrac{\cos\theta}{\rho_0 c_0} + \dfrac{i\gamma}{\rho_1\omega}}{\dfrac{\cos\theta}{\rho_0 c_0} - \dfrac{i\gamma}{\rho_1\omega}}$$

$$\frac{T}{I} = \frac{\dfrac{2\cos\theta}{\rho_0 c_0}}{\dfrac{\cos\theta}{\rho_0 c_0} - \dfrac{i\gamma}{\rho_1\omega}}.$$

The modulus of the reflection coefficient is unity, $|R/I| = 1$ and therefore all the incident energy is reflected. This is as we might expect since we have seen that the evanescent waves transport no energy away from the surfaces. All the energy is reflected and this phenomenon is called *total internal reflection*; only exponentially decaying disturbances are produced in the 'transmission' region.

It is generally true that 'subsonic' surface waves support an evanescent wave in the fluid and that these waves are always confined to the vicinity of the interface. The 'subsonic' surface waves need not necessarily be produced by sound waves incident from a medium with a lower sound speed, but could be caused by, for example, a vibrating structure.

In multilayer systems some energy can escape through the high sound speed region if it is thin enough. Consider an infinitely long slab of fluid of density ρ_1, sound speed c_1 of thickness h, surrounded by fluid with a lower sound speed c_0. An oncoming sound wave whose angle of incidence, θ, satisfies $c_0/\sin\theta < c_1$ produces evanescent waves within the slab. The magnitude of these waves decays exponentially, but if the slab is not too thick a weaker disturbance will penetrate the slab and a sound wave will escape from the other side. This phenomenon is called *barrier penetration*.

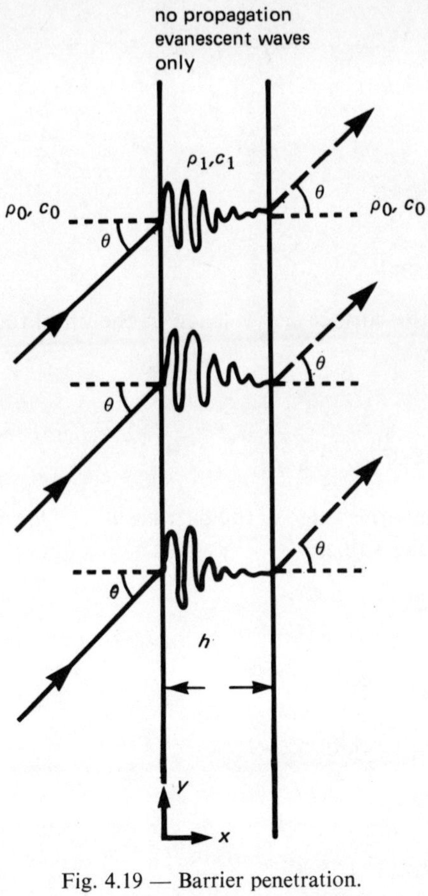

Fig. 4.19 — Barrier penetration.

Barrier penetration

Consider the system illustrated in Figure 4.19. On the irradiated side of the barrier the pressure field at $x = 0$ is the sum of the incident and reflected waves

$$p' = I\,e^{i\omega(t - y\sin\theta/c_0)} + R\,e^{i\omega(t - y\sin\theta/c_0)}$$

and the normal component of particle velocity is

$$u = \frac{\cos\theta}{\rho_0 c_0}\{I\,e^{i\omega(t - y\sin\theta/c_0)} - R\,e^{i\omega(t - y\sin\theta)/c_0)}\}.$$

These are continuous with the pressure and normal velocity within the barrier.

$$p' = A\,e^{i\omega(t - y\sin\theta/c_0)}\,e^{\gamma x} + B\,e^{i\omega(t - y\sin\theta/c_0)}\,e^{-\gamma x}$$

$$u = \frac{-\gamma A}{i\omega\rho_1}\,e^{i\omega(t - y\sin\theta/c_0)}\,e^{\gamma x} + \frac{\gamma B}{i\omega\rho_1}\,e^{i\omega(t - y\sin\theta/c_0)}\,e^{-\gamma x}$$

where

$$\gamma = \frac{\omega}{c_1}\sqrt{\frac{c_1^2}{U^2} - 1}$$

and

$$U = c_0/\sin\theta.$$

Continuity at the surface $x = 0$ leads to the amplitude relationships

$$I + R = A + B \tag{4.20}$$

$$\frac{\cos\theta}{\rho_0 c_0}(I - R) = -\frac{\gamma}{i\omega\rho_1}(A - B) \tag{4.21}$$

and continuity at the other edge of the barrier, at $x = h$, with the transmitted field that has then the values,

$$p' = T\,e^{i\omega(t - (y/c_0)\sin\theta)}$$

$$u = \frac{\cos\theta}{\rho_0 c_0}T\,e^{i\omega(t - (y/c_0)\sin\theta)}$$

leads to the amplitude relations,

$$A\,e^{\gamma h} + B\,e^{-\gamma h} = T \tag{4.22}$$

$$-\frac{\gamma}{i\omega\rho_1}(A\,e^{\gamma h} - B\,e^{-\gamma h}) = \frac{\cos\theta}{\rho_0 c_0}T. \tag{4.23}$$

The four equations are easily solved to determine T as

$$T = \frac{I}{\cosh \gamma h + \frac{1}{2} i \left(\frac{\cos\theta}{\rho_0 c_0} \frac{\rho_1 \omega}{\gamma} - \frac{\rho_0 c_0}{\cos\theta} \frac{\gamma}{\rho_1 \omega} \right) \sinh \gamma h}. \tag{4.24}$$

We see that as $\gamma h \to 0$, T tends to I, i.e. when the barrier is much thinner than γ^{-1} there is negligible impedance to the wave which then crosses the barrier unattenuated. But if the barrier is much thicker than γ^{-1} so that γh is very large,

$$T \sim \frac{2I\,\mathrm{e}^{-\gamma h}}{1 + \frac{1}{2} i \left(\frac{\cos\theta}{\rho_0 c_0} \frac{\rho_1 \omega}{\gamma} - \frac{\rho_0 c_0}{\cos\theta} \frac{\gamma}{\rho_1 \omega} \right)},$$

and the strength of the transmitted wave is exponentially small. γ^{-1} is proportional to the wavelength of sound within the barrier c_1/ω. It is the measure of the barrier thickness needed for a given degree of attenuation and can be regarded as an attenuation distance.

We have shown that there is no energy transport in a single evanescent wave. Within the barrier there are two waves A and B and it is the interaction between these two modes, that makes $\overline{p'u}$ non-zero, permitting acoustic energy to 'escape' across the barrier.

EXERCISES FOR CHAPTER 4

The sound speed and mean density of air, water and hydrogen may be taken to be 340 m/s, 1.2 kg/m^3, 1450 m/s, 10^3 kg/m^3 and 1305 m/s, 0.08 kg/m^3 respectively.

1. A steel plate of mass 400 kg/unit area separates air and water. A fluctuating normal force is applied uniformly over the plate. This force has an amplitude of 10^3 N/unit area and a frequency of 100 Hz. Determine the amplitude of vibration of the plate and the intensity of the sound radiated as plane waves into the water.

2. Find the transmission loss for a 500 Hz sound wave passing normally through a sheet of glass with a mass of 2 kg/unit area. A double glazed window is installed in the hope that it will reduce the sound transmitted through the window. If the two sheets of glass each have a mass of 2 kg/unit area and d is the distance between the two glass sheets determine the pressure transmission coefficient for a sound wave of frequency ω. Show that the normally incident 500 Hz wave is most effectively stopped by the double glazing if the gap is about 18 cm.

What is the transmission loss with such a gap? This gap is too large for the double glazing to be efficient as a heat insulator, and also many window sills are not wide enough to accommodate such a wide unit. If the windows are fitted with a gap of .014 m what is the transmission loss for the 500 Hz wave?

3. Plane sound waves in air are incident at an angle θ on a thick layer of spongy material, which can be considered to absorb all the sound entering it. The sponge has a surface impedance $Z = i\omega m + b + k/i\omega$, where ω is the frequency of the sound waves. What proportion of the incident energy is absorbed by the surface? For what frequency and incidence angle is this maximised?

4. Sound is transmitted from one medium with density ρ_0, sound speed c_0 into another with properties ρ_1, c_1. Find an expression for the angle of incidence for which all the sound energy is transmitted.

5. A sound wave is incident from air onto an infinitely long slab of hydrogen of width h. The angle of incidence is 50°. Determine an expression for the transmission loss across the hydrogen slab in terms of the thickness h. Show that, if d is greater than the wavelength of the sound in the hydrogen, the transmitted sound field is negligible. If h is 1/10th of a wavelength, what is the transmission loss?

6. A sonar receiver mounted on the surface of a moving submarine is used to monitor the sound of a distant vessel, but its performance is limited by the noise of the turbulent boundary layer that envelops the submarine. The boundary layer pressure field may be assumed to vary only in the flow direction and to convect without change with the turbulent eddies that move along the submarine's surface at 80% of the submarine's speed.

In order to improve the signal-to-noise ratio it is decided to recess the pressure sensitive surface of the sonar by a distance H into the submarine. The cavity above the sonar is filled with seawater and encased by an acoustically transparent cover which is level with the submarine surface. This modification does not change the turbulence in the boundary layer.

Determine the value of H needed to improve the signal-to-noise ratio of the sonar by 60 dB at a frequency of 1 kHz, when the submarine is moving at a speed of 15 m/s. The surface of the transparent cover may be assumed to be flat and of infinite extent, and the sonar's sensitive surface may be taken to be rigid.

Chapter 5

Ray Theory

5.1 THE RAY THEORY EQUATIONS

A plane wave has the distinctive property that its strength and direction of propagation do not vary as it propagates through a homogeneous medium while its phase increases linearly with distance:

$$p' = I\, e^{i\omega(t-\tau(x,y))}$$

where $\tau(x, y) = X/c = (x/c)\cos\theta + (y/c)\sin\theta$ and both I and θ are independent of position. The actual value of p' of course varies smoothly over a wavelength, λ.

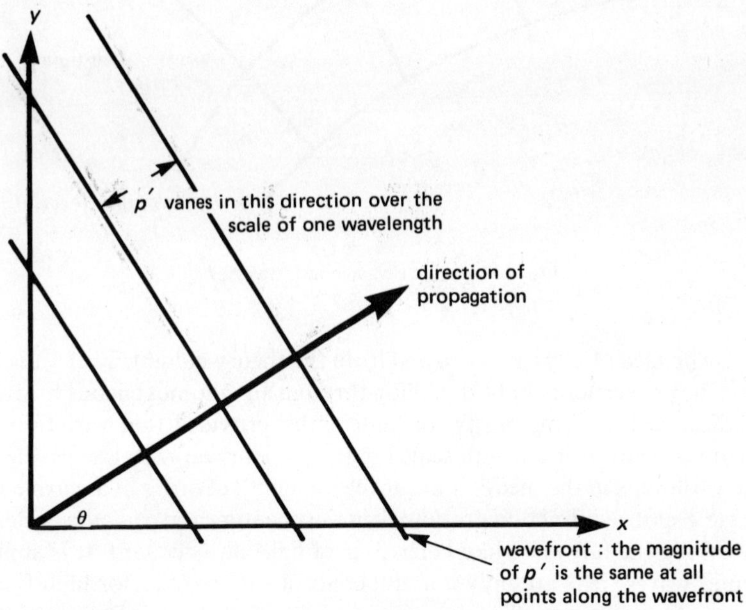

Fig. 5.1 — A plane wave.

On the other hand, a wave propagating through a region with a slowly varying sound speed will have a slightly curved wavefront, and its direction of propagation and amplitudes will vary gradually from point to point:

$$p' = I(x, y)\, e^{i\omega(t - \tau(x,y))}.$$

I varies because of gradual changes in the sound speed, and therefore an appreciable variation of I only occurs over the length scale over which the sound speed varies. The phase, τ, is a function of position and accounts for the variation of p' over the scale of wavelength. If L, the scale on which the sound speed varies, is very much greater than the wavelength λ, then I is slowly varying compared with τ and we can recognise the moving surfaces $t - \tau(x, y) = $ constant to be wavefronts. Rays are defined to be curves which are always normal to the wavefronts.

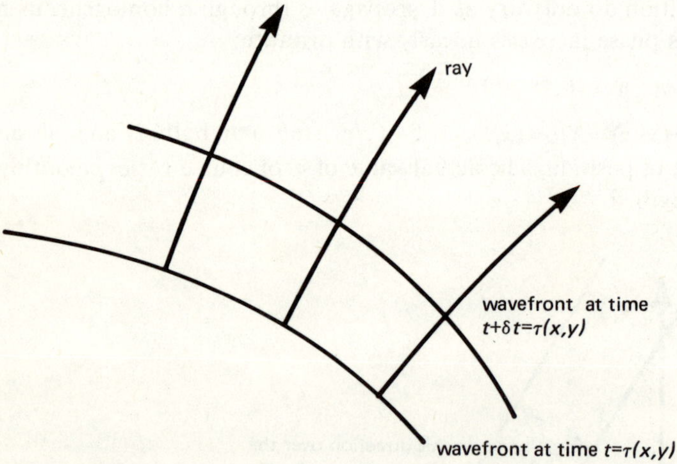

Fig. 5.2 — Short wavelength waves, $L >> \lambda$.

The idea of 'rays' is borrowed from the theory of light. There it is known that the properties of light travelling through an inhomogeneous medium can be discussed in terms of rays or light paths, provided that variations in the medium occur over a length scale long in comparison with the wavelength. If the variations in the medium occur over a length of order of a wavelength, or less (e.g. as in a diffraction grating) then the phenomenon can only be described by taking the full wave-like behaviour of light into account. It is similar for sound waves; rays are only a useful concept if $L \gg \lambda$, i.e. for high frequency waves. Then the only effect of variations in the sound speed is that the speed of propagation of the wavefront along the ray varies from point to point, and the amplitude of the wave changes slowly as the front progresses. Locally, over the scale of a wavelength the front appears plane, and so we might hope to apply some of the formulae we have derived for plane waves to the propagation of rays.

When a plane wave propagates from one medium into another with a different sound speed, we found (see equation 4.14) that $\sin\theta/c$ remained unchanged, where θ was the angle between the wavefront and the surface.

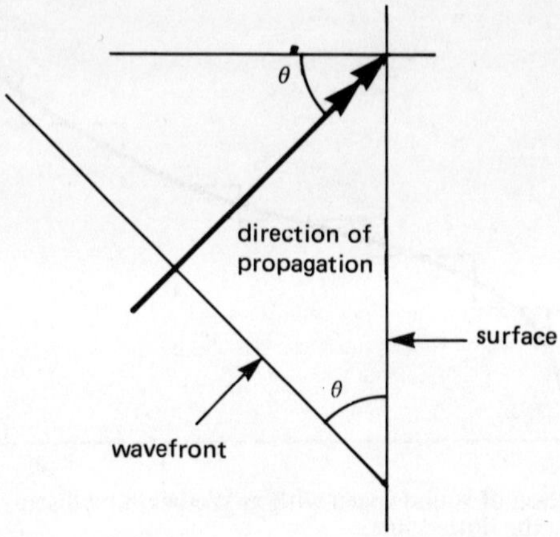

Fig. 5.3 — Plane waves.

We would like next to be able to determine what happens to the wavefronts, and hence the rays, as they pass through a stratified medium where c, the sound speed, varies continuously as a function of one variable, x say. Consider this continuous variation to be approximated by N jumps, i.e. the medium in which the sound speed varies continuously is replaced by N slabs as shown by Figures 5.4 and 5.5. The sound speed is considered to be constant within each slab, and 'jumps' at the interface between slabs. At each interface $\sin\theta/c$ remains constant for a plane wave. If we allow N, the number of slabs, to become very large we are accurately modelling the continuous sound speed profile, and so this suggests that $\sin\theta/c$ remains unchanged along a ray propagating through a stratified fluid. This result is in fact correct, although the arguments given above to derive it hardly constitute a 'proof'. A fuller derivation of this equation is given in section 5.2.

When $c = c(x)$, then Snell's law that $\sin\theta/c$ is constant along a ray gives θ, the angle between the wavefront and the y-axis, which is the same as the angle between the ray and the x-axis.

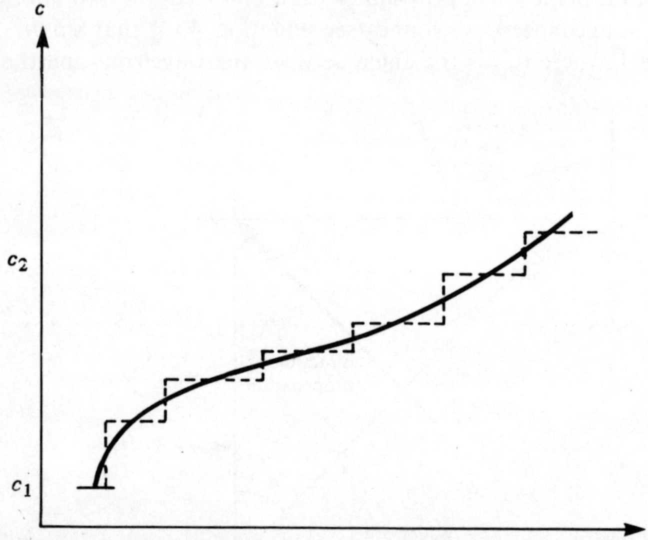

Fig. 5.4 — Variation of sound speed with x. We begin by discussing the approximation shown by the dotted line.

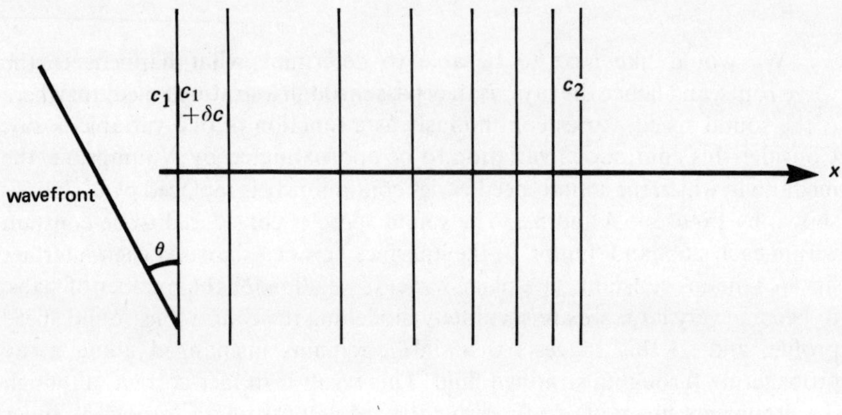

Fig. 5.5 — The sound speed is taken to be constant within each slab and to 'jump' at the interface between slabs.

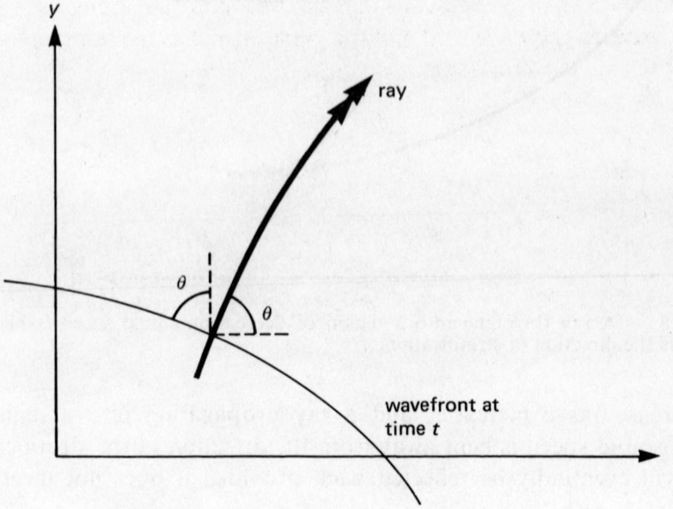

Fig. 5.6 — Rays are perpendicular to the wavefronts.

$$\sin\theta/c = \text{constant}. \tag{5.1}$$

Equation (5.1) can be used to determine the ray paths and hence to discover where sound is heard. It is immediately apparent that no sound travels into regions where $c > c_0/\sin\theta_0$, c_0 and θ_0 being the sound speed and ray angle at some reference point along the ray. In our analysis of plane waves we saw that only evanescent, or decaying, waves exist in these regions.

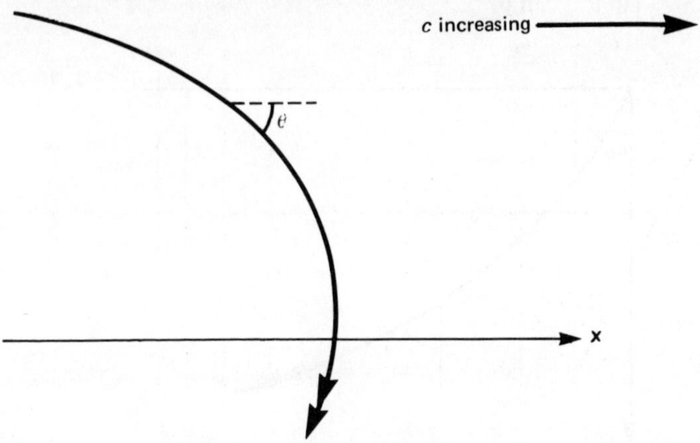

Fig. 5.7 — A ray propagating into a region of increasing sound speed is bent away from the direction of stratification.

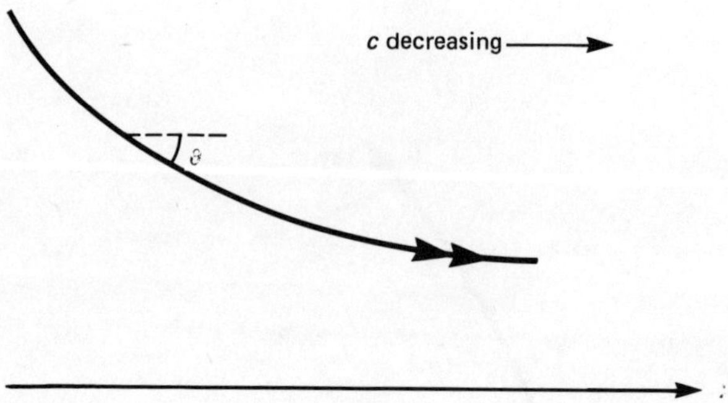

Fig. 5.8 — A ray travelling into a region of decreasing sound speed is bent towards the direction of stratification.

θ increases as c increases, and a ray propagating into a region of increasing sound speed is bent away from the direction of the stratification. The ray will eventually be reflected back provided it does not meet with obstacles in its path.

Example

In the depths of the ocean the pressure increases with depth, which causes the sound speed to increase approximately linearly with depth. What is the form of the ray paths? We will take the sound speed to be $c = c_0 + c_0\alpha x$ in $x > 0$, and we will trace the path of the ray which makes an angle θ_0 with the x-axis at (0, 0).

Let the equation of the ray path be $y = y(x)$, then along the ray

$$\frac{\sin\theta}{c} = \frac{\sin\theta_0}{c_0}$$

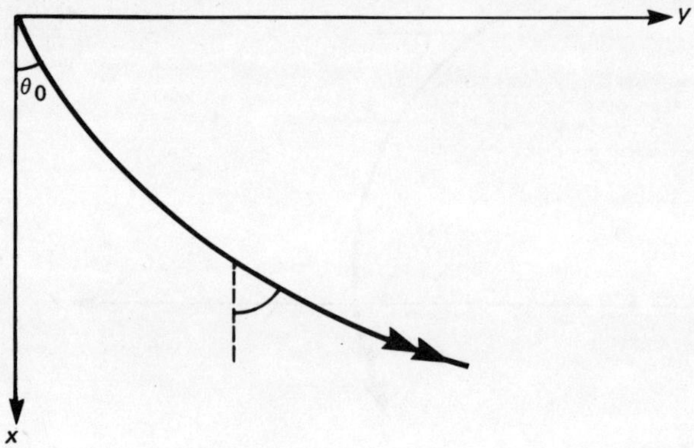

Fig. 5.9 — Ray in a stratified medium.

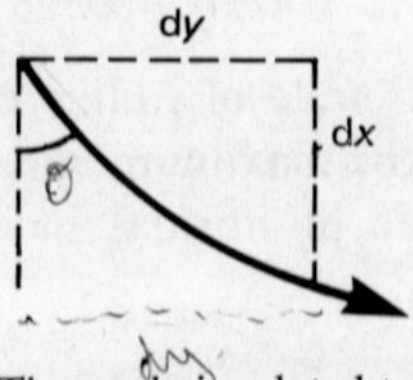

Fig. 5.10 — The angle is related to the gradient.

where

$$\sin\theta = \frac{\mathrm{d}y}{\{(\mathrm{d}y)^2 + (\mathrm{d}x)^2\}^{\frac{1}{2}}}$$

$$= \frac{\dfrac{\mathrm{d}y}{\mathrm{d}x}}{\left\{\left(\dfrac{\mathrm{d}y}{\mathrm{d}x}\right)^2 + 1\right\}^{\frac{1}{2}}},$$

and so Snell's law can be rewritten as

$$\frac{\dfrac{\mathrm{d}y}{\mathrm{d}x}}{\left\{\left(\dfrac{\mathrm{d}y}{\mathrm{d}x}\right)^2 + 1\right\}^{\frac{1}{2}}} = \sin\theta_0(1 + \alpha x).$$

We will now rearrange this equation into an integrable form. First we square it,

$$\left(\frac{\mathrm{d}y}{\mathrm{d}x}\right)^2 = \left(\frac{\mathrm{d}y}{\mathrm{d}x}\right)^2 \sin^2\theta_0(1 + \alpha x)^2 + \sin^2\theta_0(1 + \alpha x)^2.$$

When we bring all the terms involving the derivatives $\mathrm{d}y/\mathrm{d}x$ onto the left hand side and then take the square root of the equation, we obtain

$$\frac{\mathrm{d}y}{\mathrm{d}x} = \pm\frac{\sin\theta_0(1 + \alpha x)}{\{1 - \sin^2\theta_0(1 + \alpha x)^2\}^{\frac{1}{2}}}.$$

Conveniently the right-hand side is a perfect differential.

$$y = \pm\int_0^x \frac{\sin\theta_0(1 + \alpha x)\,\mathrm{d}x}{\{1 - \sin^2\theta_0(1 + \alpha x)^2\}^{\frac{1}{2}}}$$

$$= \frac{\mp 1}{\alpha\sin\theta_0}\{1 - \sin^2\theta_0(1 + \alpha x)^2\}^{\frac{1}{2}} \pm \frac{\cos\theta}{\alpha\sin\theta_0}.$$

This curve takes on a more recognizable form if we rearrange and square the equation

$$(y - \cot \theta_0/\alpha)^2 + (x + 1/\alpha)^2 = \operatorname{cosec}^2 \theta_0/\alpha^2. \tag{5.2}$$

The ray path is an arc of a circle of radius cosec θ_0/α, with its centre at $(-1, \cot \theta_0)/\alpha$. We note that the maximum value of x is $(\operatorname{cosec} \theta_0 - 1)/\alpha$ and that all rays (except the one at normal incidence when $\sin \theta_0 = 0$) are eventually reflected upwards.

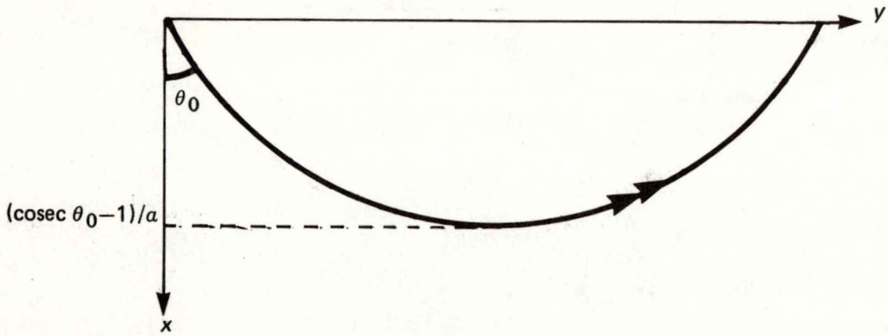

Fig. 5.11 — When the sound speed varies linearly with depth the ray paths are arcs of circles.

Ray theory can be used to determine more than just where the sound is heard. It will also give information about the sound level. To see this we will again divide our continuously varying medium up into N slabs, and consider the sound speed to be constant within each slab and to jump at the interfaces between slabs. Suppose c_{MIN} and c_{MAX} are respectively the minimum and maximum values of the sound speed in the region. Then the change in the sound speed at each interface is of the order of $(c_{\text{MAX}} - c_{\text{MIN}})/N$. From equation (4.15) the reflection coefficient for plane waves at an interface between two regions with sound speed c_0, c_1 and densities ρ_0, ρ_1 is given by

$$\frac{R}{I} = \frac{\dfrac{\rho_1 c_1}{\cos \varphi} - \dfrac{\rho_0 c_0}{\cos \theta}}{\dfrac{\rho_1 c_1}{\cos \varphi} + \dfrac{\rho_0 c_0}{\cos \theta}},$$

where $\cos \varphi = (1 - \sin^2 \theta c_1^2/c_0^2)^{\frac{1}{2}}$. Here the sound speed only changes slightly at each interface and

$$\left|\frac{R}{I}\right| \sim \frac{1}{N}.$$

Of course there are also many multiply reflected waves as sound reverberates in our laminate model of the continuous profile. If the layers are

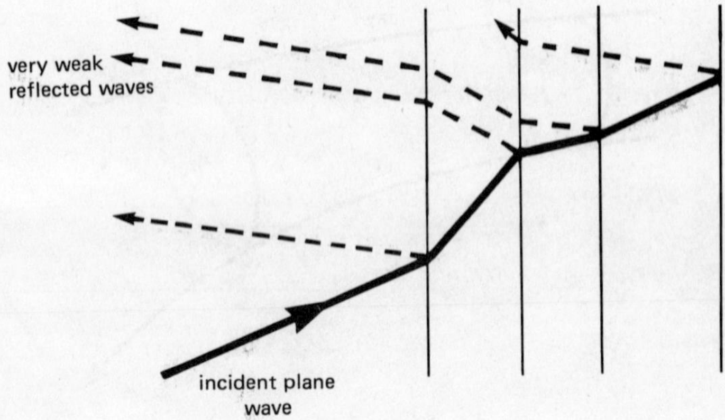

Fig. 5.12 — The continuously varying medium in divided into N slabs. In each slab the sound speed is considered to be constant and to 'jump' at the interface between slabs. By taking N to be large we see that no energy is reflected in a continuously varying medium.

thin enough for those multiply reflected waves to arrive back at an interface while still in phase with the primary wave, those reverberant elements are very important indeed. In fact it is easy then to show by considering a simple layered medium (two layers is an adequate number) that the transmission and reflection properties of 'compact' layers are determined only by the net change in acoustic properties across the layer and not in any way by the distribution of that change within the layer. But ray theory describes the opposite case where the layer is grossly 'non-compact'; each elementary slab is much thicker than the acoustic wavelength and there is a large variation in the phase of a wave between its first arrival at an interface and subsequent arrivals following multiple reflection. The energy in a sum of many wavelets of highly variable phase is negligible and that is the reason why ray theory can be understood without mention of the multiply reflected waves in our many layered approximation of the continuous profile. The energy reflected from each interface is therefore order $1/N^2$ smaller than the incident energy, and since there are N similar interfaces the total reflected energy $\sim N|I|^2/N^2 = |I|^2/N$. As N becomes very large we are accurately modelling the actual continuous variation in sound speed and no energy is reflected. This suggests there is no reflection of energy for a high frequency sound ray propagating through a medium in which the sound speed varies continuously, and from energy conservation we conclude that $\overline{p'uA}$ is constant along a ray tube, where A is the cross-sectional area of the tube. Since the wave-front is locally plane, we can write u in terms of p' to conclude that $\overline{p'^2}\,A/\rho_0 c$ remains constant along a ray tube of sound propagating through a medium in which ρ_0 and c vary slowly.

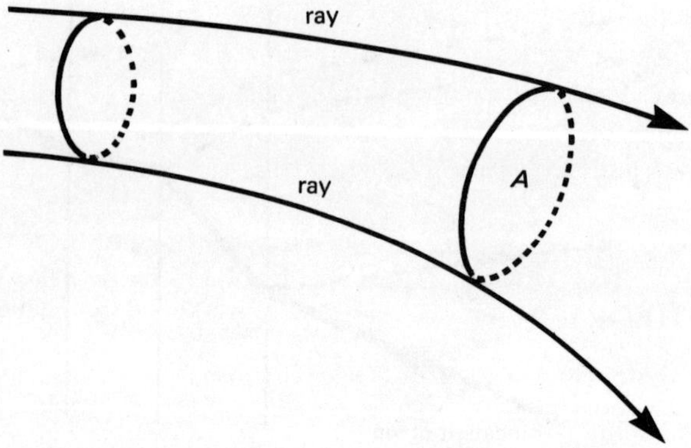

Fig. 5.13 — A ray tube.

$$\frac{A}{\rho_0 c}\,\overline{p'^2} = \text{constant.} \tag{5.3}$$

The area, A, may be calculated by tracing the paths of adjacent rays. If A decreases the pressure tends to increase as the rays focus the sound.

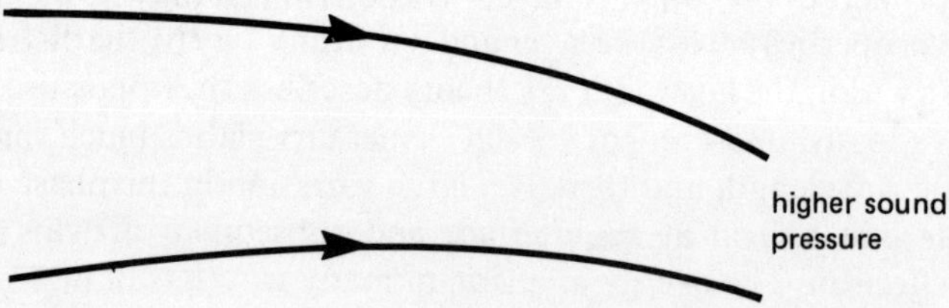

Fig. 5.14 — Converging rays.

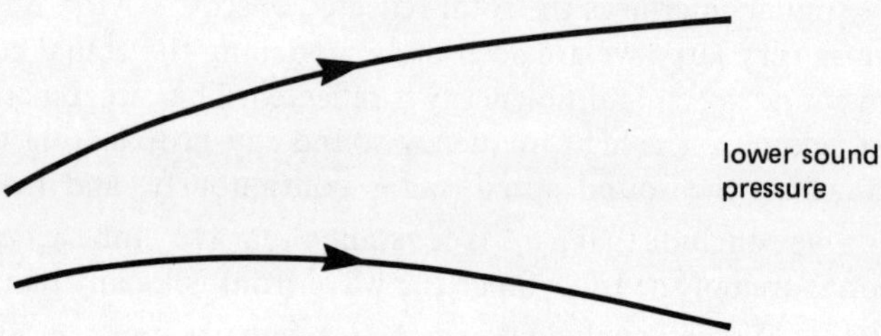

Fig. 5.15 — Diverging rays.

Conversely as rays diverge the acoustic energy is spread out over a larger area, and the amplitude of the pressure disturbance is decreased.

(5.1) and (5.3) are the main equations of ray theory. The arguments given in this section suggest that these equations are reasonable, but do not fully justify them. A full derivation of the ray theory equations is given below.

5.2 A MORE RIGOROUS DERIVATION OF RAY THEORY

The foregoing development of ray theory was based on intuitive arguments, and is not really very satisfactory. We could not, for example, easily extend those arguments to determine the next order term in the ray theory approximations. We will now give a more convincing derivation of the theory working from the equations of motion.

We must begin by extending the two dimensional equations of fluid flow discussed in section (2.1) to include the case where the sound speed and mean density vary. Conservation of mass states that

$$\frac{\partial \rho}{\partial t} = -\frac{\partial}{\partial x}(\rho_0 u) - \frac{\partial}{\partial y}(\rho_0 v), \tag{5.4}$$

where u and v are the components of particle velocity in the x and y directions respectively. The density at a point can change over a time interval not only because of the pressure perturbations in this compressible fluid but also because fluid with a different mean density may be convected into view:

$$\frac{\partial \rho}{\partial t} = \frac{1}{c^2}\frac{\partial p}{\partial t} - u\frac{\partial \rho_0}{\partial x} - v\frac{\partial \rho_0}{\partial y}.$$

If we use this equation to eliminate ρ from (5.4) we obtain

$$\frac{1}{c^2}\frac{\partial p}{\partial t} = -\rho_0\frac{\partial u}{\partial x} - \rho_0\frac{\partial v}{\partial y}.$$

The equation of x-momentum is, as before

$$\frac{\partial u}{\partial t} = -\frac{1}{\rho_0}\frac{\partial p}{\partial x}.$$

Similarly

$$\frac{\partial v}{\partial t} = -\frac{1}{\rho_0}\frac{\partial p}{\partial y}.$$

These two equations can be used to substitute for u and v in (5.4):

$$\frac{1}{c^2}\frac{\partial^2 p'}{\partial t^2} = \rho_0\frac{\partial}{\partial x}\left(\frac{1}{\rho_0}\frac{\partial p'}{\partial x}\right) + \rho_0\frac{\partial}{\partial y}\left(\frac{1}{\rho_0}\frac{\partial p'}{\partial y}\right). \tag{5.5}$$

We will look for a harmonic solution of this equation, where p' is proportional to $e^{i\omega t}$, and we will express the solution as a power series in inverse powers of ω of the form

$$p'(x, y, t) = e^{i\omega(t-\tau(x,y))} \sum_{n=0}^{\infty} (i\omega)^{-n} I_n(x, y). \tag{5.6}$$

Such a series is called a ray series. The advantage of looking for a solution of (5.5) of this form is that for high frequencies it is sufficient to consider only the first few terms of the series. We will see later that the first term satisfies the ray equations described in section 5.1.

$\tau(x, y)$ is called the phase function and surfaces $t = \tau(x, y)$ are wavefronts. Rays are defined to be curves which are everywhere normal to the wavefronts; e.g. let a ray be denoted by the parametric curve

$$x = X(s), \quad y = Y(s),$$

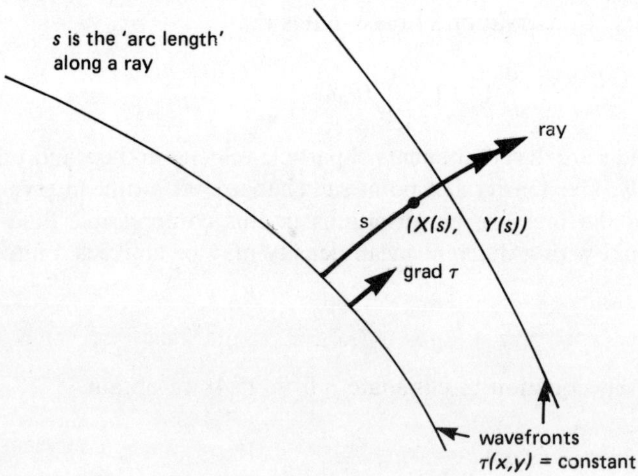

Fig. 5.16 — Ray path co-ordinates.

where s is the distance travelled along the ray path. The tangent to a ray, $(dX/ds, dY/ds)$, is always perpendicular to the surfaces $\tau(x, y) = $ constant. It is therefore always parallel to grad τ and we can write

$$\left(\frac{dX}{ds}, \frac{dY}{ds}\right) = \eta \operatorname{grad} \tau \tag{5.7}$$

for some function $\eta(x, y)$. Now $|dX/ds, dY/ds| = 1$, since $dX^2 + dY^2 = ds^2$, and η can be found by evaluating the magnitude of (5.7):

$$\left(\frac{\mathrm{d}X}{\mathrm{d}s}, \frac{\mathrm{d}Y}{\mathrm{d}s}\right) = \frac{\operatorname{grad} \tau}{|\operatorname{grad} \tau|}. \tag{5.8}$$

It is simple to evaluate the derivatives of the form of the pressure given in (5.6):

$$\frac{\partial^2 p'}{\partial t^2} = (i\omega)^2 \, \mathrm{e}^{i\omega(t-\tau)} \sum_{n=0}^{\infty} (i\omega)^{-n} I_n$$

$$\frac{\partial p'}{\partial x} = \mathrm{e}^{i\omega(t-\tau)} \sum_{n=0}^{\infty} (i\omega)^{-n} \left\{ -i\omega \frac{\partial \tau}{\partial x} I_n + \frac{\partial I_n}{\partial x} \right\}$$

$$\rho_0 \frac{\partial}{\partial x}\left(\frac{1}{\rho_0}\frac{\partial p'}{\partial x}\right) = \mathrm{e}^{i\omega(t-\tau)} \sum_{n=0}^{\infty} (i\omega)^{-n} \left[(i\omega)^2 \left(\frac{\partial \tau}{\partial x}\right)^2 I_n - i\omega \left\{ \rho_0 \frac{\partial}{\partial x}\left(\frac{1}{\rho_0}\frac{\partial \tau}{\partial x} I_n\right) + \frac{\partial \tau}{\partial x}\frac{\partial I_n}{\partial x} \right\} + \rho_0 \frac{\partial}{\partial x}\left(\frac{1}{\rho_0}\frac{\partial I_n}{\partial x}\right) \right]$$

The y derivatives are of course similar. Substitution of the correct form for the derivatives into equation (5.5) leads to

$$\sum_{n=0}^{\infty} (i\omega)^{-n} \left[(i\omega)^2 \left\{ \left(\frac{\partial \tau}{\partial x}\right)^2 + \left(\frac{\partial \tau}{\partial y}\right)^2 - \frac{1}{c^2} \right\} I_n - i\omega \left\{ \rho_0 \frac{\partial}{\partial x}\left(\frac{1}{\rho_0}\frac{\partial \tau}{\partial x} I_n\right) + \rho_0 \frac{\partial}{\partial y}\left(\frac{1}{\rho_0}\frac{\partial \tau}{\partial y} I_n\right) + \frac{\partial \tau}{\partial x}\frac{\partial I_n}{\partial x} + \frac{\partial \tau}{\partial y}\frac{\partial I_n}{\partial y} \right\} + \rho_0 \frac{\partial}{\partial x}\left(\frac{1}{\rho_0}\frac{\partial I_n}{\partial x}\right) + \rho_0 \frac{\partial}{\partial y}\left(\frac{1}{\rho_0}\frac{\partial I_n}{\partial y}\right) \right] = 0.$$

If this equation is to be true for all values of ω, the coefficient of each power of ω must vanish. Setting the coefficient of the highest power of ω, ω^2, to zero gives

$$\left(\frac{\partial \tau}{\partial x}\right)^2 + \left(\frac{\partial \tau}{\partial y}\right)^2 - \frac{1}{c^2} = 0. \tag{5.9}$$

This is called the *eikonal* equation. It determines the wavefronts and the ray paths.

The coefficient of ω must also vanish and this condition leads to the equation

$$\rho_0 \frac{\partial}{\partial x}\left\{\frac{1}{\rho_0}\frac{\partial \tau}{\partial x} I_0\right\} + \rho_0 \frac{\partial}{\partial y}\left\{\frac{1}{\rho_0}\frac{\partial \tau}{\partial y} I_0\right\} + \frac{\partial \tau}{\partial x}\frac{\partial I_0}{\partial x} + \frac{\partial \tau}{\partial y}\frac{\partial I_0}{\partial y} = 0. \tag{5.10}$$

This is a differential equation for I_0, the leading amplitude term. Equating the coefficients of the other powers of ω to zero produces equations satisfied by $I_1, I_2 \ldots$ etc. However for high frequencies only the I_0 amplitude is important, because I_1 is divided by ω, I_2 by ω^2 etc. and are negligible for high enough ω.

We will now solve the two equations (5.9) and (5.10). We begin by manipulating (5.9) into a form where it gives information about the ray paths. We will consider again the case where c is a function of only one variable, $c = c(x)$, say. Then the ray paths have a particularly simple form. The eikonal equation, (5.9) states that

$$|\text{grad } \tau| = 1/c. \tag{5.11}$$

We use this to substitute for $|\text{grad } \tau|$ in the y-component of equation (5.8) and obtain

$$\frac{1}{c}\frac{\mathrm{d}Y}{\mathrm{d}s} = \frac{\partial \tau}{\partial y}.$$

If we differentiate this with respect to s we find that

$$\frac{\mathrm{d}}{\mathrm{d}s}\left(\frac{1}{c}\frac{\mathrm{d}Y}{\mathrm{d}s}\right) = \frac{\mathrm{d}}{\mathrm{d}s}\frac{\partial \tau}{\partial y}. \tag{5.12}$$

We will now show that the right-hand side of this equation is zero

$$\frac{\mathrm{d}}{\mathrm{d}s}\frac{\partial \tau}{\partial y} = \left(\frac{\partial X}{\partial s}\frac{\partial}{\partial x} + \frac{\partial Y}{\partial s}\frac{\partial}{\partial y}\right)\frac{\partial \tau}{\partial y} \quad \text{by the Chain Rule.}$$

$$= c\left(\frac{\partial \tau}{\partial x}\frac{\partial}{\partial x} + \frac{\partial \tau}{\partial y}\frac{\partial}{\partial y}\right)\frac{\partial \tau}{\partial y} \quad \text{from equations (5.8) and (5.11).}$$

After a simple rearrangement this leads to

$$\frac{\mathrm{d}}{\mathrm{d}s}\frac{\partial \tau}{\partial y} = c\left\{\frac{\partial \tau}{\partial x}\frac{\partial}{\partial y}\left(\frac{\partial \tau}{\partial x}\right) + \frac{\partial \tau}{\partial y}\frac{\partial}{\partial y}\left(\frac{\partial \tau}{\partial y}\right)\right\}$$

$$= \frac{c}{2}\frac{\partial}{\partial y}\left\{\left(\frac{\partial \tau}{\partial x}\right)^2 + \left(\frac{\partial \tau}{\partial y}\right)^2\right\}$$

$$= \frac{c}{2}\frac{\partial}{\partial y}\left(\frac{1}{c^2}\right) \quad \text{from equation (5.9)}$$

$$= 0, \quad \text{since } c \text{ is independent of } y.$$

Hence equation (5.12) states that

$$\frac{1}{c}\frac{\mathrm{d}Y}{\mathrm{d}s} = \text{constant along ray.}$$

A diagram quickly shows that

$$\frac{\mathrm{d}Y}{\mathrm{d}s} = \sin\theta,$$

where θ is the angle between the ray and the x-axis. We have therefore shown

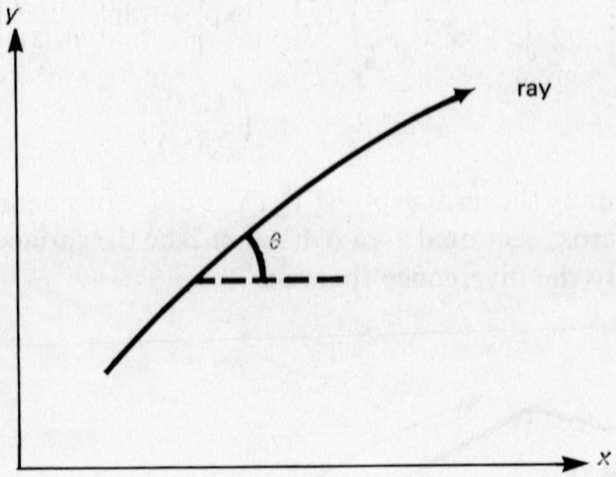

Fig. 5.17 — $\dfrac{\text{Sin}\,\theta}{c}$ is a constant.

that $\sin\theta/c$ is constant along a ray. This is the Snell's law relationship given in equation (5.1).

The intensity I_0 must satisfy equation (5.10)

$$2\left(\frac{\partial\tau}{\partial x}\frac{\partial I_0}{\partial x}+\frac{\partial\tau}{\partial y}\frac{\partial I_0}{\partial y}\right)+I_0\left(\frac{\partial^2\tau}{\partial x^2}+\frac{\partial^2\tau}{\partial y^2}+\rho_0\frac{\partial\tau}{\partial x}\frac{\partial}{\partial x}\left(\frac{1}{\rho_0}\right)+\rho_0\frac{\partial\tau}{\partial y}\frac{\partial}{\partial y}\left(\frac{1}{\rho_0}\right)\right)=0.$$

First note that

$$\frac{\partial\tau}{\partial x}\frac{\partial}{\partial x}+\frac{\partial\tau}{\partial y}\frac{\partial}{\partial y}=\frac{1}{c}\left(\frac{\partial X}{\partial s}\frac{\partial}{\partial x}+\frac{\partial Y}{\partial s}\frac{\partial}{\partial y}\right)\quad\text{from equations (5.8) and (5.11)}$$

$$=\frac{1}{c}\frac{\partial}{\partial s}\quad\text{by the Chain Rule.}$$

This enables us to rewrite the equation for I_0 in a much simpler form,

$$2\frac{\partial I_0}{\partial s}+I_0\left(c\nabla^2\tau+\rho_0\frac{\partial}{\partial s}\left(\frac{1}{\rho_0}\right)\right)=0.$$

The solution to this equation is

$$I_0(s)=I_0(s_0)\exp\left[-\frac{1}{2}\int_{s_0}^{s}\left\{c\nabla^2\tau+\rho_0\frac{\partial}{\partial s}\left(\frac{1}{\rho_0}\right)\right\}\mathrm{d}s\right]\tag{5.13}$$

where s_0 is some reference point on the ray.

We first observe that

$$\exp\left[-\frac{1}{2}\int_{s_0}^{s}\rho_0\frac{\partial}{\partial s}\left(\frac{1}{\rho_0}\right)\mathrm{d}s\right]=\exp\left[-\tfrac{1}{2}\ln\left(\frac{\rho_0(s_0)}{\rho_0(s)}\right)\right]$$

$$=\left(\frac{\rho_0(s)}{\rho_0(s_0)}\right)^{\frac{1}{2}}.$$

We can also simplify the term $\exp[-\frac{1}{2}\int_{s_0}^{s}c\nabla^2\tau\,\mathrm{d}s]$. Consider a short length δs of a ray tube of cross-sectional area $\Delta A(s)$. Let Σ be the surface of the ray tube, V its volume. By the divergence theorem

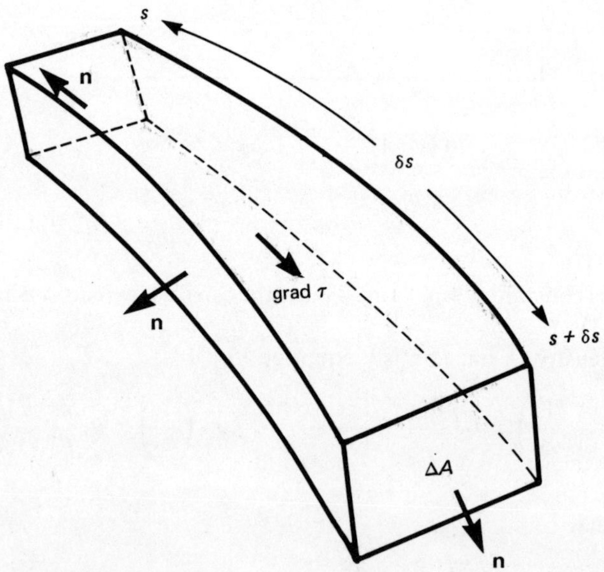

Fig. 5.18 — A ray tube.

$$\int_V \nabla^2\tau\,\mathrm{d}V=\int_\Sigma \mathbf{n}\cdot\text{grad}\,\tau\,\mathrm{d}\Sigma. \tag{5.14}$$

Now on the sides of the ray tube $\mathbf{n}$ is perpendicular to grad τ:

$$\mathbf{n}\cdot\text{grad}\,\tau=0.$$

At the ends of the tube $\mathbf{n}=\pm c\,\text{grad}\,\tau$, and $\mathbf{n}\cdot\text{grad}\,\tau=\pm c\,\text{grad}\,\tau^2=\pm 1/c$ using the eikonal equation. Therefore (5.14) states that

$$\nabla^2\tau\,\delta s\Delta A=\frac{\Delta A(s+\delta s)}{c(s+\delta s)}-\frac{\Delta A(s)}{c(s)}=\frac{\partial}{\partial s}\left(\frac{\Delta A}{c}\right)\delta s.$$

Thus

$$c\nabla^2\tau = \frac{c}{\Delta A}\frac{\partial}{\partial s}\left(\frac{\Delta A}{c}\right)$$

and

$$\exp\left[-\frac{1}{2}\int_{s_0}^{s} c\nabla^2\tau \, ds\right] = \exp\left[-\frac{1}{2}\log\frac{\Delta A(s)c(s_0)}{\Delta A(s_0)c(s)}\right]$$

$$= \left(\frac{c(s)\Delta A(s_0)}{\Delta A(s)c(s_0)}\right)^{\frac{1}{2}}.$$

Returning to equation (5.13) we see that the solution can be expressed as

$$I_0(s) = I_0(s_0)\left(\frac{\rho_0(s)c(s)}{\Delta A(s)}\frac{\Delta A(s_0)}{\rho_0(s_0)c(s_0)}\right)^{\frac{1}{2}}$$

i.e. $I_0^2(s)A(s)/\rho_0(s)c(s)$ is constant along the ray tube. This is precisely the form given for the pressure amplitude in equation (5.3).

We have therefore shown that the ray theory equations are satisfied by the leading term for the pressure in an expansion in inverse powers of ω. For high frequencies it is sufficient to consider just this first term. But the complete ray theory can, if need be, be extended by including further terms to cope with lower frequencies and regions where the first term vanishes.

5.3 UNDERWATER SOUND PROPAGATION

Water transmits sound waves far better than it does optical, radio or magnetic waves, and so sound is used extensively for underwater communication. The water is warm near the surface of the ocean but the sea temperature decreases with depth. This temperature variation means that the sound speed decreases also, stabilising at a depth of about 1000 m. Beyond that depth the sea is approximately isothermal, and a variation in sound occurs only because of pressure changes. There the increase of pressure with depth leads to an increase in the sound speed. The minimum value of the sea water sound speed is about 1480 m/s and occurs at a depth of 1000 m, but these values vary slightly from ocean to ocean, and also have seasonal variations.

We can use equation (5.1) to discuss what happens to a sound ray propagating through such a sound speed profile. Firstly we note that any ray propagating downwards will eventually come into a region where the sound speed increases with depth, and will be bent back. Similarly a ray propagating upwards will also find itself travelling into a region where the sound speed increases, and will also be refracted back (provided it does not reach the surface of the sea first). We see that a ray can thus become 'trapped' or confined within certain levels in the ocean. These regions are called *sound channels*. In them the sound can only spread out in a circle rather than over a sphere as it would in a homogeneous medium. As a result the sound decays more slowly and can be detected at far greater distances. For example, the sound produced

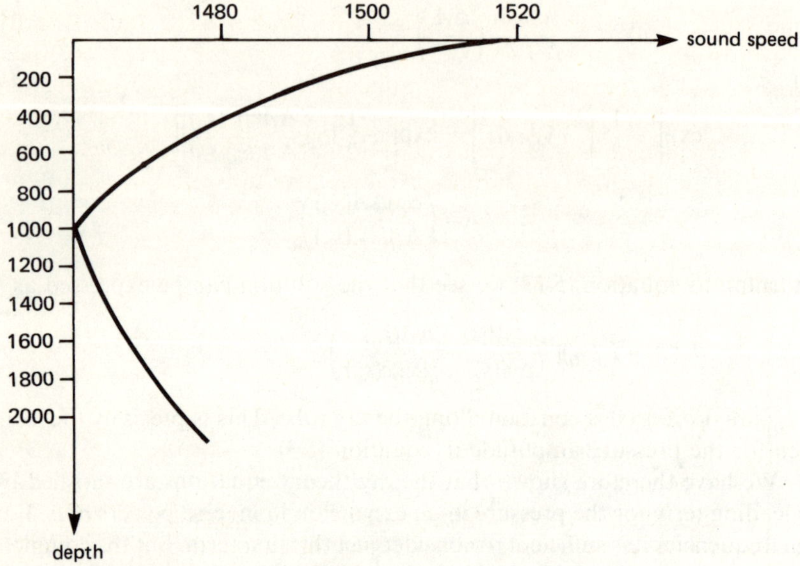

Fig. 5.19 — A typical sound speed variation with depth in the ocean.

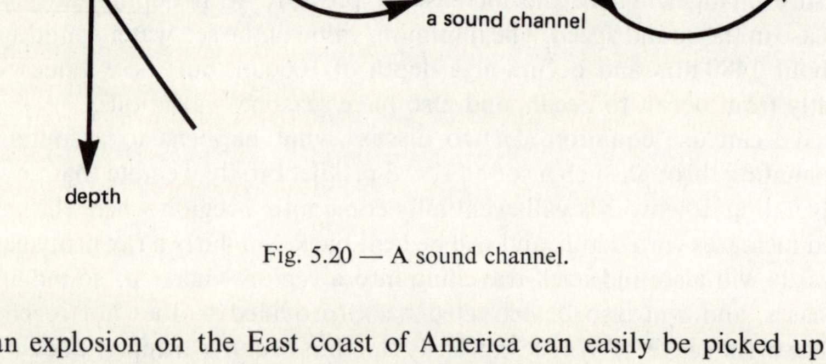

Fig. 5.20 — A sound channel.

by an explosion on the East coast of America can easily be picked up on hydrophones off the west coast of Scotland. It has also been shown that whales can detect sounds over a range of at least 2000 km.

The refraction of sound rays by anomalies of the sound speed profile in the ocean can lead to the formation of *shadow zones*. Consider for example a sound source in the ocean near a position of maximum in the sound velocity. Rays propagating upwards move into regions where the sound speed decreases and are refracted upwards. Similarly rays moving downwards are refracted downwards more steeply. It is evident from Figure 5.21 that there is a region into which no rays can penetrate. Such quiet regions are called shadow zones, and are useful to submarines avoiding acoustic detection.

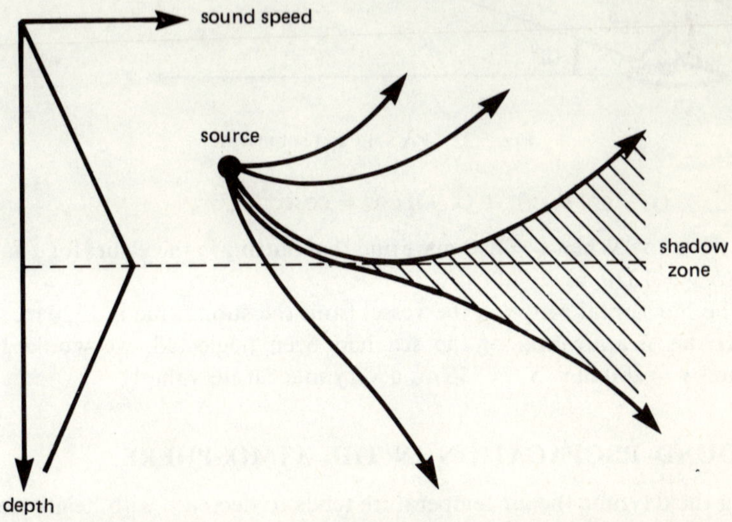

Fig. 5.21 — The formation of a shadow zone.

Example

A submarine is at a depth of 200 m. Its sonar transducer detects a surface vessel at an upward angle of 15° with the horizontal. What is the horizontal range of the vessel from the submarine?

The sound speed can be considered to vary linearly with depth and is 1500 m/s on the surface and 1450 m/s at a depth of 200 m. We will choose the origin of our co-ordinate system to be centred on the submarine and we will take the x-axis to be vertically upwards. Then

$$c = 1450 \text{ m/s} + 0.25x \text{ s}^{-1}$$

$$= c_0(1 + \alpha x), \quad \text{where } \mathbf{c}_0 = 1450, \alpha = 1.724 \times 10^{-4} \text{ m}^{-1}.$$

We wish to follow the ray which makes an angle of 75° with the x-axis at (0, 0); and in particular to determine the value of y when $x = +200$ m. We have already traced the ray paths for this sound velocity profile and from equation (5.2) we see that the ray follows the path:

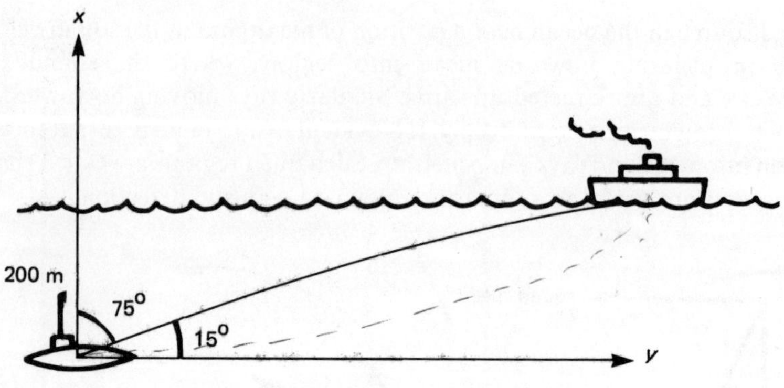

Fig. 5.22 — A ship and submarine.

$$(y - \cot \theta_0/\alpha)^2 + (x + 1/\alpha)^2 = \operatorname{cosec}^2 \theta_0/\alpha^2,$$

where $\theta_0 = 75°$. When $x = 200$ m we find that putting in the values for α and θ_0 leads to $y = 1320$ m.

The horizontal range to the vessel from the submarine is 1320 m.

[If the stratification of the sea had been neglected, we would have obtained $y = 200 \tan 75° = 746$ m; a very inaccurate value!]

5.4 SOUND PROPAGATION IN THE ATMOSPHERE

During the daytime the air temperature tends to decrease with height above the earth, and hence the sound speed decreases upwards. Rays are then bent up, and most of the sound will pass over the head of a distant observer.

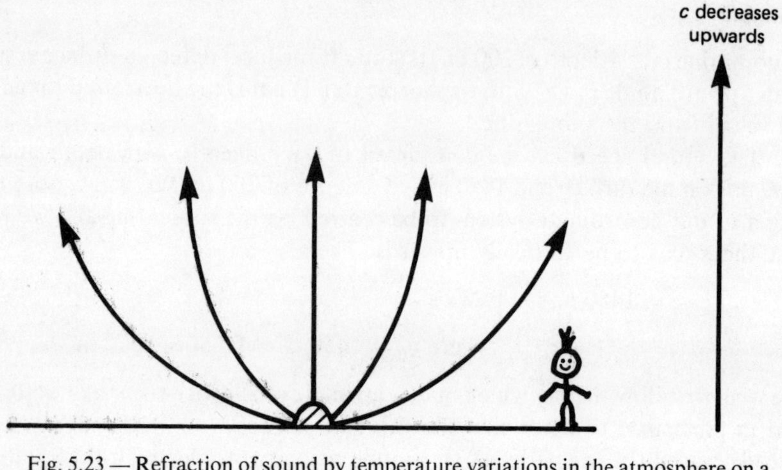

Fig. 5.23 — Refraction of sound by temperature variations in the atmosphere on a typical day.

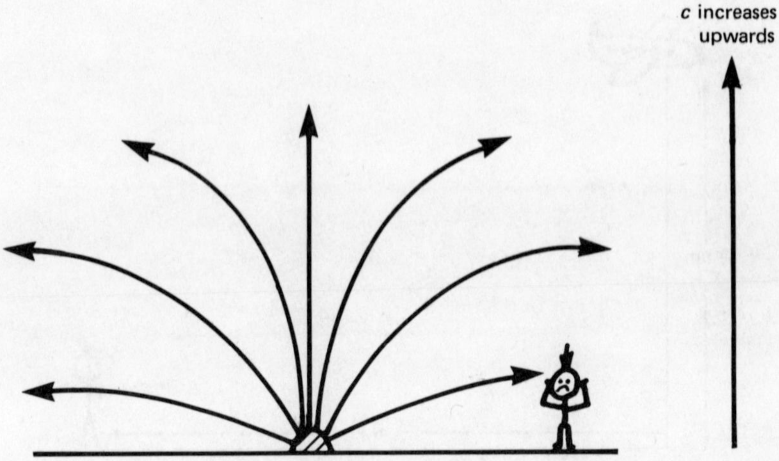

Fig. 5.24 — Refraction of sound by temperature variations in the atmosphere on a clear night following a warm day.

Sometimes on a clear night following a warm day, the ground cools more quickly than the air, and the lower strata of the atmosphere are then colder than the air at higher levels. Then the rays are bent back to the ground and are heard more distinctly by an observer. Strong gradients in wind will refract sound in a similar way because sound travels relative to the moving air at speed c, and has therefore a variable speed relative to the ground. Atmospheric effects need to be considered and accounted for when estimating or measuring noise at large distances from factories and aircraft.

Example

An observer measures the sound pressure level from an aircraft at a height of 2 km and a horizontal distance of 5 km to be 80 dB. Calculate the sound power output of the aircraft if

(1) the sound speed is assumed to be constant between the aircraft and the ground

(2) the sound speed decreases linearly with height at a rate of 2.10^{-2} m/s per metre.

[Take the aircraft to be an omnidirectional acoustic source.]

The sound pressure level is 80 dB. I.e.

$$80 = 20\log_{10}\left(\frac{p'_{rms}}{2.10^{-5}\ \text{N/m}^2}\right) \quad \text{and} \quad p'_{rms} = .2\ \text{N/m}^2.$$

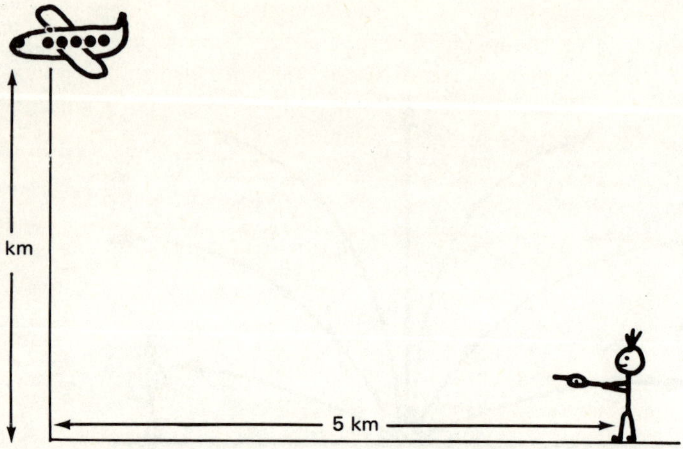

Fig. 5.25 — Measuring the sound of an aircraft.

Let W be the total sound power output of the aircraft. We will choose to use a spherical co-ordinate system (R, θ, φ) centred on the aircraft.

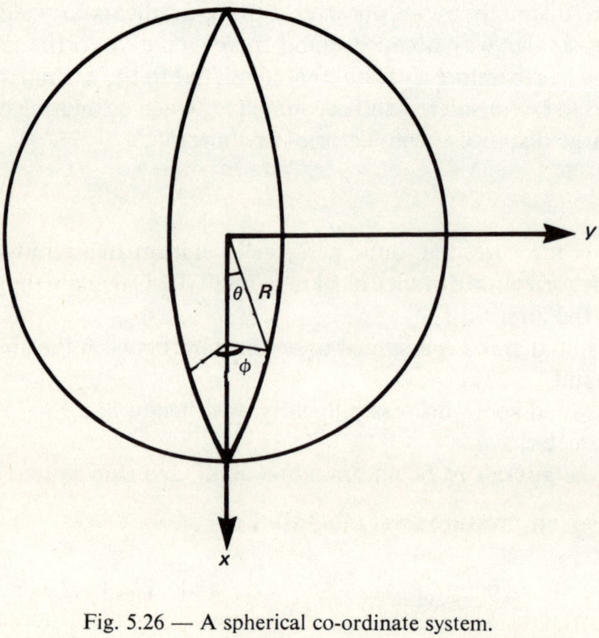

Fig. 5.26 — A spherical co-ordinate system.

Consider a sphere $R = \varepsilon$, where ε is small enough for the sound speed to be constant within the sphere. The total energy flux through the surface of the sphere is W, and since the source is omnidirectional the energy flux per unit area is $W/4\pi\varepsilon^2$. Hence the energy flux into a ray tube, which consists of the bundle of rays with directions between (θ_0, φ_0) and $(\theta_0 + \delta\theta, \varphi_0 + \delta\varphi)$, is

$$\frac{W\varepsilon\,\delta\theta\varepsilon \sin\theta_0\,\delta\varphi}{4\pi\varepsilon^2} = \frac{W \sin\theta_0\,\delta\theta\,\delta\varphi}{4\pi}.$$

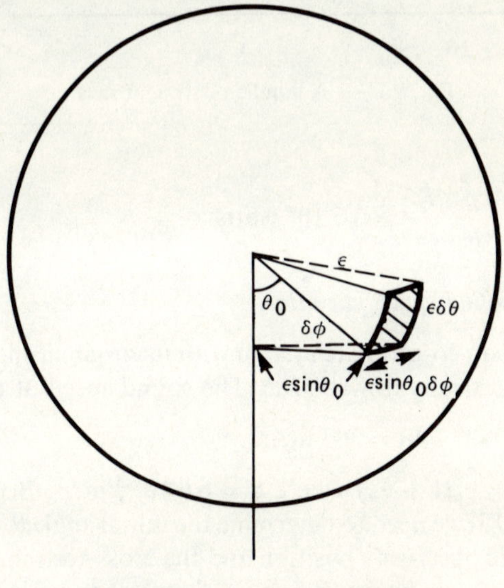

Fig. 5.27 — Ray tube co-ordinates.

If we trace the ray tube that passes through the observer and calculate A, the cross-sectional area of the tube at the observer we can use the energy equation

$$\frac{|p'_{\mathrm{rms}}|^2 A}{\rho_1 c_1} = \frac{W \sin\theta_0\,\delta\theta\,\delta\varphi}{4\pi} \tag{5.15}$$

to calculate W.

Case (1) Uniform sound speed

The rays travel in straight lines and

$$A = R\,\delta\theta R \sin\theta_0\,\delta\varphi$$

where R is the distance of the aircraft from the observer. Hence

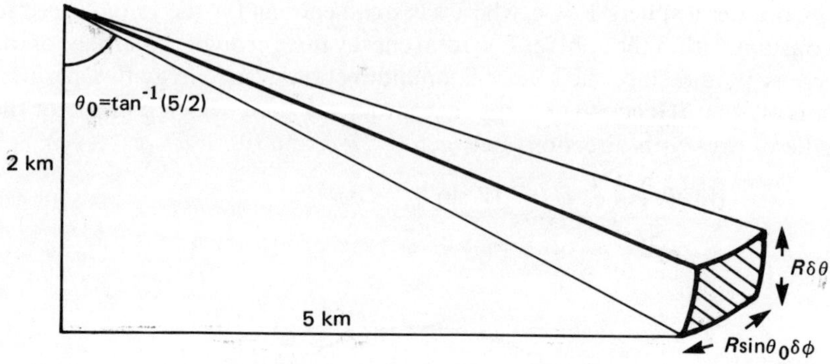

Fig. 5.28 — A bundle of straight rays.

$$W = \frac{4\pi p'^2_{\text{rms}} R^2}{\rho_1 c_1} = 3.5\ 10^4 \text{ watts.}$$

Case (2) Linear sound speed variation

Consider a cartesian co-ordinate system with its origin at the aircraft and the x-axis pointing vertically downwards. The sound speed at the origin is

$$c_0 = 343 - 40 = 303 \text{ m/s},$$

and elsewhere $c = c_0(1 + \alpha x)$ where $\alpha = 6.6\ 10^{-5}\ \text{m}^{-1}$. Before we can calculate W from (5.15) we need to determine the initial angle θ_0 of the ray which passes through the observer's position and the cross-sectional area, A, of this ray tube. From (5.2) we know that the ray paths are circles

$$(y - \cot\theta_0/\alpha)^2 + (x + 1/\alpha)^2 = \operatorname{cosec}^2\theta_0/\alpha^2$$

i.e.

$$y^2 - 2y\cot\theta_0/\alpha + (x + 1/\alpha)^2 = 1/\alpha^2.$$

The ray that passes through $(2, 5)10^3$ has $\cot\theta_0 = .59$, $\theta_0 = 59.4°$. The area, A, of this ray tube can be found by tracing the ray $\theta_0 + \delta\theta$ and finding where it meets the ground. If the distance along the ground between the two rays is ΔD, the perpendicular distance between rays is $\Delta D \cos\theta_1$, and

$$A = \Delta D \cos\theta_1\ 5 \times 10^3\ \delta\varphi \text{ m}^2.$$

θ_1, the ray angle at ground level, can be calculated from Snell's law:

$$\frac{\sin\theta_1}{343} = \frac{\sin 59.4}{303}, \quad \theta_1 = 77°.$$

D, the horizontal range of the ray, is the value of y when $x = 2 \times 10^3$ i.e.

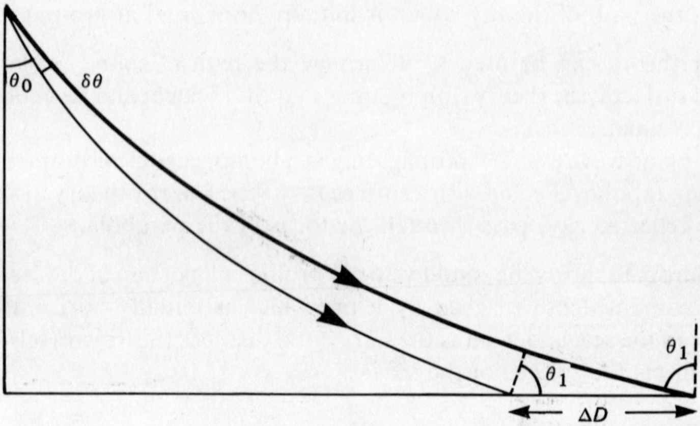

Fig. 5.29 — Two adjacent rays.

$$D^2 - 2D \cot \theta_0/\alpha + (2 \times 10^3 \text{ m} + 1/\alpha)^2 = 1/\alpha^2.$$

ΔD can be calculated by evaluating $D(\theta_0 + \delta\theta)$ and $D(\theta_0)$ and subtracting the two values. But since this difference is $\delta\theta \, \mathrm{d}D/\mathrm{d}\theta_0$ it is quickest to differentiate

$$\frac{\mathrm{d}D}{\mathrm{d}\theta}(D - \cot \theta_0/\alpha) + D \operatorname{cosec}^2 \theta_0/\alpha = 0.$$

Therefore

$$\Delta D = \delta\theta \frac{\mathrm{d}D}{\mathrm{d}\theta} = \delta\theta \frac{D \operatorname{cosec}^2 \theta_0}{\cot \theta_0 - \alpha D} = \delta\theta \, 2.6 \times 10^4 \text{ m}.$$

Hence in this stratified case the power output

$$W = \frac{4\pi p'^2_{\text{rms}} A}{\rho_1 c_1 \sin \theta_0} = 4.1 \times 10^4 \text{ watts},$$

and is more than when the same sound level is heard beneath an unstratified atmosphere.

EXERCISES FOR CHAPTER 5

1. During the day the air temperature in the lower strata of the atmosphere decreases linearly with height, and can be written as $T = T_0(1 - \alpha x)$ where x is measured vertically upwards from the ground. Show that the path of a ray, which makes an angle θ_0 with the vertical at the origin, can be expressed as

$$x = \frac{\cos 2\theta - \cos 2\theta_0}{2\alpha \sin^2 \theta_0}, \quad y = \frac{2(\theta_0 - \theta) + \sin 2\theta - \sin 2\theta_0}{2\alpha \sin^2 \theta_0}.$$

Sketch the path of the ray which is initially horizontal at ground level.

2. Ray theory can be used to determine the path of sound reflected from curved surfaces; the theory simply states that the incident and reflection angles must be equal.

A plane wave $e^{i\omega(t+x/c)}$ propagating in a homogeneous medium is incident onto a parabolic reflector with a surface $r^2 = 4ax$. Use ray theory to show that all the reflected rays pass through the focus of the parabola.

3. Figure 5.30 shows the sound velocity profile in a portion of the sea. What is the maximum depth reached by a ray which is initially horizontal at the surface of the sea, and what is the horizontal distance this ray travels before it again meets the surface of the sea?

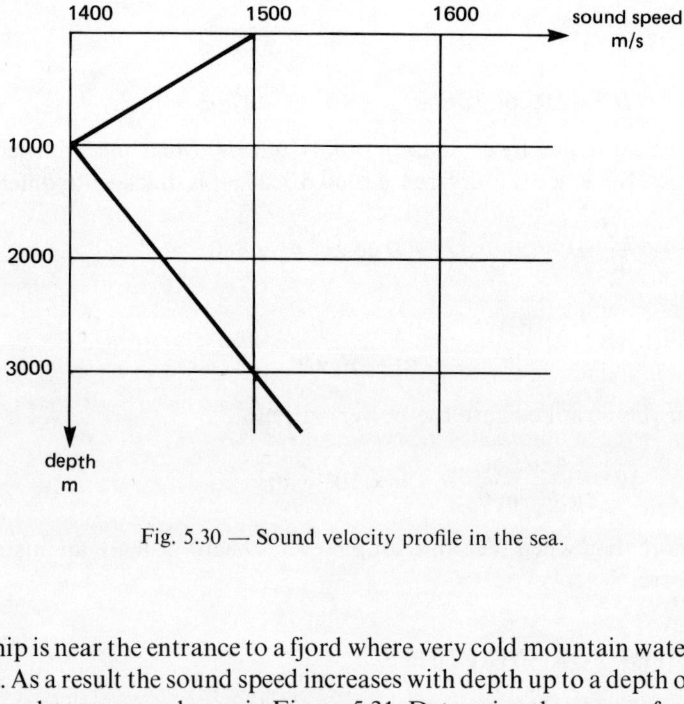

Fig. 5.30 — Sound velocity profile in the sea.

4. A ship is near the entrance to a fjord where very cold mountain waters enter the sea. As a result the sound speed increases with depth up to a depth of 200 m and then decreases as shown in Figure 5.31. Determine the range of positions from which a submarine could not be detected by the ship's sonar.

5. A ship transmits a sonar ray which is horizontal at sea level. The ray meets a shoal of fish and is reflected back along the same path. The time delay between the transmission and reception of the beam is 2 seconds. What is the depth of the shoal and what is its horizontal range from the ship?

[Assume that the sound speed decreases linearly with depth at a rate of .1 m/s per metre and takes a value of 1500 m/s on the surface.]

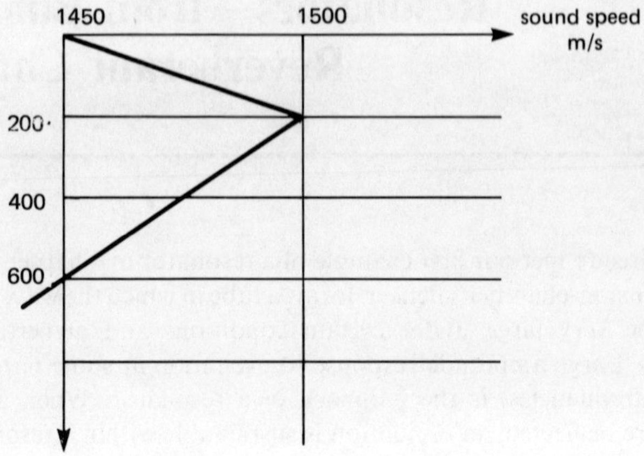

Fig. 5.31 — Sound speed in a fjord.

6. An explosion occurs underwater at a depth of 200 m. The sound pressure level on the sea surface at a horizontal distance of 1 km from the site of the explosion is found to be 100 dB. What is the sound pressure level at the sea surface vertically above the site of the explosion?

[Assume that the sound speed decreases linearly with depth at a rate of .1 m/s per metre, and takes a value of 1520 m/s on the surface, and that the explosion is an omnidirectional sound source.]

Chapter 6

Resonators—from Bubbles to Reverberant Chambers

We have already met our first example of a resonator in Chapter 3 where the single expansion-chamber 'silencer' forms a tube in which the wave amplitudes can become very large under certain conditions and at certain discrete frequencies. Large amplitude response to excitation at some pure tones (the resonance frequencies) is the hallmark of a resonator. When all forms of damping are neglected, an oscillation is sustainable within a resonator in the absence of any excitation, again at the same discrete resonance frequency. If an undamped resonator were to be excited by some external source the response level would grow as the resonant motion accumulated more and more energy from the driver. But most resonators do have damping, or losses, and their peak response to external driving is limited at the resonance frequencies by that damping.

Figure 6.1 illustrates the single expansion chamber silencer for which we found in Chapter 3 (equation 3.6) that the strength of the reflected wave vanished whenever $\sin \omega l/c$ was zero and that then there was perfect transmission of sound across the expansion chamber. We note now that this is true regardless of how severe the constriction of the (negative) expansion might be and deduce that the acoustic activity within the narrow tube must be very large indeed to convey the energy flux across it without any evident

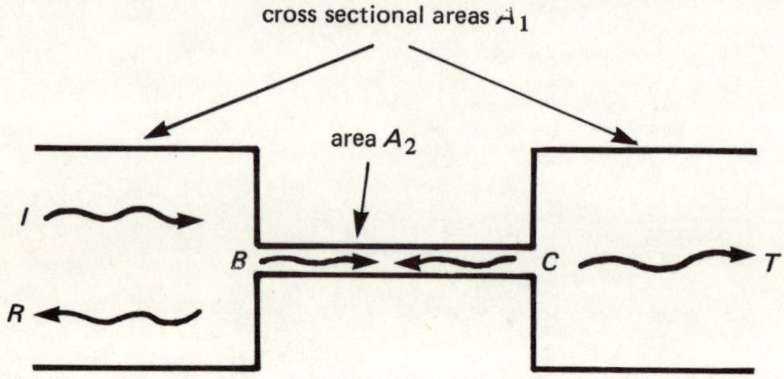

Fig. 6.1 — A single (negative) expansion chamber.

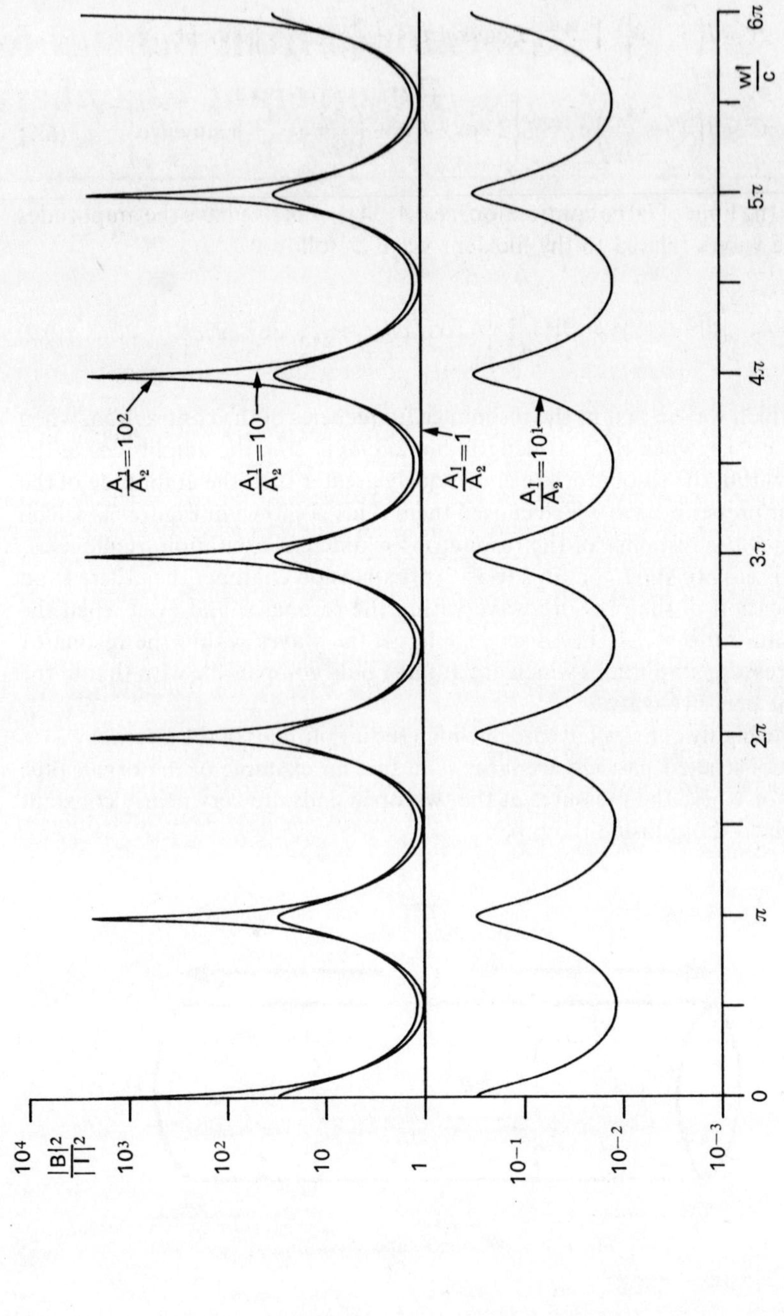

Fig. 6.2 — The response of a resonator excited at different harmonic frequencies according to equation (6.1)

obstruction. The strengths of the waves B and C are easily deduced from the equations that led to (3.6) to be

$$B = I\left(1 + \frac{A_1}{A_2}\right) e^{i\omega l/c}\left\{2\cos\omega l/c + \left(\frac{A_1}{A_2} + \frac{A_2}{A_1}\right) i \sin\omega l/c\right\}^{-1}$$

$$C = I\left(1 - \frac{A_1}{A_2}\right) e^{-i\omega l/c}\left\{2\cos\omega l/c + \left(\frac{A_1}{A_2} + \frac{A_2}{A_1}\right) i \sin\omega l/c\right\}^{-1}. \quad (6.1)$$

In the limit of large contraction, i.e. $(A_1/A_2) \to \infty$, we have the amplitudes of these waves related to the incident wave as follows:

$$|B|^2 = |C|^2 = |I|^2 \frac{A_1^2}{A_2^2}\left\{4\cos^2\omega l/c + \frac{A_1^2}{A_2^2}\sin^2\omega l/c\right\}^{-1} \quad (6.2)$$

from which we see that at the resonance frequencies of the contraction, when $\omega l/c = n\pi$, i.e. when the wavelength $\lambda = 2\pi c/\omega$ is $2l/n$, the amplitudes of the waves within the resonator are enormously greater than the amplitude of the incident pressure wave which caused them. This is shown in Figure 6.2 which illustrates the response of the resonator for different excitation frequencies.

It is easy to show that in a (positive) expansion chamber then there is no amplification of the pressure wave within the resonator and even when the expansion ratio A_2/A_1 becomes very large, the waves within the resonator have pressure amplitudes which are at most only comparable with that in the incident pressure wave.

The highly constricted case in which the amplitudes of the internal waves in the constricted passage are large is in fact an example of the organ pipe resonator where the pressures at the two open ends are very nearly constant (relative to those inside the pipe).

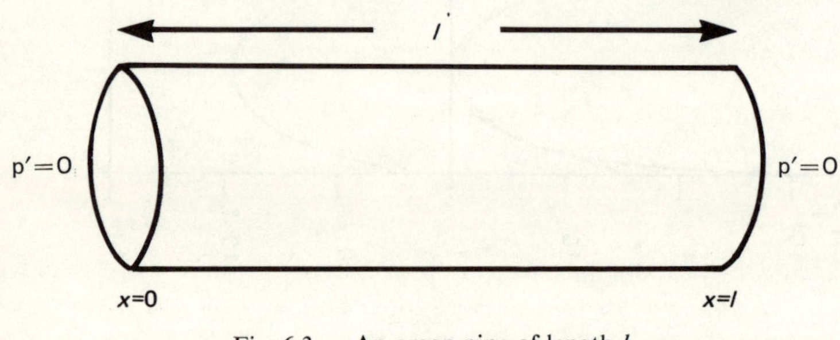

Fig. 6.3 — An organ pipe of length l.

6.1 ORGAN PIPES

If an organ pipe is open at both ends, the pressure at the ends will be (very nearly) atmospheric, and $p' = 0$ there. For one-dimensional waves of frequency ω

$$p' = B\,e^{i\omega(t-x/c)} + C\,e^{i\omega(t+x/c)}, \quad \text{within the pipe.} \tag{6.3}$$

But since $p' = 0$ at $x = 0$,

$$B = -C.$$

The second condition of no pressure perturbation at $x = l$ then shows that

$$e^{-i\omega l/c} - e^{i\omega l/c} = 0$$

or

$$\sin\frac{\omega l}{c} = 0.$$

The organ pipe frequencies are at

$$\omega = \frac{\pi c}{l}, \frac{2\pi c}{l}, \frac{3\pi c}{l} \text{ etc.,}$$

and the wavelengths of these modes are $2\pi c/\omega = 2l, l, \frac{2}{3}l \ldots$ etc.

At the fundamental frequency $\omega = \pi c/l$, putting $B = -C$ in equation (6.3) shows that

$$p' = -2iB\,e^{i\omega t}\sin\pi x/l \tag{6.4}$$

and

$$u = \frac{2B}{\rho_0 c}\,e^{i\omega t}\cos\pi x/l$$

where we have used the linear momentum equation to relate p' and u. The particle displacement can easily be calculated because

$$u = \frac{\partial\eta}{\partial t}.$$

Integration shows that

$$\eta = \frac{-2Bi}{\rho_0 c\omega}\,e^{i\omega t}\cos\pi x/l. \tag{6.5}$$

From a comparison of (6.4) and (6.5)

$$\frac{p'}{\rho_0 c^2} = \frac{\pi\eta}{l}\tan\pi x/l. \tag{6.6}$$

The pressure and displacement are in phase if $\tan\pi x/l$ is positive i.e. $0 < x < l/2$, and they are diametrically out of phase if $l/2 < x < l$. This will be used in the discussion of the Rijke tube.

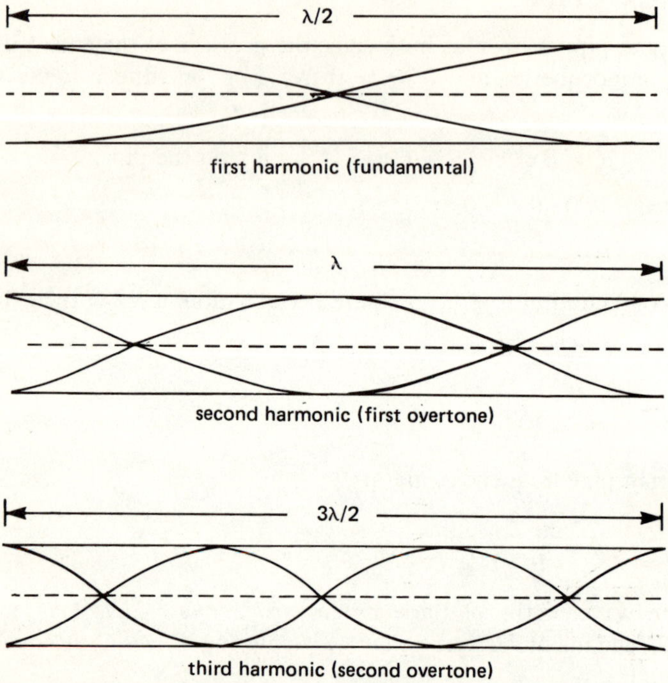

Fig. 6.4 — Harmonics of an open ended organ pipe.

6.2 THE RIJKE TUBE

A Rijke tube consists of a pipe open at both ends with a heated grid. The pipe is held vertically. When the grid is in the lower half of the pipe it excites the fundamental pipe resonance, but when the grid is in the upper half of the pipe no sound is heard. Rayleigh (Lord Rayleigh—The Theory of Sound (1894)) explained this phenomenon. He began by discussing how the addition of heat affects sound waves. He noted that if heat is added to a sound wave when it is at the high temperature phase of its cycle, energy is fed into the acoustic disturbance, but if heat is added at the low temperature limit the sound wave loses energy. Since the compression in a sound wave is adiabatic this means that pressure and temperature fluctuations are in phase and heat addition during a positive pressure disturbance increases the amplitude of the sound waves, but it has the opposite effect when the pressure disturbance is negative. (This is similar to the effect of heat addition in thermodynamic cycles. When heat is added at the high pressure part of a cycle, the system gains energy. This energy is then usually used to do work as in, for example, the Joule and Carnot cycles.)

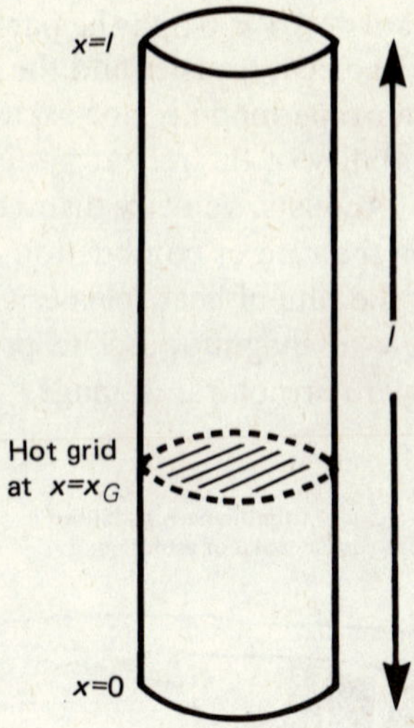

Fig. 6.5 — A Rijke tube.

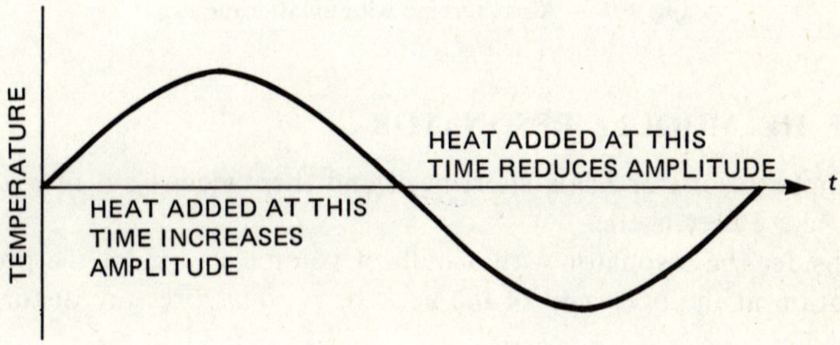

Fig. 6.6 — Rayleigh's criterion.

In the Rijke tube the grid heats the air around it and this hot air rises. The particle displacement due to the sound wave is superimposed on this natural convection. When the acoustic particle displacement is positive upwards more fresh colder air from below the grid passes through the grid to be heated, but when the acoustic particle displacement is negative relatively little heat transfer takes place. Hence we conclude that the maximum heat transfer occurs when the particle displacement η is positive. From (6.6) the particle displacement is in phase with the pressure at the grid if $x_G < l/2$; i.e. when the grid is in the lower half of the pipe the heat transfer coincides with the high pressure and energy is fed into the acoustic disturbances. When the grid is in the upper half at the pipe $\tan \pi x_G/l < 0$ and the particle displacement and the pressure are out of phase. The heat transfer and the pressure are therefore out of phase too and the organ pipe mode is not excited.

Reheat buzz, an instability of the reheat system of a jet engine, can be triggered in a similar way. Acoustic velocity disturbances in the pipe disturb the reheat flame and alter the rate of combustion. If the phase relationship between the pressure and the rate of heat release is suitable the disturbances can gain energy and grow in magnitude. The pressure perturbations can become so large that they do structural damage.

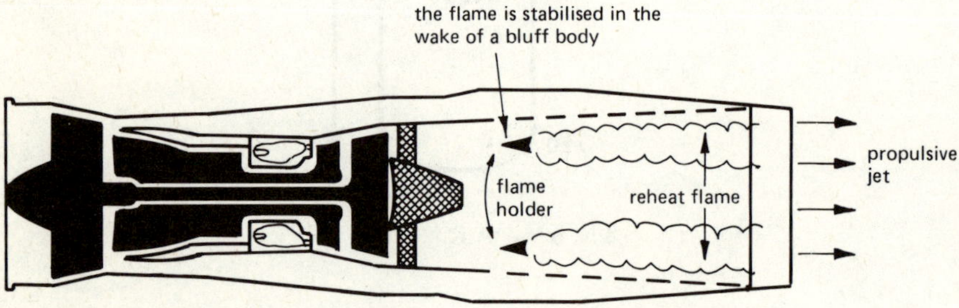

Fig. 6.7 — A gas turbine with an afterburner.

6.3 THE HELMHOLTZ RESONATOR

A Helmholtz resonator has a short neck and then widens out into a large volume (like a beer bottle).

Consider the resonator with a bulb of volume V, and let the pressure perturbation at the open end of the neck be p'_1. This pressure disturbance

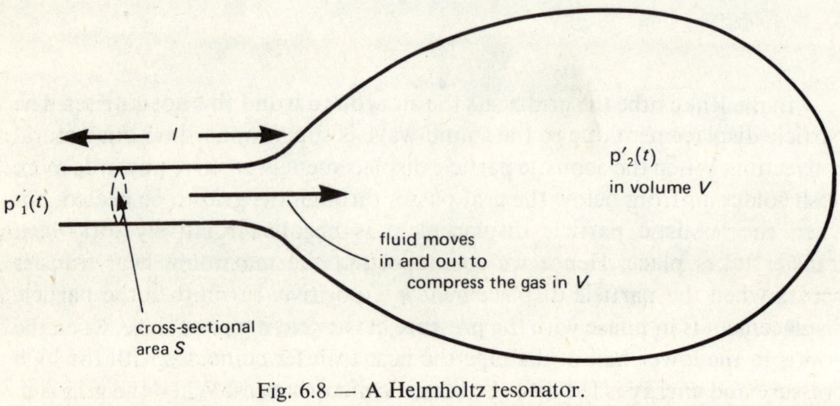

Fig. 6.8 — A Helmholtz resonator.

causes mass to flow into the bulb at a rate of Q kg/s. It is possible to derive a simple relationship between p'_1 and Q.

As mass flows into the bulb the density there must rise:

$$V\frac{\partial \rho}{\partial t} = Q$$

i.e.

$$\rho' = \frac{Q}{Vi\omega} \quad \text{for harmonic disturbances.}$$

Associated with this density change is a corresponding rise in the pressure in the bulb

$$p'_2 = c^2\rho' = \frac{c^2 Q}{Vi\omega}. \tag{6.7}$$

In the neck, from momentum balance, the pressure difference between the ends equals the rate of momentum change

$$p'_1 - p'_2 = \rho_0 l \frac{\partial u}{\partial t}.$$

This can be rewritten because $u = Q/\rho_0 S$ and hence

$$p'_1 - p'_2 = \frac{i\omega l Q}{S}.$$

After substituting for p'_2 from (6.7) we find

$$p'_1 = \left\{\frac{c^2}{Vi\omega} + \frac{i\omega l}{S}\right\}Q = \left\{\omega^2 - \frac{c^2 S}{Vl}\right\}\frac{ilQ}{\omega S}. \tag{6.8}$$

The system resonates at a frequency $\omega = \sqrt{c^2 S/Vl}$. Near this frequency very small pressure disturbances can lead to large mass variations.

For a 'typical' beer bottle $V = 5.10^{-4}$ m^3, $S = 2.10^{-4}$ m^2, $l = 5.10^{-2}$ m and the resonance frequency is $\omega = 1000$ rad/s, about 150 Hz. A tuneful note! Higher notes are of course obtained from partly filled bottles when V is smaller.

Helmholtz resonators also have a practical use. Their characteristic behaviour of small pressures at the neck producing large mass fluctuations can be exploited to produce an efficient muffler. To attenuate sound travelling along a duct a Helmholtz resonator is connected to the side of a duct. We will consider an incident wave of strength I and frequency ω. The Helmholtz resonator causes a wave of strength R, to be reflected back along the duct and lets a wave of strength T be transmitted on down the duct. We choose a co-

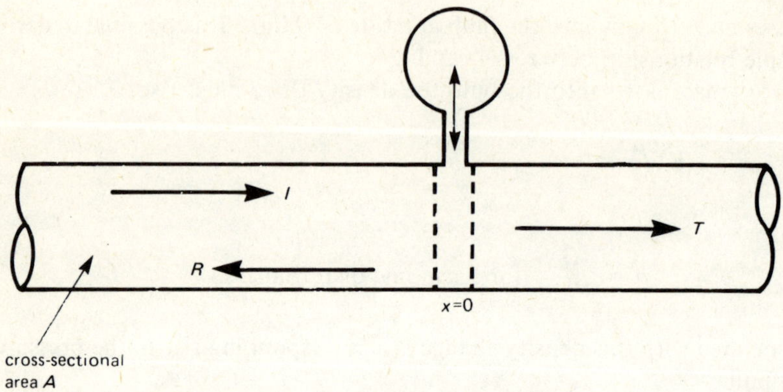

Fig. 6.9 — A volume-resonator silencer.

ordinate system so that the resonator is connected at $x = 0$. Then

$$\begin{aligned} p' &= I\,\mathrm{e}^{i\omega(t-x/c)} + R\,\mathrm{e}^{i\omega(t+x/c)} \quad \text{in } x < 0 \\ &= T\,\mathrm{e}^{i\omega(t-x/c)} \quad \text{in } x > 0. \end{aligned}$$

The pressure at the neck of the resonator at $x = 0$ is p'_1 and therefore

$$\left.\begin{aligned} p'_1 &= (I+R)\,\mathrm{e}^{i\omega t} \\ p'_1 &= T\,\mathrm{e}^{i\omega t} \end{aligned}\right\} \tag{6.9}$$

Now consider the mass flux into the control volume outlined by the dotted line. The mass flux into this volume from the left is $A(I - R)\,\mathrm{e}^{i\omega t}/c$ where A is the cross-sectional area of the duct, and we have again used the linear plane wave relations to relate the velocity to the pressure. This mass flux into the control volume must be equal to the sum of the mass flux into the resonator and into the right-hand duct, i.e.

$$\frac{A}{c}(I-R)\,\mathrm{e}^{i\omega t} = Q + \frac{AT}{c}\,\mathrm{e}^{i\omega t}. \tag{6.10}$$

Equations (6.8)–(6.10) are four algebraic equations involving the four unknowns Q, R, T, p'_1 and so can be solved to find T. After some algebra it can be shown that

$$T = \frac{I}{1 + \dfrac{1}{2A}\left\{\dfrac{c}{i\omega V} + \dfrac{i\omega l}{cS}\right\}^{-1}}. \tag{6.11}$$

The transmission loss is

$$L_T = 10\log_{10}\left\{\frac{|I|^2}{|T|^2}\right\},$$

i.e.

$$L_T = 10 \log_{10}\left\{1 + \frac{1}{4A^2}\left(\frac{c}{\omega V} - \frac{\omega l}{cS}\right)^{-2}\right\}. \tag{6.12}$$

The agreement between this theoretical expression and experimental results is good. A practical muffler incorporating a Helmholtz resonator often uses the duct walls as part of the surface of the bulb and this principle of attenuation is used in some car 'silencers' as shown in figure 6.11.

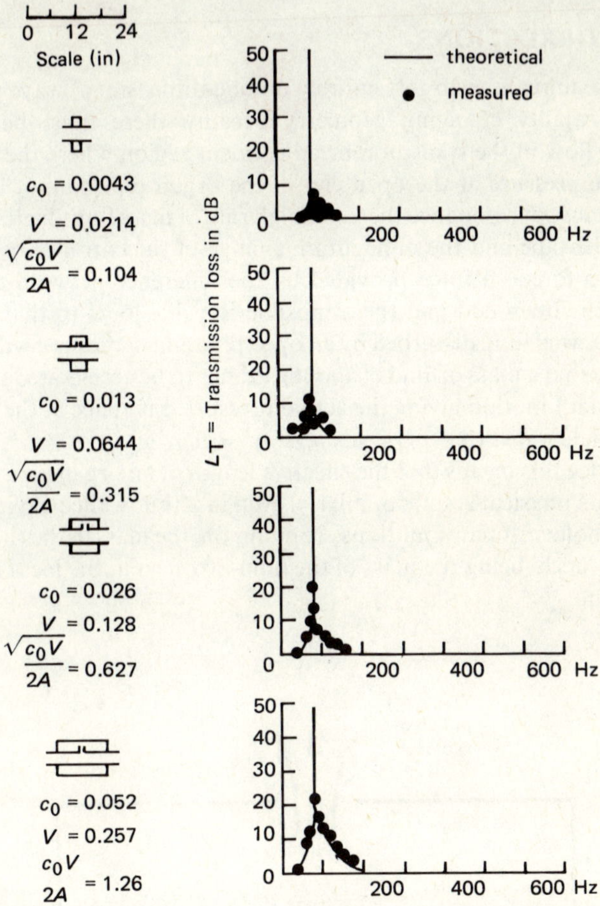

Fig. 6.10 — Graphs for different resonators with the same resonance frequency. $c_0 = S/l$. (from Davies *et al.* NACA report 1192 (1954)). The theoretical curve is from equation (6.12).

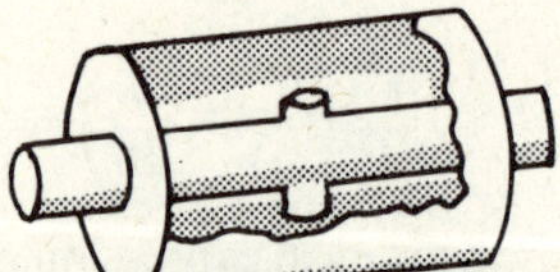

Fig. 6.11 — A practical muffler incorporating a Helmholtz resonator.

6.4 END CORRECTIONS

It is of course unrealistic to rely entirely on one-dimensional wave motion in ducts with rapidly changing geometry because there must be a three-dimensional flow in the transition or adjustment region where the geometry changes. The pressure at the open end of the organ pipe cannot be exactly constant because at resonance there is a high rate of mass flow fluctuating into and out of the pipe and the momentum change of the entrant flow must be supplied by a force—a force provided by the difference between the actual pressure at the open end and the atmospheric value local to that end. The volume of external fluid disturbed by an open pipe end of radius a will be of the order of a^3 so that a mass of fluid of about $\rho_0 a^3$ has to be accelerated to velocity v in the resonant motion giving the actual pressure difference at the open end an approximate value $(p - p_0) = i\omega v \rho_0 a^3/a^2 = i\omega\rho_0 av$.

In practice this means that the effective length of an organ pipe is slightly longer than its physical length. Similarly, even in a thin-walled cavity with an orifice Helmholtz resonator motions are common, the mass in the then almost non-existent 'neck' being the mass of the fluid involved in the local motion, a mass of about $\rho_0 a^3$.

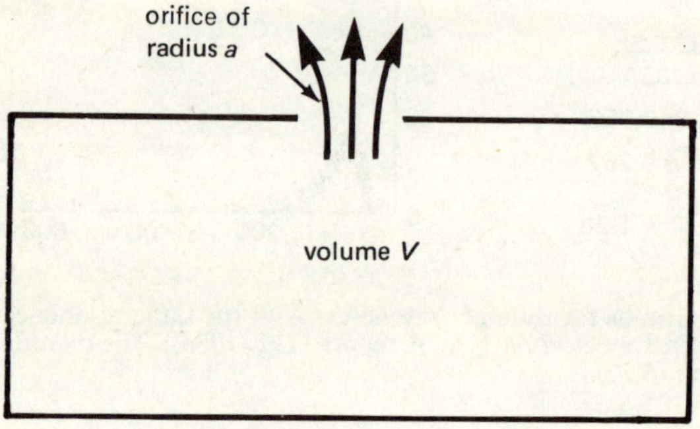

Fig. 6.12 — Mass of about $\rho_0 a^3$ is involved in the 'virtual' neck of this Helmholtz resonator.

6.5 THE ISOLATED BUBBLE AS A RESONANT SCATTERER

We return now to consider more carefully the sound generated by an isolated gas bubble in otherwise uniform liquid. Again we suppose that the environment of the bubble is exposed to an unsteady 'incident' pressure fluctuation $p_i'(t)$ that is uniform on the scale of the bubble. The bubble responds in a spherically symmetric way, its volume pulsating to accommodate the imposed external pressures. The incident pressure is therefore supplemented by the vibration induced pressure $p'(r, t)$ which is the sound field scattered by the bubble's volumetric pulsation. Suppose that the incident pressure field is harmonic of angular frequency ω. In that case we found in equation (2.29) that the scattered surface pressure field is related to the bubble's surface velocity u_a by

$$\hat{p}(a) = \frac{a_0 i\omega\rho_0\hat{u}_a}{1 + i\omega a_0/c}. \tag{6.13}$$

a_0 is the mean radius of the bubble and $\hat{p}$ and $\hat{u}_a$ again denote the complex amplitudes of pressure perturbation and surface velocity. The total surface pressure on the bubble's external surface is

$$p_0 + p_i' + \frac{a_0 i\omega\rho_0\hat{u}_a}{1 + i\omega a_0/c}\,\mathrm{e}^{i\omega t} \tag{6.14}$$

where p_0 is the mean pressure in the external fluid.

Surface tension T in the surface of the bubble will supplement this pressure by $2T/a$ so that the interior gas in the bubble, which is assumed small enough to be at the spacewise uniform pressure p_g, is

$$p_g = p_0 + \frac{2T}{a} + p_i' + \frac{a_0 i\omega\rho_0\hat{u}_a}{1 + i\omega a_0/c}\,\mathrm{e}^{i\omega t}. \tag{6.15}$$

If the gas inside the bubble is perfect and in an adiabatic state, then

$$p_g a^{3\gamma} = p_{g0} a_0^{3\gamma} \tag{6.16}$$

where a is the time dependent radius of the pulsating bubble and the suffix zero indicates the mean value. γ is the ratio of the principal specific heats of the bubble gas.

Providing the motion is weak enough that linear equations describe the small vibrations, the relationship between the internal pressure fluctuations and the surface velocity can be found by differentiating equation (6.16) with respect to time

$$\frac{\partial p_g}{\partial t} a_0^{3\gamma} + p_{g0} 3\gamma a_0^{(3\gamma-1)} \frac{\partial a}{\partial t} = \frac{\partial p_g}{\partial t} a_0^{3\gamma} + p_{g0} 3\gamma a_0^{(3\gamma-1)} u_a = 0$$

i.e.

$$\frac{\partial}{\partial t} p_g = \frac{-3p_{g0}\gamma}{a_0} u_a. \tag{6.17}$$

This equation, when combined with (6.15), then shows the bubble surface velocity to be determined in terms of the incident pressure field

$$\hat{u}_a = \frac{i\omega}{\rho_0 a_0} \hat{p}_i \left\{ \frac{\omega^2}{(1 + i\omega a_0/c)} - \left(\frac{3\gamma p_{g0}}{\rho_0 a_0^2} - \frac{2T}{\rho_0 a_0^3} \right) \right\}^{-1}. \tag{6.18}$$

The internal mean gas pressure p_{g0} exceeds the external mean pressure in the liquid p_0 by the surface tension induced mean pressure $2T/a_0$ so that (6.18) is alternatively written as

$$\hat{u}_a = \frac{i\omega}{\rho_0 a_0} \hat{p}_i \left\{ \frac{\omega^2}{(1 + i\omega a_0/c)} - \left(\frac{3\gamma p_0}{\rho_0 a_0^2} + (3\gamma - 1) \frac{2T}{\rho_0 a_0^3} \right) \right\}^{-1}$$

or

$$\hat{u}_a = \frac{i\omega}{\rho_0 a_0} \hat{p}_i \left\{ \frac{\omega^2}{1 + i\omega a_0/c} - \omega_0^2 \right\}^{-1}. \tag{6.19}$$

For air and vapour bubbles in water the Helmholtz number $\omega a_0/c$ is usually extremely small so that the bubble response is seen to be that of a lightly damped oscillator which resonates at the frequency ω_0 such that

$$\omega_0^2 = \left(\frac{3\gamma p_0}{\rho_0 a_0^2} + (3\gamma - 1) \frac{2T}{\rho_0 a_0^3} \right). \tag{6.20}$$

Surface tension acts to raise the resonance frequency but not usually by very much, for example by only some 5% for a bubble of mean radius 10^{-3} cm. The surface tension term is negligible for bubbles resonant below 30 kHz ($a_0 > 10^{-2}$ cm). For air bubbles in water at 20°C at mean pressure of one atmosphere (10^5 N/m^2) equation (6.20) is effectively

$$\omega_0 a_0 = \sqrt{\frac{3\gamma p_0}{\rho_0}} = 20.5 \text{ m/sec} \tag{6.21}$$

which makes the Helmholtz number at resonance

$$\frac{\omega_0 a_0}{c} = 0.014. \tag{6.22}$$

Low frequency bubble vibrations can be subject to sufficient heat transfer from the gas to the liquid that the gas motion is effectively isothermal in which case the foregoing analysis can be made to apply by formally setting γ equal to unity in the various formulae.

Having determined the surface velocity of the pulsating bubble it is a straightforward matter to write the scattered sound field by combining equations (2.30) and (6.19).

$$\hat{p}(r) = \frac{-a}{r}\hat{p}_i\,\mathrm{e}^{-i\omega(r-a)/c}\left\{1 - \frac{\omega_0^2}{\omega^2}\left(1 + \frac{i\omega a}{c}\right)\right\}^{-1}. \tag{6.23}$$

As we anticipated in Chapter 2 the simplified treatment of the problem that we gave there (cf. equation 2.27) is the limiting high frequency behaviour of this more general result which incorporates the effects due to the finite compressibility of the internal gas. It is this finite compressibility that gives the bubble its resonant character. At resonance, where $\omega = \omega_0$, the scattered field is extremely strong

$$\hat{p}(r) \underset{\omega\to\omega_0}{\sim} \frac{-ic}{\omega r}\hat{p}_i = \frac{-i}{2\pi}\frac{\lambda}{r}\hat{p}_i. \tag{6.24}$$

It is evident from this result that the sound field is established as if the incident field were applied over a sphere of radius $\lambda/2\pi$ in place of the physically much smaller radius of the bubble.

Sonar observations of reflections, or scatter, from single bubbles, that exist in the swim bladder by which fishes remain neutrally buoyant, indicate a great deal. They are strong reflectors at the resonance frequency, which immediately from equation (6.22) indicates the radius of the bubble and therefore the probable size of the fish!

6.6 RESONANT BOXES

We have discussed in Chapter 3, section 3.2, the higher order modes that propagate as sound in a square sectional hard-walled duct of side length a. Equation (3.14) gives the general field in such a duct as the sum of upstream and downstream travelling wave elements. If the duct is only of length L and is terminated by hard walls at $x_3 = 0$ and $x_3 = L$, then the coefficients A_{mn} and B_{mn} in that equation can be related by the requirement for the velocity to vanish, i.e. $\partial p/\partial x_3 = 0$ at $x_3 = 0$ and $x_3 = L$.

$$-ik_{mn}A_{mn} + ik_{mn}B_{mn} = 0$$

and (6.25)

$$-ik_{mn}A_{mn}\,\mathrm{e}^{-ik_{mn}L} + ik_{mn}B_{mn}\,\mathrm{e}^{ik_{mn}L} = 0$$

i.e.

$$A_{mn} = B_{mn}$$

and

$$\sin k_{mn}L = 0; \quad k_{mn} = \frac{q\pi}{L} \tag{6.26}$$

q being an integer. The general expression for the pressure field in the box

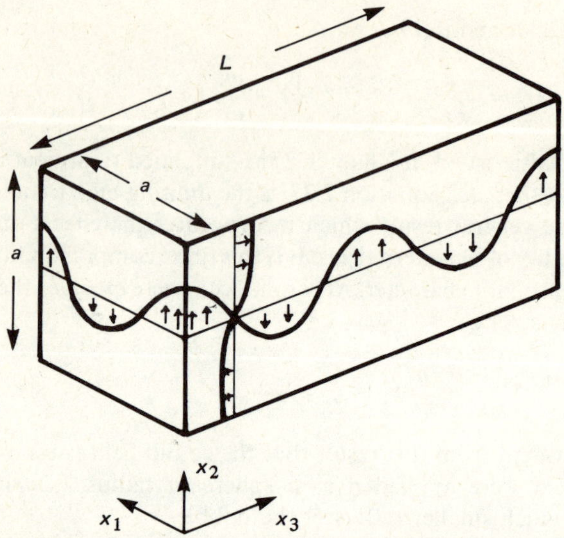

Fig. 6.13 — Box shaped resonator with higher order modes depicted for the case $m = 3, n = 1, q = 5$.

shaped space depicted in Figure 6.13 is then

$$p'(\mathbf{x}, t) = \sum_m \sum_n \sum_q A_{mnq} \cos\frac{m\pi x_1}{a} \cos\frac{n\pi x_2}{a} \cos\frac{q\pi x_3}{L} \mathrm{e}^{i\omega t}. \tag{6.27}$$

The axial wave number k_{mn} is determined from (3.13) and (6.26):

$$k_{mn} = \frac{\pi q}{L} = \sqrt{\frac{\omega^2}{c^2} - \frac{\pi^2}{a^2}(m^2 + n^2)},$$

i.e. the resonance frequencies of the room are given by

$$\omega = c\pi\sqrt{\frac{(m^2 + n^2)}{a^2} + \frac{q^2}{L^2}}. \tag{6.28}$$

When there is a sound source in a room the room acts like a resonator and responds strongly to excitation by the source at frequencies in the immediate vicinity of any of its resonant frequencies. A source emitting low frequency sound will only excite a few of these resonant modes, but for very high frequencies, or more strictly very high values of the Helmholtz number $(\omega a/c)$, there are many resonant modes with frequencies near ω. Then highly intricate multimodal fields are generated, and some statistical way of accounting for the many modes becomes appropriate. In the simple box depicted in Figure 6.13, the method of multiple images could be used to illustrate the field, the field of each image source corresponding to one mode, the waves reflecting off the plane surfaces back into the space in an everlasting reverberant sound. But in practice, some absorption takes place upon reflection and the geometry is rarely perfect, so that some approximate description of the complex reverberant field is more appropriate. That is the technique of room acoustics which we will now briefly describe.

6.7 ROOM ACOUSTICS

When a sound source is in a room, sound waves emanating from the source will propagate until they strike the walls. Some energy will be absorbed by a real wall and a weaker wave will be reflected back. Again this propagates until it reaches another wall where it is again reflected with partial absorption. This process continues until all the sound energy is eventually absorbed. The overall array of randomly criss-crossing rays is called a *reverberant* sound field.

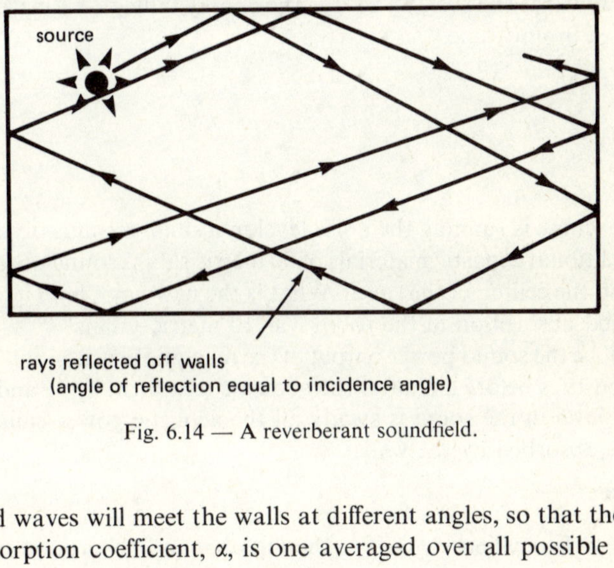

Fig. 6.14 — A reverberant soundfield.

Sound waves will meet the walls at different angles, so that the relevant global absorption coefficient, α, is one averaged over all possible incidence angles.

The wall and furnishings in a room will in general have different absorption coefficients. If we wished to determine the details of the actual sound absorption we would have to trace the progress of the individual sound rays from their source and discover on which surfaces they impinged and compute the sound absorbed at each impact. A difficult and intricate calculation! Fortunately there is a great simplification when the sound field is diffuse. A diffuse sound field is one in which there is no preferred direction of sound propagation and the different criss-cossing rays travel in all directions with equal probability. Then the number of rays hitting a particular wall in unit time depends only on the wall area. If a room has surfaces $S_1, S_2, \ldots, S_n$ with absorption coefficients $\alpha_1, \alpha_2, \ldots, \alpha_n$, the mean absorption coefficient, $\bar{\alpha}$, is defined to be the area-weighted average of the local absorption coefficients.

$$\bar{\alpha} = \frac{1}{S}(\alpha_1 S_1 + \alpha_2 S_2 + \cdots + \alpha_n S_n) \tag{6.29}$$

where S is the total surface area $S_1 + S_2 + \cdots + S_n$.

An open space (in the ray theory approximation) produces no reflected wave and is sometimes taken as a standard of unity absorption. An open window of area $a = S\bar{\alpha}$ is like a perfect absorber of area a and will absorb as much sound as a room with total area S and mean absorption coefficient $\bar{\alpha}$. If S is in square metres a is called the sound absorption of the room measured in metric *sabins* (it has units m^2). The sound power absorbed in a room is aI, where I is the (scalar) intensity of those elements in the diffuse sound that are travelling towards the surface, i.e. I is the energy falling on (one side of) unit surface area in unit time.

Example

When a machine is running the noise level in a small room is observed to be 80 dB. Additional acoustic materials of 50 metric sabins sound absorption are mounted to the ceiling of the room. What is the new noise level if initially the initial sound absorption in the room was 10 metric sabins?

Let W be the sound power output of the machine in watts, and denote the sound intensities before and after the acoustic treatment by I_1 and I_2. When the sound level in the room is steady all the acoustic power emitted by the machine is absorbed by the walls.

Hence

$W = a_1 I_1$, where $a_1 = 10$, the initial sound absorption of the room in metric sabins

and

$W = a_2 I_2$, where $a_2 = 60$, the final sound absorption of the room in metric sabins.

Elimination of W from these two equations shows

$$I_2 = I_1 \frac{a_1}{a_2}$$

and so the new sound pressure level

$$= 20 \log_{10}\left(\frac{p_{2\text{rms}}}{2 \times 10^{-5} \text{ N/m}^2}\right)$$

$$= 20 \log_{10}\left(\frac{p_{1\text{rms}}}{2 \times 10^{-5} \text{ N/m}^2}\right) + 10 \log_{10}\left(\frac{a_1}{a_2}\right) = 72 \text{ dB}.$$

The new noise level is 72 dB.

Reverberation time

The reverberation time, T, was a concept introduced by W. C. Sabine. The reverberation time is used to characterise the acoustic behaviour of a room and is defined to be the time in which the sound of a previous source drops from 60 dB to the threshold of hearing. Typical values of reverberation times are about .3 sec for living rooms or up to 10 sec for larger churches. Most large rooms have reverberation times between .7 and 2 s. If the reverberation time is too short sounds appear 'dead' as the lack of echo produces a very clipped sound. If the reverberation time is too long speech becomes incoherent, and echoes drown the speaker.

The reverberation time in a room can be calculated from the sound absorption if certain simplifying assumptions are made. The sound produced by a short duration source is assumed to travel in rays and those rays are reflected with partial absorption each time they strike the walls of a room. After a large number of successive reflections the sound in the room is assumed to become diffuse and must decay as its energy is absorbed by the walls. (If the sound absorption in a room is sufficiently large all the sound may be absorbed

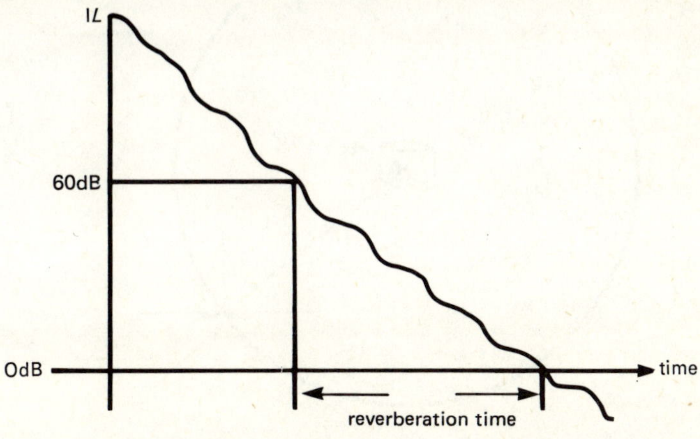

Fig. 6.15 — The definition of reverberation time.

after the first couple of reflections and a diffuse sound field is never obtained. Such rooms are called 'dead' rooms and their reverberation time must be calculated by a different method. See, for example, Eyring, 1930, *J. Acoust. Soc. Am.*, **1**, 217). Once a diffuse field has been obtained the (short-) time-averaged energy density of the sound field, e, is the same throughout the room. Hence the total energy in the room is eV, where V is the volume of the room. If we denote the sound intensity at the walls by I, the rate of energy absorption is Ia, where a is the total absorption of the room in metric sabins. The energy balance equation is therefore

$$V\frac{\mathrm{d}e}{\mathrm{d}t} = -Ia. \tag{6.30}$$

We will now use ray theory to relate e and I. Because these arguments are based on ray theory they are accurate only when the wavelengths of the sound are much shorter than the typical length scales in the room.

The energy in a small volume δV is $e\,\delta V$, since e is the energy density. When the sound field is diffuse all directions of propagation of rays are equally probable. The rays travel at speed c and carry the energy with them and so after a time interval r/c the energy which was initially concentrated within δV is now uniformly distributed over a spherical shell of radius r, centred on the region δV. The energy incident on unit area of the shell is therefore $e\,\delta V/4\pi r^2$, and hence the energy falling on a surface element of area δS making an angle θ with the radius is $e\,\delta V/4\pi r^2\;\delta S\cos\theta$, because $\delta S\cos\theta$ is the projection of δS

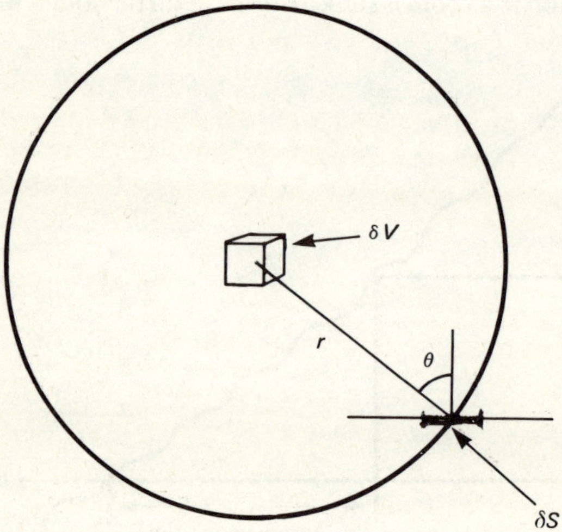

Fig. 6.16 — After a time interval r/c the energy which was initially concentrated within δV is uniformly distributed over the spherical shell of radius r.

onto the sphere. This is the part of the energy initially in δV that subsequently strikes the surface element δS. The intensity, I, is defined as the energy incident on unit area in unit time and so in order to calculate the intensity we wish to find the total energy that strikes δS in a known time interval δt. Since sound travels at speed c this energy must initially have been within a hemisphere of radius $c\,\delta t$ of the surface. The total energy, δE, impinging on area δS in time δt is

$$\delta E = \frac{e\,\delta S}{4\pi} \int_{\substack{\text{hemisphere}\\ \text{of radius } c\,\delta t}} \frac{\delta V \cos\theta}{r^2}.$$

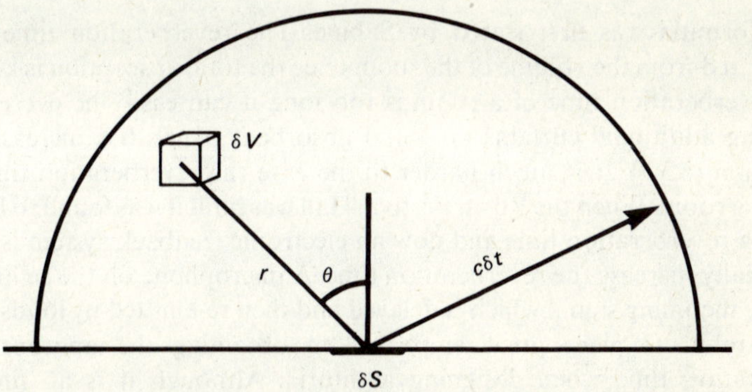

Fig. 6.17 — Sound energy incident on δS in time δt lies initially within a hemisphere of radius $c\delta t$.

In spherical co-ordinates (r, θ, φ), $\mathrm{d}V = r^2 \sin\theta \,\mathrm{d}\varphi\,\mathrm{d}\theta\,\mathrm{d}r$ and this integral is

$$\delta E = \frac{e\,\delta S}{4\pi} \int_0^{c\,\delta t} \mathrm{d}r \int_0^{\pi/2} \sin\theta\cos\theta\,\mathrm{d}\theta \int_0^{2\pi} \mathrm{d}\varphi$$

$$= \frac{e\,\delta S}{4}\, c\,\delta t.$$

The intensity is given by

$$I = \frac{\delta E}{\delta S\,\delta t} = \frac{ec}{4}. \tag{6.31}$$

We see that for diffuse sound fields when each direction of propagation is equally likely $I = ec/4$. We can compare this with the relationship for a plane unidirectional wave given in section 1.8; there $I = ec$.

Substitution of e from (6.31) into equation (6.30) gives

$$\frac{4V}{c}\frac{\partial I}{\partial t} = -Ia,$$

which has the solution

$$I = I_0\,\mathrm{e}^{-act/4V},$$

where I_0 is the initial value of the intensity.

The reverberation time T is defined to be the time at which $I/I_0 = 10^{-6}$, i.e. $e^{-(ac/4V)T} = 10^{-6}$ or

$$T = \frac{V}{ac}\,55.3. \tag{6.32}$$

This formula was first stated by Sabine. The reverberation time can be calculated from the volume of the room once the total absorption is known. If the reverberation time of a room is too long it can easily be decreased by hanging additional curtains or sound absorbent panels (i.e. increasing a in equation (6.32)). It is much harder to increase the reverberation time of an existing room. When the Royal Festival Hall was built it was found to have too short a reverberation time and now an electronic feedback system is used to artificially increase the reverberation time. A microphone on the ceiling picks up the incoming signal which is delayed and then re-emitted by loudspeakers.

Architects place great emphasis on obtaining the appropriate reverberation time when designing auditoria. Although it is an important parameter related to a room's acoustic behaviour it is by no means the only one (it is however the easiest one to measure!). Two halls with the same reverberation time will not necessarily be equally pleasant rooms in which to hear music, because other parameters, like, for example, the time delay between the direct sound field and the first reflection, also affect our enjoyment of sounds.

EXERCISES FOR CHAPTER 6

1. Determine the normal modes of vibration of the air inside a rectangular room 5 m × 6 m × 3 m. What are the resonance frequencies of the room below 50 Hz?

2. An oboe can be considered as a tube of conical cross-section of length l. At the mouth-piece the cross-sectional area is effectively zero and the pressure is finite and at the open end the pressure vanishes. By considering pressure disturbances of the form

$$p'(\mathbf{x}, t) = \frac{f(t - r/c)}{r} + \frac{g(t + r/c)}{r}$$

where r is the distance from the mouth-piece, show that $\omega = n\pi c/l$, with n taking integer values, are a set of resonant frequencies of the air inside the oboe.

3. A dynamic loudspeaker having a cone of 0.2 m diameter and 0.01 kg mass is mounted in the opening of a rigid-walled back-enclosed cabinet whose inside dimensions are 0.3 m × 0.5 m × 0.4 m. (a) What is the Helmholtz resonator frequency of the cabinet alone? (b) If the suspension system of the speaker cone has stiffness of 1 kN/m, what is the cone's resonant frequency when mounted in the cabinet? (c) The cone is then vibrated at a frequency ω by an applied fluctuating force which has a peak value of 20 N. The cone radiates a plane acoustic wave. Find the frequency at which the radiated acoustic power is maximised, and calculate the value of the maximum radiated power. Assume that the 'effective' length of the neck of the Helmholtz resonator is 0.1 m.

4. Determine the resonant frequencies of a duct of length L which has a uniform cross-sectional area and has one open and one closed end. For each resonant frequency, determine how the relationship between the particle velocity and the pressure perturbation varies with the distance x from the closed end.

An exit pipe in a jet engine can be considered to have one acoustically closed end at the turbine and one end open to the atmosphere. The distance between the turbine and the open end is L. Sound waves propagating in the pipe cause unsteadiness in the rate of combustion. Assume that the change in the heat release rate, q, caused by the acoustic waves lags the acoustic particle velocity, u, by an amount α; i.e. $q = k\,\mathrm{e}^{-i\alpha}u$, where k is a positive constant. Show that the reheat system is unstable at the fundamental organ pipe frequency when $-\pi < \alpha < 0$. What would determine the onset of this instability in practice?

5. A bubble of radius $a(t)$ pulsates adiabatically in a fluid with mean density ρ_0, sound speed c and ambient pressure p_0. Show that

$$\rho_0 \frac{\mathrm{d}^2 a}{\mathrm{d}t^2} + \frac{3\gamma p_0}{a_0 c}\frac{\mathrm{d}a}{\mathrm{d}t} + \frac{3\gamma p_0}{a_0^2}(a - a_0) = 0,$$

where a_0 is the mean radius of the bubble and γ is the ratio of specific heats of the gas inside the bubble.

What is the resonance frequency of an air bubble of radius 1 mm in water at atmospheric pressure?

6. Twenty machines are to be installed in a workshop 20 m × 10 m × 5 m. The reverberation time of the empty workshop is 1.5 s. Estimate the maximum sound power output of each machine if the sound pressure level in the workshop is to be less than 90 dB when all the machines are running. Assume that each machine has an absorption of 1 metric sabin.

Chapter 7

Sources of Sound

The problem of identifying sources of sound in practical situations and of determining the radiated sound field is not easy. It is a problem that has challenged, and continues to challenge, many engineers and mathematicians. In Chapters 1 and 2 we investigated the simple sound fields generated by the vibration of boundaries. There the source was definite and easily recognizable, but in many cases the sound sources may be less tangible. Turbulence in vigorous flows for example can generate sound, and in order to describe these more complex distributed sources correctly we need to develop a careful definition of acoustic sources. We take as our starting point the fact that while the identification of sources might be difficult, the total absence of a sound source is easy to recognize.

7.1 SILENCE IS THE ONLY HOMOGENEOUS SOUND FIELD IN UNBOUNDED SPACE!

We ask the question of what sound field can occupy all of space in the absence of boundaries and in doing so insist that there be no incoming field from infinity. The one-dimensional fields $f(x_1 \pm ct)$ certainly cannot satisfy that constraint; they come in from one side and disappear without change on the other. In equation (2.21) we seem to have established that

$$\left(\frac{1}{c^2}\frac{\partial^2}{\partial t^2} - \nabla^2\right)\frac{f(t - r/c)}{r} = 0 \tag{7.1}$$

so that $f(t - r/c)/r$, which certainly satisfies the radiation condition, appears to be a solution of the homogeneous wave equation. But closer inspection of the procedure used in establishing (2.21) will reveal the qualification that the point $r = 0$ must not be considered since the operator,

$$\nabla^2 p' = \frac{1}{r^2}\frac{\partial}{\partial r}\left\{r^2\frac{\partial p'}{\partial r}\right\}, \quad \text{used in equation (2.21),}$$

is a pretty meaningless thing there. We now go on to ask the question of precisely what value the function generated by the left hand side of (7.1) has in the vicinity of the co-ordinate origin where $r = 0$? $f(t - r/c)/r$ has a singularity at the origin, and we observed in Chapter 2 that near a singularity the wave equation reduces to Laplace's equation. So that

$$\left(\frac{1}{c^2}\frac{\partial^2}{\partial t^2} - \nabla^2\right)\frac{f(t - r/c)}{r}$$

is effectively $-\nabla^2(f(t - r/c)/r)$ close enough to the point $r = 0$. $\nabla^2(f(t - r/c)/r)$ is not an ordinary function. It has the remarkable property that its integral over a vanishingly small volume surrounding the origin is finite. That integral is

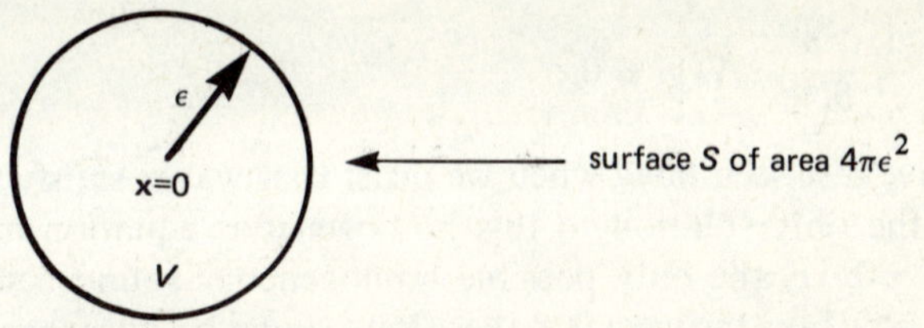

Fig. 7.1 — V is the volume of a small sphere of radius ϵ.

$$\int_V \nabla^2\left(\frac{f(t - r/c)}{r}\right)\mathrm{d}V = \int_S \frac{\partial}{\partial n}\left(\frac{f(t - r/c)}{r}\right)\mathrm{d}S \tag{7.2}$$

$$= 4\pi\varepsilon^2\left\{-\frac{f(t - \varepsilon/c)}{\varepsilon^2} + \frac{1}{\varepsilon}\frac{\partial f}{\partial r}(t - \varepsilon/c)\right\}$$

$$= -4\pi f(t) \quad \text{as } \varepsilon \to 0. \tag{7.3}$$

The function that integrates to unity over a vanishingly small volume surrounding the origin is defined to be the three-dimensional delta function $\delta(\mathbf{x})$.

$$\int_V \delta(\mathbf{x})\,\mathrm{d}V = 1 \quad \text{if } V \text{ includes } \mathbf{x} = 0$$

$$= 0 \quad \text{otherwise} \tag{7.4}$$

i.e. $\delta(\mathbf{x}) = 0$ when $\mathbf{x} \neq 0$, but near $\mathbf{x} = 0$ it is sufficiently large that its integral over any volume enclosing the origin is unity. When (7.4) is multiplied by $4\pi f(t)$ and compared with (7.3) we see that

$$\left(\frac{1}{c^2}\frac{\partial^2}{\partial t^2} - \nabla^2\right)\frac{f(t - r/c)}{r} = 4\pi f(t)\,\delta(\mathbf{x}). \tag{7.5}$$

This is an important result, which will be used later in this chapter as a building block in determining the solution of practical problems of sound generation. For our present purposes we observe that the only way of ensuring that the wave equation is homogeneous, i.e. that both sides of (7.5) are zero everywhere is to set $f = 0$. That suppresses the sound and shows the reasonableness of the somewhat startling heading of this section.

7.2 THE DEFINITION OF A SOUND SOURCE

We have defined sound to be very weak disturbances to a material at rest and determined that such disturbances must conform with the wave equation

$$\frac{1}{c^2}\frac{\partial^2 p'}{\partial t^2} - \nabla^2 p' = 0. \tag{7.6}$$

We have just seen how, when we insist that waves satisfy the radiation condition, the only solution to this homogeneous equation in unbounded space is $p' = 0$, i.e. the only possible homogeneous sound field is absolute silence. We now take the view that there is no sound because there is no source and, somewhat arbitrarily, define the left hand side of equation (7.6) to be that source; we call it $q(\mathbf{x}, t)$.

$$\frac{1}{c^2}\frac{\partial^2 p'}{\partial t^2} - \nabla^2 p' = q. \tag{7.7}$$

In the sound field, q must be zero because sound waves satisfy the homogeneous wave equation. The source region, V, where $q(\mathbf{x}, t)$ is non-zero, is thus clearly separated from the sound field where q must vanish. We have said that this definition of a sound source is somewhat arbitrary, and indeed knowledge of the sound field is not sufficient to uniquely determine the source. To illustrate this non-uniqueness consider the sound field $p' + q$. It is identical to the sound field p', since by definition $q = 0$ in the sound field. Indeed the sound field $p' + f(q)$ is identical to the sound field p' for a vast variety of functions f given only that $f(0) = 0$. But the sources of these three identical sound fields are all different. Since

$$\frac{1}{c^2}\frac{\partial^2 p'}{\partial t^2} - \nabla^2 p' = q,$$

q is the source of p'. But

$$\frac{1}{c^2}\frac{\partial^2 (p' + q)}{\partial t^2} - \nabla^2 (p' + q) = q + \frac{1}{c^2}\frac{\partial^2 q}{\partial t^2} - \nabla^2 q,$$

and so we must conclude that $p' + q$, which is equal to p' in the sound field, is generated by the source field $q + (1/c^2)(\partial^2 q/\partial t^2) - \nabla^2 q$. Similarly it follows from

$$\frac{1}{c^2}\frac{\partial^2(p' + f(q))}{\partial t^2} - \nabla^2(p' + f(q)) = q + \frac{1}{c^2}\frac{\partial^2 f(q)}{\partial t^2} - \nabla^2 f(q)$$

that $q + (1/c^2)(\partial^2 f(q)/\partial t^2) - \nabla^2 f(q)$ is the source of $p' + f(q)$. This demonstrates that the wavefield does not contain sufficient information for its source to be identified. This has important practical consequences. It is no good listening to a sound, or even analysing its structure with the most sophisticated techniques and equipment, if the aim is to describe its source with certainty. That aim is not realisable.

It is however a fact that the sound field generated by a source field q is unique. Once the source is known so is the sound, even though the converse is not true. We can determine that sound field by solving equation (7.7). Formally we rewrite the right-hand side as

$$q(\mathbf{x}, t) = \int_V q(\mathbf{y}, t)\, \delta(\mathbf{x} - \mathbf{y})\, \mathrm{d}^3\mathbf{y}.$$

Equation (7.5) shows that

$$\left(\frac{1}{c^2}\frac{\partial^2}{\partial t^2} - \nabla^2\right)\frac{q(\mathbf{y}, t - r/c)}{r} = 4\pi q(\mathbf{y}, t)\, \delta(\mathbf{x} - \mathbf{y}),$$

where $r = |\mathbf{x} - \mathbf{y}|$. Hence by superposition, i.e. by integrating this equation with respect to $\mathbf{y}$ and comparing the result with equation (7.7), we see that the solution of (7.7) is

$$p'(\mathbf{x}, t) = \int_V \frac{q(\mathbf{y}, t - |\mathbf{x} - \mathbf{y}|/c)}{4\pi|\mathbf{x} - \mathbf{y}|}\, \mathrm{d}^3\mathbf{y}. \tag{7.8}$$

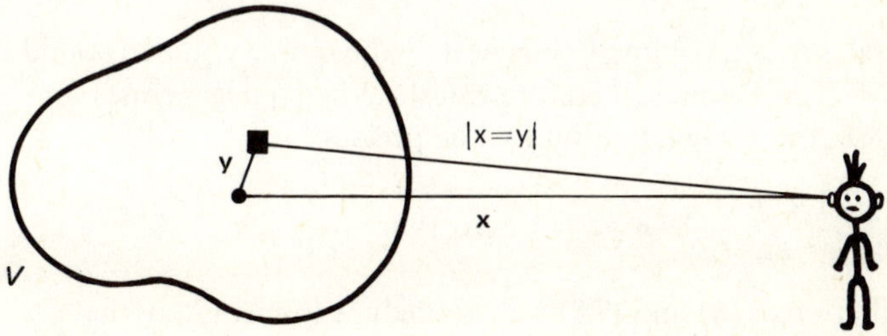

Fig. 7.2 — Sound emitted at $\mathbf{y}$ is heard by the observer at $\mathbf{x}$ at a time $|\mathbf{x} - \mathbf{y}|/c$ after emission.

Each source element at $\mathbf{y}$ radiates sound some of which travels towards the observer at $\mathbf{x}$. The sound arriving at $\mathbf{x}$ at time t must have been launched from the source at the previous time, $t - |\mathbf{x} - \mathbf{y}|/c$, since it travels at speed c a distance $|\mathbf{x} - \mathbf{y}|$ from source to observer. $t - |\mathbf{x} - \mathbf{y}|/c$ is usually called the retarded time. Whenever the region V where q is non-zero is compact i.e. small in comparison with the wavelength the variation in retarded time over the source region is negligible for a distant observer, and (7.8) reduces to the approximate simple form

$$p'(\mathbf{x}, t) = \frac{Q(t - |\mathbf{x}|/c)}{4\pi|\mathbf{x}|} \tag{7.9}$$

where $|\mathbf{x}| \gg |\mathbf{y}|$ and $Q(t) = \int_V q(\mathbf{y}, t)\, \mathrm{d}^3\mathbf{y}$.

Monopole and dipole source distributions

The sound field generated by the source distribution $q(\mathbf{x}, t)$ is an integral (7.8) representing the accumulated sum of wavelets generated by all elements of the source field. Now it may happen that those source elements are particularly arranged so that their sum vanishes. It would be important to recognize that fact in any sound field generated by their net effect because such sound fields might then be extremely weak. The net sound field will only be non-zero because sources at different positions radiate sound to the point $(\mathbf{x}, t)$ at different retarded times, and this de-emphasises the significance of the instantaneous sum of all sources being zero.

The monopole source distribution $q(\mathbf{x}, t)$ is said to degenerate into a dipole distribution if q can be expressed as $-\operatorname{div} \mathbf{f}$ for a function $\mathbf{f}$ that vanishes in the sound field. Such fields are ones in which different elements of q are in destructive interference, their instantaneous sum being zero.

$$q(\mathbf{x}, t) = -\operatorname{div} \mathbf{f}(\mathbf{x}, t)$$

$$Q(t) = \int_{\bar{V}} q(\mathbf{x}, t)\, \mathrm{d}V = -\int_{\bar{V}} \operatorname{div} \mathbf{f}\, \mathrm{d}V = -\int_S f_n(\mathbf{x}, t)\, \mathrm{d}S = 0;$$

the integral is over a volume $\bar{V}$ (bigger than the source volume) bounded by the surface S on which $\mathbf{f}$ must therefore vanish. When q degenerates into a dipole distribution, the wave equation for the pressure is

$$\frac{1}{c^2}\frac{\partial^2 p'}{\partial t^2} - \nabla^2 p' = -\operatorname{div} \mathbf{f}(\mathbf{x}, t) = -\frac{\partial}{\partial x_i} f_i(\mathbf{x}, t). \tag{7.10}$$

From equations (7.7) and (7.8) we can deduce immediately that

$$\left(\frac{1}{c^2}\frac{\partial^2}{\partial t^2} - \nabla^2\right) \int_V \frac{f_i(\mathbf{y}, t - |\mathbf{x} - \mathbf{y}|/c)}{4\pi|\mathbf{x} - \mathbf{y}|}\, \mathrm{d}^3\mathbf{y} = f_i(\mathbf{x}, t).$$

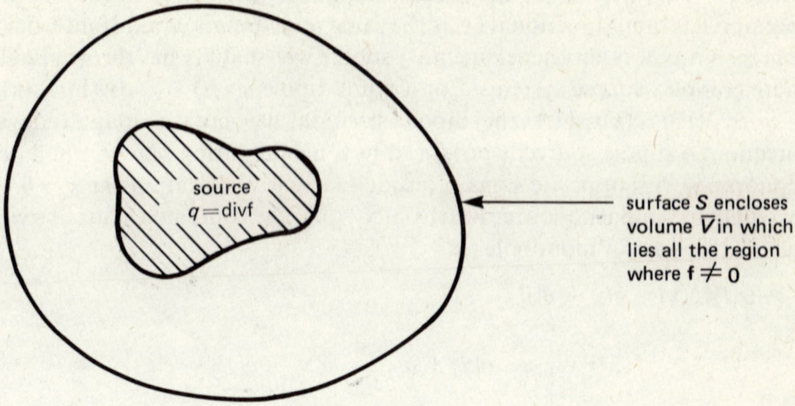

Fig. 7.3 — Volume $\bar{V}$ encloses the source region and $\int_{\bar{V}} q(\mathbf{x}, t)\mathrm{d}V = 0$.

After differentiation with respect to x_i this shows that

$$\left(\frac{1}{c^2}\frac{\partial^2}{\partial t^2} - \nabla^2\right)\frac{\partial}{\partial x_i}\int_V \frac{f_i(\mathbf{y}, t - |\mathbf{x} - \mathbf{y}|/c)}{4\pi|\mathbf{x} - \mathbf{y}|}\mathrm{d}^3\mathbf{y} = \frac{\partial f_i}{\partial x_i} \tag{7.11}$$

and so

$$p'(\mathbf{x}, t) = -\frac{\partial}{\partial x_i}\int_V \frac{f_i(\mathbf{y}, t - |\mathbf{x} - \mathbf{y}|/c)}{4\pi|\mathbf{x} - \mathbf{y}|}\mathrm{d}^3\mathbf{y} \tag{7.12}$$

is the solution of (7.10).

Of course the source q could be regarded as a monopole distribution even when it is expressible as $-\mathrm{div}\,\mathbf{f}$. But since the structure of the sound field is dominated by the destructive interference between adjoining monopole elements it is as well to recognize that from the outset; that is effectively what is done in writing the sound generated by the source distribution of equation (7.10) in the form of equation (7.12) rather than the fully equivalent monopole formula:

$$p'(\mathbf{x}, t) = -\int_V \frac{\mathrm{div}\,\mathbf{f}(y, t - |\mathbf{x} - \mathbf{y}|/c)}{4\pi|\mathbf{x} - \mathbf{y}|}\mathrm{d}^3\mathbf{y}.$$

Point monopoles and dipoles

A source which is concentrated at a point and whose field can be expressed as $q(\mathbf{x}, t) = Q(t)\,\delta(\mathbf{x})$ is called a monopole point source. For such a source (7.8) reduces to

$$p'(\mathbf{x}, t) = \frac{Q(t - r/c)}{4\pi r} \tag{7.13}$$

where $r = |\mathbf{x}|$. $Q(t)$ is called the source strength. The sound field due to a point monopole is omnidirectional i.e. is the same at all points equidistant from the source. A dipole is another elementary source whose study has thrown light on more complex source systems. For a point dipole $q(\mathbf{x}, t) = -\text{div}(\mathbf{F}(t)\,\delta(\mathbf{x})) = -(\partial/\partial x_i)(F_i(t)\,\delta(\mathbf{x}))$. $\mathbf{F}(t)$, the dipole strength, has both a magnitude and direction. A dipole source is produced by a neighbouring pair of equal point monopoles with opposite signs. Consider a negative monopole at $\mathbf{x} = \mathbf{0}$, and an equal and opposite source nearby at $\mathbf{x} = \boldsymbol{\delta}\mathbf{l}$. The combined source strength due to this pair of monopoles is

$$
\begin{aligned}
-Q(t)\{\delta(\mathbf{x}) - \delta(\mathbf{x} - \boldsymbol{\delta}\mathbf{l})\} \\
&= -Q(t)\left\{\delta l_1 \frac{\partial}{\partial x_1}\,\delta(\mathbf{x}) + \delta l_2 \frac{\partial}{\partial x_2}\,\delta(\mathbf{x}) + \delta l_3 \frac{\partial}{\partial x_3}\,\delta(\mathbf{x})\right\} \quad \text{as } \boldsymbol{\delta}\mathbf{l} \to \mathbf{0} \\
&= -\text{div}\{Q(t)\,\boldsymbol{\delta}\mathbf{l}\,\delta(\mathbf{x})\} \quad \text{as } \boldsymbol{\delta}\mathbf{l} \to \mathbf{0}.
\end{aligned}
$$

This is a dipole source with strength $\mathbf{F}(t) = Q(t)\,\boldsymbol{\delta}\mathbf{l}$. A dipole is the limiting source formed as two opposite monopoles are brought together, if the monopoles' strength increases in such a way that the product $\boldsymbol{\delta}\mathbf{l}Q$ remains finite, while $\boldsymbol{\delta}\mathbf{l}$ tends to zero. The dipole pressure field, p'_d, say, is given by (7.12) with $\mathbf{f}(\mathbf{x}) = \mathbf{F}(\mathbf{x})\,\delta(\mathbf{x})$.

$$
p'_d(x, t) = -\frac{\partial}{\partial x_i}\left[\frac{F_i(t - r/c)}{4\pi r}\right]. \tag{7.14}
$$

The pressure at distance r and at an angle θ to the dipole axis is

$$
\begin{aligned}
p'_d(r, \theta, t) &= -\frac{\partial r}{\partial x_i}\frac{\partial}{\partial r}\left\{\frac{F_i(t - r/c)}{4\pi r}\right\} \\
&= \frac{\cos\theta}{4\pi}\left\{\frac{1}{cr}\frac{\partial F}{\partial t} + \frac{F}{r^2}\right\}.
\end{aligned} \tag{7.15}
$$

The structure of this dipole field can be easily understood by considering the dipole in terms of its constituent cancelling monopoles. The sound field has a $\cos\theta$ angular dependence. There are no disturbances at 90° to the dipole axis because an observer there is equidistant from the two monopoles and their sound field exactly cancels. Similarly the pressure disturbances have a maximum magnitude on the dipole axis because in these directions the observer is nearer to one source than the other thereby reducing the cancellation. We see that radial variation of the dipole field has two terms, one falling off with the inverse square of distance and the other linearly. The term with the inverse square dependence clearly dominates the field near the source and for this reason it is called the 'near field'. The other, in which the amplitude falls off inversely with increasing distance, is obviously dominant at large distances and for that reason is called the 'far field'. For monopoles we found

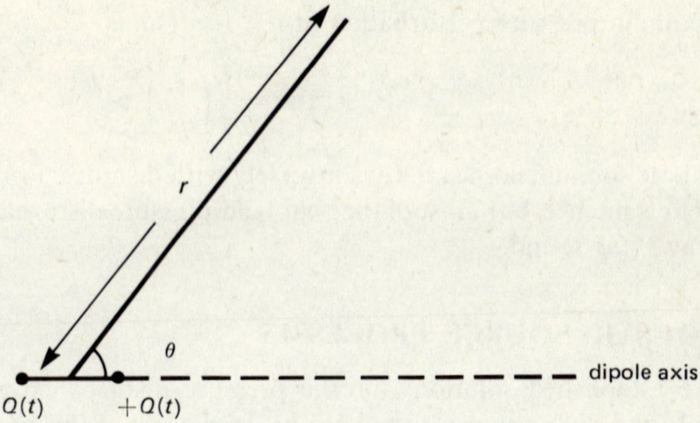

Fig. 7.4 — A dipole can be formed by adjacent, opposite monopole sources.

that the sound field everywhere decays like r^{-1}. Dipoles have a near field, monopoles do not. Because a dipole involves sources of opposite sign there is some cancellation of the acoustic field and dipoles are less efficient radiators than monopoles.

Example

A (i) monopole and (ii) dipole point source at $r = 0$ radiates sound of frequency ω. Determine the ratio of the maximum pressure disturbance at $r = \varepsilon(\ll c/\omega)$ to that at $r = l(\gg c/\omega)$ for the two sources.

The pressure perturbation generated by a monopole has the form

$$p'(r, t) = \frac{A}{4\pi r}\,e^{i\omega(t - r/c)}.$$

The maximum pressure perturbation at $r = \varepsilon$ is $A/4\pi\varepsilon$
The maximum pressure perturbation at $r = l$ is $A/4\pi l$

$$\text{ratio} = \frac{l}{\varepsilon}.$$

The pressure perturbation generated by a dipole is

$$p'(\mathbf{x}, t) = \frac{\cos\theta f}{4\pi}\left\{\frac{i\omega}{rc} + \frac{1}{r^2}\right\} e^{i\omega(t - r/c)}.$$

The maximum pressure perturbation at $r = \varepsilon \ll c/\omega$ is $\dfrac{f}{4\pi\varepsilon^2}$

The maximum pressure perturbation at $r = l \gg c/\omega$ is $\dfrac{\omega f}{4\pi l c}$

$$\text{ratio} = \frac{l}{\varepsilon}\left(\frac{c}{\omega\varepsilon}\right) \gg \frac{l}{\varepsilon}.$$

A monopole sound field just decays inversely with distance. For a dipole the 'near' field is intense, but most of the near field pressure disturbance does not radiate away as sound.

7.3 ACOUSTIC SOURCE PROCESSES

We have established solutions for the pressure disturbance generated by monopole and dipole source distributions. In the rest of this chapter we will investigate mechanisms which produce these source distributions in nature. We saw in Chapter 2 that the homogeneous wave equation is an expression of the conservation laws of mass and momentum in a weakly disturbed homogeneous material in which $p = p(\rho)$. A source process is one in which there is a forcing term in the wave equation and therefore occurs when the criteria leading to the homogeneous equation are violated in some way. For example if a weakly disturbed fluid is bounded by some surface, then sound can be generated at that surface, and we have already seen an example of that in the case of the pulsating sphere in Chapter 2. Many commonly encountered sources of sound are due to the vibration of boundary surfaces. We now show that a source will also result from a breakdown of the state equation $p = p(\rho)$, when, for example, heat is added unsteadily to a fluid.

Combustion noise

When heat is added to a fluid the density of the fluid is a function of two variables, the pressure and the heat supplied to unit mass of fluid, $h(\mathbf{x}, t)$, say:

$$\rho = \rho(p, h).$$

Hence

$$\begin{aligned} \mathrm{d}\rho &= \left.\frac{\partial\rho}{\partial p}\right|_h \mathrm{d}p + \left.\frac{\partial\rho}{\partial h}\right|_p \mathrm{d}\boldsymbol{h} \\ &= \frac{1}{c^2}\,\mathrm{d}p + \left.\frac{\partial\rho}{\partial h}\right|_p \mathrm{d}\boldsymbol{h}. \end{aligned}$$

For a perfect gas $\rho = p/RT$, so that

$$\begin{aligned} \left.\frac{\partial\rho}{\partial h}\right|_p &= -\frac{p}{RT^2}\left.\frac{\partial T}{\partial h}\right|_p = -\frac{p}{RT^2 c_p} \\ &= -\frac{\rho_0(\gamma-1)}{c^2} \end{aligned}$$

giving

$$\mathrm{d}\rho = \frac{1}{c^2}\,\mathrm{d}p - \frac{\rho_0(\gamma - 1)}{c^2}\,\mathrm{d}h.$$

The linearised equation describing motion of a perfect gas with unsteady heat addition is a combination of this with the equation that follows from (2.1) and (2.3)

$$\frac{\partial^2 \rho'}{\partial t^2} - \nabla^2 p' = 0$$

i.e.

$$\frac{1}{c^2}\frac{\partial^2 p'}{\partial t^2} - \nabla^2 p' = \frac{\rho_0(\gamma - 1)}{c^2}\frac{\partial^2 h}{\partial t^2}. \tag{7.16}$$

The solution of this equation is

$$p'(\mathbf{x}, t) = \frac{\rho_0(\gamma - 1)}{4\pi c^2}\frac{\partial^2}{\partial t^2}\int \frac{h(\mathbf{y}, t - |\mathbf{x} - \mathbf{y}|/c)}{|\mathbf{x} - \mathbf{y}|}\,\mathrm{d}^3\mathbf{y}. \tag{7.17}$$

The sound field is forced by the rate of change in the heat addition rate. Steady heating is silent.

Actually the sound of the monopole source caused by unsteady heating is purely due to the unsteady expansion of the fluid at essentially constant pressure. To see this we first recognize that

$$\frac{\rho_0(\gamma - 1)}{c^2}\frac{\partial^2 h}{\partial t^2} = \left.\frac{-\partial^2 \rho}{\partial t^2}\right|_p.$$

Now $-(1/\rho_0)(\partial\rho/\partial t)$ is equal to div $\mathbf{v}$ by mass continuity, and when we recall that the divergence of the velocity field is identically equal to the rate of volumetric growth of unit material volume, it follows that the monopole strength density

$$\frac{\rho_0(\gamma - 1)}{c^2}\frac{\partial^2 h}{\partial t^2},$$

is precisely ρ_0 times the rate of increase in the fractional rate of volumetric growth. We will see below, in equation (7.18), that this is also the strength of the monopole that would be induced by the spontaneous creation of mass.

Sound generation by the linear creation of matter and externally applied forces

Acoustic sources would also arise if the conservation statements were false in some way. Let us suppose that somehow mass is created at a rate $\boldsymbol{m}$ per unit volume and that the newly created mass occupies a fraction β of unit volume from which it has displaced the original fluid of mean density ρ_0. Then the mass conservation equation would be

$$\frac{\partial \rho}{\partial t} + \rho_0 \operatorname{div} \mathbf{v} = m.$$

A source would be similarly induced if the momentum balance was upset by applying to every volume of fluid an external force $\mathbf{f}(\mathbf{x}, t)$, say. The linearised momentum equation would then be

$$\rho_0 \frac{\partial \mathbf{v}}{\partial t} + \operatorname{grad} p' = \mathbf{f}.$$

The density may not now be a function of pressure alone because the newly created mass, occupying a volume fraction β might have density ρ_m that is different from that in the rest of the fluid, ρ_f say, that occupies the fraction $(1 - \beta)$ of unit volume, even though the two materials co-exist at the same pressure. The density ρ, i.e. the total mass contained in unit volume, is consequently

$$\rho = \beta\rho_m + (1 - \beta)\rho_f$$

and m the volumetric mass creation rate must be equal to

$$m = \frac{\partial}{\partial t}(\beta\rho_m)$$

so that

$$\frac{\partial \rho}{\partial t} = m + \frac{\partial \rho_f}{\partial t} - \frac{\partial}{\partial t}(\beta\rho_f) = m + \frac{1}{c^2}\frac{\partial p'}{\partial t} - \frac{\partial}{\partial t}(\beta\rho_0)$$

and

$$\frac{1}{c^2}\frac{\partial^2 p'}{\partial t^2} - \frac{\partial^2 \rho}{\partial t^2} = \frac{\partial^2}{\partial t^2}(\beta\rho_0) - \frac{\partial m}{\partial t}.$$

When these equations are combined to produce a wave equation we see the structure of the sources generated by externally imposed effects.

$$\frac{1}{c^2}\frac{\partial^2 p'}{\partial t^2} - \nabla^2 p' = \rho_0 \frac{\partial^2 \beta}{\partial t^2} - \operatorname{div} \mathbf{f}. \tag{7.18}$$

Mass creation per se is of no acoustical consequence, it is only the volume occupied by that newly created mass that drives a sound field; mass created in a very dense form hardly disturbs the surrounding fluid. We see also that just as we found previously for the case of steady heating, the steady creation of mass (actually volume) is silent and that it is only the unsteady displacement of fluid by an accelerating growth in β that gives rise to sound. That acceleration is a very effective source, being a monopole of strength $\rho_0(\partial^2\beta/\partial t^2)$ per unit volume.

The application of an external force gives rise to the less effective dipole source, the strength of the dipole being equal to the force applied to unit volume of this linearly disturbed homogeneous fluid.

The solution to equation (7.18) is

$$p'(\mathbf{x}, t) = \rho_0 \frac{\partial^2}{\partial t^2} \int_V \frac{\beta(\mathbf{y}, t - |\mathbf{x} - \mathbf{y}|/c)}{4\pi|\mathbf{x} - \mathbf{y}|} \mathrm{d}^3\mathbf{y}$$

$$- \frac{\partial}{\partial x_i} \int_V \frac{f_i(\mathbf{y}, t - |\mathbf{x} - \mathbf{y}|/c)}{4\pi|\mathbf{x} - \mathbf{y}|} \mathrm{d}^3\mathbf{y}. \tag{7.19}$$

The unsteady displacement of mass, just like the unsteady supply of heat, induces a monopole source distribution while the unsteady applied force supports a dipole. If, for example, 'volume' is injected at the origin at a rate $(\partial\beta/\partial t)(t)$, then

$$p'(\mathbf{x}, t) = \rho_0 \frac{\partial^2}{\partial t^2} \left\{ \frac{\beta(t - r/c)}{4\pi r} \right\}, \quad \text{with } r = |\mathbf{x}|.$$

By comparison with the expression (7.13) for a monopole field we see that this volume displacement corresponds to a point monopole source of strength $\rho_0(\partial^2\beta/\partial t^2)$. The pulsating sphere of section 2.2 is acoustically similar to the unsteady creation of mass and the oscillating sphere of constant volume can be considered to induce dipole sound partly because of the unsteady force it applies to the surrounding fluid and partly because of its surface displacement.

Acoustic sources also arise if the flow velocities are too large to comply with linear equations. Then we might collect all the non-linear terms together on the right hand side of the wave equation, a procedure that makes clear the fact that non-linearities in the motion of matter do definitely act as sources of sound which can radiate away from those regions of vigorous activity. That process is the essence of aerodynamic sound production which we deal with in detail in the next section.

7.4 SOUND GENERATION BY FLOW—LIGHTHILL'S ACOUSTIC ANALOGY

We have defined sound to be weak motion about a homogeneous state of rest and seen in the last section how deviations from the assumed state of quiescence induce acoustic source processes. All that introduction was somewhat artificial because we know how impossible it is to create mass and induce external forces, and to do so at a point would undoubtedly violate the 'weak perturbation' assumption that allows the equations to be linearised. The question of how precisely to identify the real origins of a sound wave was not successfully addressed until Lighthill, in 1951, developed his theory of aerodynamic sound in response to the then emerging need to control the noise of jet propelled aircraft. Lighthill defined the source of sound to be the difference between the exact statements of natural laws and their acoustical approximations.

The exact statement of mass conservation equates the rate at which the mass in unit volume increases to the convergence of the mass flux vector $\rho\mathbf{v}$.

$$\frac{\partial \rho}{\partial t} + \frac{\partial}{\partial x_i}(\rho v_i) = 0. \tag{7.20}$$

The exact momentum density is $\rho\mathbf{v}$ and that will increase because of the force that acts on unit volume, the negative gradient of the stress field, p_{ij} say, and because a convergence of moving material will bring with it into unit volume additional momentum. The exact momentum equation is

$$\frac{\partial}{\partial t}(\rho v_i) + \frac{\partial}{\partial x_j}(p_{ij} + \rho v_i v_j) = 0 \tag{7.21}$$

where $p_{ij} = p'\,\delta_{ij} - \tau_{ij}$, δ_{ij} is the Kronecker δ-function ($\delta_{ij} = 1$ if $i = j$, $\delta_{ij} = 0$ if $i \neq j$), and τ_{ij} is the viscous stress.

Now subtract $\partial/\partial x_i$ (7.21) from $\partial/\partial t$ (7.20) to show that

$$\frac{\partial^2 \rho}{\partial t^2} = \frac{\partial^2 \rho'}{\partial t^2} = \frac{\partial^2}{\partial x_i\,\partial x_j}(\rho v_i v_j + p_{ij}),$$

and then subtract from both sides,

$$c^2\nabla^2\rho' = c^2\,\frac{\partial^2 \rho'\,\delta_{ij}}{\partial x_i\,\partial x_j}.$$

Lighthill's equation then follows the definition of Lighthill's stress tensor

$$T_{ij} = \rho v_i v_j + p_{ij} - c^2\rho'\,\delta_{ij}$$

$$\frac{\partial^2 \rho'}{\partial t^2} - c^2\nabla^2\rho' = \frac{\partial^2 T_{ij}}{\partial x_i\,\partial x_j}. \tag{7.22}$$

No approximation is made in this equation. Real material motion can be thought of as an acoustic field in which waves propagate at constant speed c and the source field for those waves is a quadrupole distribution, the strength of the quadrupole in unit volume being Lighthill's stress tensor T_{ij}. That the source distribution is quadrupole is revealed by the double divergence structure of the right-hand side of equation (7.22). There is a double tendency for source elements to cancel. Just as we saw in (7.10) and (7.12) how the dipole field is a divergence of a monopole field, so the quadrupole field is the divergence of a dipole field or the double divergence of a monopole. The solution to the exact equation (7.22) for the sound field ρ' is

$$\rho'(\mathbf{x}, t) = \frac{\partial^2}{\partial x_i\,\partial x_j}\int_V \frac{T_{ij}(\mathbf{y}, t - |\mathbf{x} - \mathbf{y}|/c)}{4\pi c^2|\mathbf{x} - \mathbf{y}|}\,\mathrm{d}V \tag{7.23}$$

where the integration is to range over all $\mathbf{y}$ in the volume V where T_{ij} is non-zero; it is of course zero in the sound field proper according to our basic definition of sound and we take it for granted that most useful sounds do involve motions that accord with the linearised equations where $T_{ij} = 0$. That is in practice a most useful definition.

Every quadrupole element at $\mathbf{y}$ generates a field that travels out at the speed of sound to reach the observer at $\mathbf{x}$ at a time $|\mathbf{x} - \mathbf{y}|/c$ later. The effect of each source element falls off inversely as the distance travelled.

In linear inviscid flow variations in T_{ij} vanish. $\rho v_i v_j$, the unsteady Reynolds stress, is second order in the small fluctuating velocity. $p_{ij} = p' \, \delta_{ij}$ in inviscid flow and $p' = c^2 \rho'$ is the definition of the speed of sound; $\mathrm{d}p/\mathrm{d}\rho = c^2$.

But in turbulent flow T_{ij} does not vanish and turbulence generates sound as a quadrupole source distribution. To estimate the magnitude of the turbulence generated field we have to resort to dimensional analysis because we have pathetically little useful information about turbulence; it is one of nature's best guarded secrets.

Jet noise

We consider a jet flow of velocity U and diameter D. If there is a characteristic frequency in that flow it will have to be U/D, for dimensional consistency. The source frequency U/D will also be the frequency of the sound it generates, so that the time scale of the sound field is D/U. During this time the sound travels, at speed c, one characteristic wave scale, or one wavelength, λ, say,

$$\lambda = c\frac{D}{U}, \quad \lambda = DM^{-1}. \tag{7.24}$$

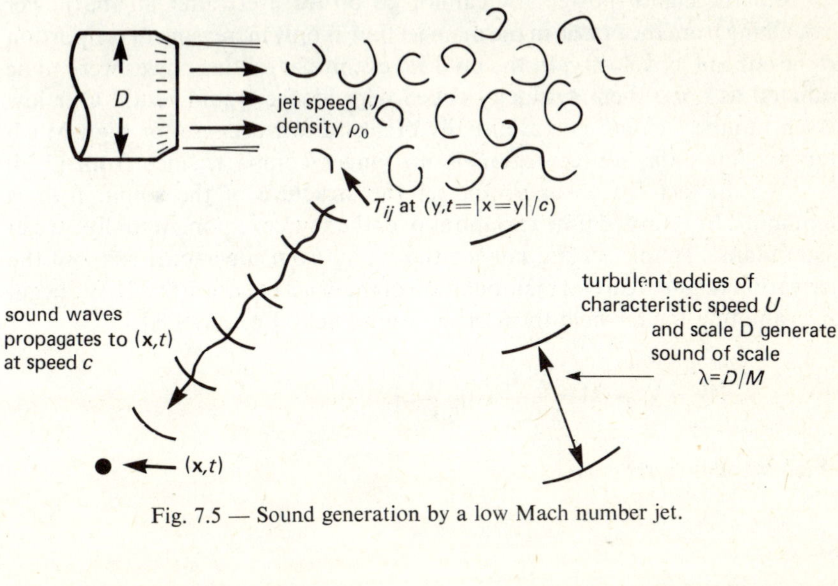

Fig. 7.5 — Sound generation by a low Mach number jet.

M, the jet Mach number, is evidently equal to the ratio of the source dimension to the acoustic wavelength, which we call the compactness ratio. For low Mach number flows $\lambda \gg D$, and the source region is compact. We can therefore ignore the variation of retarded time as $\mathbf{y}$ ranges over all the source volume in the integral in (7.23). The magnitude of the radiated sound field given in equation (7.23) can then be estimated in a straightforward manner:

$$\frac{\partial}{\partial x_i} \text{ is equivalent to } \frac{1}{\lambda} \text{ or } \frac{M}{D}$$

$$\int \mathrm{d}V \text{ is equivalent to } D^3$$

$$T_{ij} \text{ is dimensionally the same as } \rho_0 U^2. \tag{7.25}$$

Together these give the estimate

$$\rho' \sim \rho_0 M^4 \frac{D}{|\mathbf{x}|}$$

and

$$\overline{\rho'^2} \sim \rho_0^2 M^8 \frac{D^2}{|\mathbf{x}|^2} \quad \text{when } M \ll 1. \tag{7.26}$$

This law, that the mean square value of the radiated sound field from cold subsonic jets increases in proportion to the eighth power of the jet velocity, is the celebrated result due to Lighthill. It is extremely well confirmed experimentally. Low Mach number jets are acoustically very inefficient, as are all compact sources.

But the eighth power law cannot go on for ever; after all the power exhausting from the nozzle in the main jet flow is only increasing in proportion to the cube of jet velocity. In the unlikely event that all this power were to be radiated as noise then a velocity cubed dependence would result. Our low Mach number scaling law eventually breaks down because for high Mach number flows the source region is no longer compact, since from (7.24) $D = M\lambda$. In fact for a supersonic jet the structure of the sound field is dominated by retarded time variations over the source region. Actually we can also obtain a simple scaling law for the sound from supersonic jets, but the derivation requires careful manipulation of the exact equation (7.23). We begin by evaluating the far-field form of the sound field. For large $|\mathbf{x}|$

$$|\mathbf{x} - \mathbf{y}| = |\mathbf{x}| - \frac{\mathbf{y} \cdot \mathbf{x}}{|\mathbf{x}|} + 0\left(\frac{1}{|\mathbf{x}|^2}\right),$$

which we will rewrite as

$$|\mathbf{x} - \mathbf{y}| \sim r - y_r,$$

where $r = |\mathbf{x}|$, and $y_r = \mathbf{y}\cdot\mathbf{x}/|\mathbf{x}|$ is the component of $\mathbf{y}$ in the direction in which we consider the observer to lie with respect to the source.

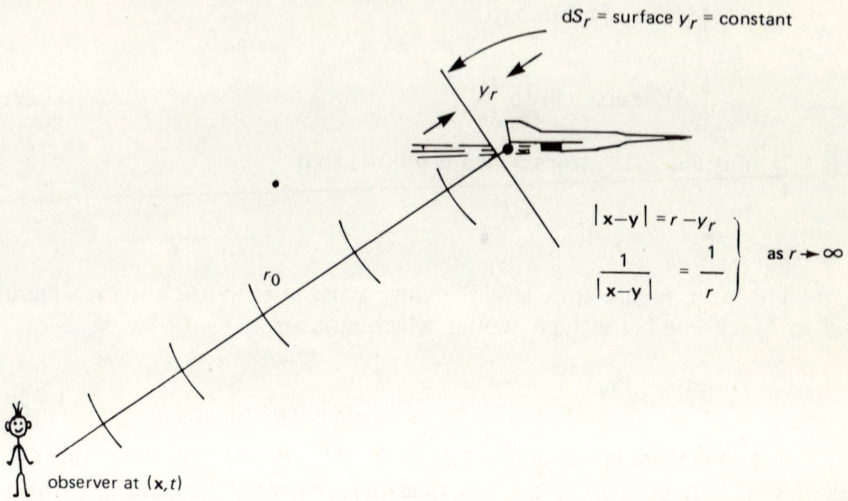

Fig. 7.6 — The geometry for calculating the sound generated by a supersonic jet.

Using this far-field approximation (7.23) simplifies to

$$\rho' = \frac{1}{4\pi c^2 r}\frac{\partial^2}{\partial r^2}\int T_{rr}(\mathbf{y}, \tau)\,\mathrm{d}V \tag{7.27}$$

where $T_{rr} = (x_i x_j/|\mathbf{x}|^2)T_{ij}$ is the component of T_{ij} in the observer's direction, and τ is the retarded time, $\tau = t - r/c + y_r/c$. The repetition of the r suffices does *not* imply tensor summation. The integration variable $\mathrm{d}V$ could be written as $\mathrm{d}y_r\,\mathrm{d}S_r$, where S_r is the surface y_r = constant. In order to bring out the retarded time variations explicitly we will change the integration variable from y_r to τ. Since $\tau = t - r/c + y_r/c$, $\mathrm{d}y_r = c\,\mathrm{d}\tau$ and (7.27) becomes

$$\rho' = \frac{1}{4\pi cr}\frac{\partial^2}{\partial r^2}\int T_{rr}(\mathbf{y}, \tau)\,\mathrm{d}S_r\,\mathrm{d}\tau, \tag{7.28}$$

with $\mathbf{y} = (\mathbf{y}_s, y_r) = (\mathbf{y}_s, c(\tau - t) + r)$, where $\mathbf{y}_s$ is a vector lying in the surface S_r. Now the variation in y_r as τ ranges over the source time scale D/U is of order D/M. For highly supersonic jets this is negligible, and so any variation of y_r can be neglected in estimating the integral in (7.28). Also the operator $\partial/\partial r$ is seen to act only through a variation in y_r so that it induces a magnitude change of order $1/D$. The characteristic magnitude of the density perturbation in the sound induced by supersonic jets is therefore estimated by setting:

$$\frac{\partial}{\partial r} \text{ equivalent to } \frac{1}{D}$$

$$\int_S \mathrm{d}S \text{ equivalent to } D^2$$

$$\int_\tau \mathrm{d}\tau \text{ equivalent to } \frac{D}{U} \tag{7.29}$$

and T_{ij} of order $\rho_0 U^2$, from which it follows that

$$\rho' \sim \rho_0 \frac{D}{r} M.$$

The mean square noise level therefore scales in proportion to the square of jet Mach number at high enough Mach number

$$\overline{\rho'^2} \sim \rho_0^2 M^2 \frac{D^2}{r^2}. \tag{7.30}$$

Actually, there is one other effect to do with the fact that the number of eddies that can be heard at any one time increases with Mach number which

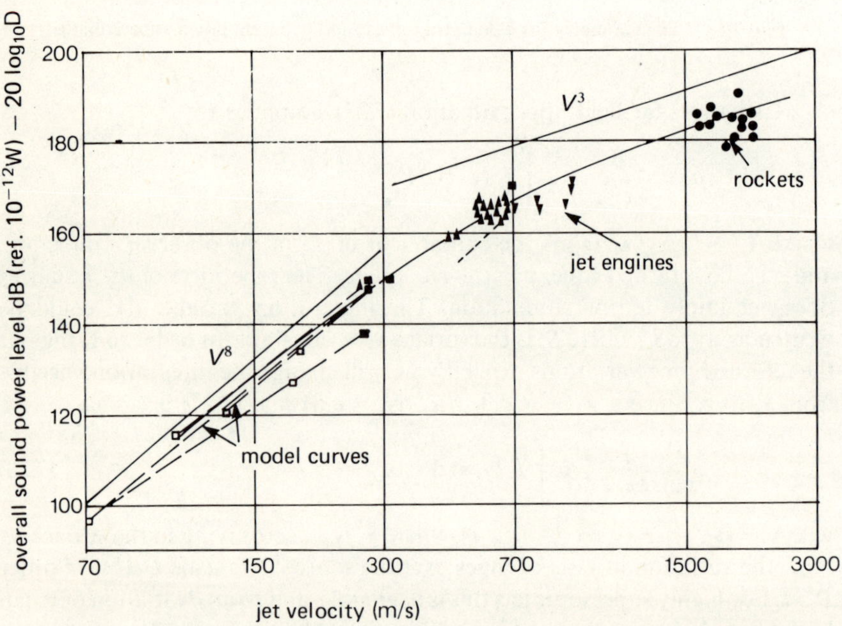

Fig. 7.7 — Variation of acoustic power levels (from Chobotov & Powell: 1957 Ramo Woolridge Corp. Rep. EM-7-7).
● rocket; ▼ turbojet (afterburning); ▲ turbojet (military power); ■ exit velocity > M = 0·8; □ air model (exit velocity < M = 0·8). D is the jet diameter in metres.

changes this law to the velocity cubed dependence that is in agreement with the experimental observations of rocket noise.

7.5 THE SOUND OF FOREIGN BODIES

Because the Lighthill theory is exact it is a convenient way of finding the sound generated by foreign bodies. The Lighthill theory shows that sound generation can be described in terms of an integral of T_{ij}, the Lighthill quadrupole source. When there are foreign bodies in the flow it is sometimes convenient to lump together the quadrupoles distributed over the interior of foreign bodies as boundary sources which can sometimes be specified without need for detailed knowledge of the stress and vibration field within such bodies. Our aim is therefore to express the integral,

$$\frac{\partial^2}{\partial x_i\,\partial x_j}\int \frac{T_{ij}(\mathbf{y}, t - |\mathbf{x}-\mathbf{y}|/c)}{|\mathbf{x}-\mathbf{y}|}\,\mathrm{d}V,$$

over the interior of a foreign body in terms of integrals over the body's surface. The steps taken to achieve this are formal. We will first rewrite the derivatives $\partial/\partial x_i$ with respect to the observer's position in terms of derivatives with respect to source position $\mathbf{y}$. Then application of the divergence theorem will convert the volume integrals into surface integrals. We begin by noting that since $|\mathbf{x}-\mathbf{y}|$ is a symmetric function of $\mathbf{x}$ and $\mathbf{y}$

$$\frac{\partial}{\partial x_i}\left\{\frac{T_{ij}\left(\mathbf{y}, t - \dfrac{|\mathbf{x}-\mathbf{y}|}{c}\right)}{|\mathbf{x}-\mathbf{y}|}\right\} = \frac{\dfrac{\partial T_{ij}}{\partial y_i}}{|\mathbf{x}-\mathbf{y}|} - \frac{\partial}{\partial y_i}\left\{\frac{T_{ij}\left(\mathbf{y}, t - \dfrac{|\mathbf{x}-\mathbf{y}|}{c}\right)}{|\mathbf{x}-\mathbf{y}|}\right\} \tag{7.31}$$

We now write $|\mathbf{x}-\mathbf{y}| = r$ and adopt the notation that the square brackets [ ] imply always that the function they enclose is to be evaluated at position $\mathbf{y}$ at the retarded time $t - r/c$. Using equation (7.21) and the definition of T_{ij}, (7.31) can be re-arranged as,

$$\frac{\partial}{\partial x_i}\left[\frac{T_{ij}}{r}\right] = -\frac{\partial}{\partial y_i}\left[\frac{\rho v_i v_j + p_{ij}}{r}\right] - c^2\frac{\partial}{\partial x_j}\left[\frac{\rho'}{r}\right] - \left[\frac{1}{r}\frac{\partial}{\partial t}(\rho v_j)\right]. \tag{7.32}$$

Differentiation of (7.32) with respect to x_j gives

$$\frac{\partial^2}{\partial x_i\,\partial x_j}\left[\frac{T_{ij}}{r}\right] = -\frac{\partial^2}{\partial y_i\,\partial x_j}\left[\frac{\rho v_i v_j + p_{ij}}{r}\right] - c^2\frac{\partial^2}{\partial x_j\,\partial x_j}\left[\frac{\rho'}{r}\right] - \frac{\partial}{\partial x_j}\left[\frac{1}{r}\frac{\partial}{\partial t}(\rho v_j)\right].$$

The procedure described above shows that

$$\frac{\partial}{\partial x_j}\left[\frac{1}{r}\frac{\partial}{\partial t}(\rho v_j)\right] = \left[\frac{1}{r}\frac{\partial^2}{\partial t\,\partial y_j}(\rho v_j)\right] - \frac{\partial}{\partial y_j}\left[\frac{1}{r}\frac{\partial}{\partial t}(\rho v_j)\right]$$

$$= \left[-\frac{1}{r}\frac{\partial^2 \rho'}{\partial t^2}\right] - \frac{\partial}{\partial y_j}\left[\frac{1}{r}\frac{\partial}{\partial t}(\rho v_j)\right],$$

after using the equation of mass conservation. Hence it follows that

$$\frac{\partial^2}{\partial x_i\,\partial x_j}\left[\frac{T_{ij}}{r}\right] = -\frac{\partial^2}{\partial x_j\,\partial y_i}\left[\frac{\rho v_i v_j + p_{ij}}{r}\right] + \frac{\partial}{\partial y_j}\left[\frac{1}{r}\frac{\partial}{\partial t}(\rho v_j)\right]$$

$$+\left\{\frac{\partial^2}{\partial t^2} - c^2\frac{\partial^2}{\partial x_i\,\partial x_i}\right\}\left[\frac{\rho'}{r}\right]. \tag{7.33}$$

The last term in this equation is zero because, for constant $\mathbf{y}$ the function $[\rho(\mathbf{y}, t - r/c)/r]$ is a solution of the homogeneous wave equation in $(\mathbf{x}, t)$ provided we do not attempt to evaluate it at the point $r = 0$, i.e. at $\mathbf{x} = \mathbf{y}$. We assume that $\mathbf{x}$ and $\mathbf{y}$ are clearly separated i.e. the observer is not on top of the body. Then integrating the remaining terms of equation (7.33) over a stationary volume V_0, which is large enough to contain the foreign body shows, by using the divergence theorem, that

$$\frac{\partial^2}{\partial x_i\,\partial x_j}\int_{V_0}\left[\frac{T_{ij}}{r}\right]\mathrm{d}V = \frac{-\partial}{\partial x_j}\int_S n_i\left[\frac{\rho v_i v_j + p_{ij}}{r}\right]\mathrm{d}S$$

$$+\frac{\partial}{\partial t}\int_S\left[\frac{\rho\mathbf{v}\cdot\mathbf{n}}{r}\right]\mathrm{d}S \tag{7.34}$$

where $\mathbf{n}$ is the direction cosine of the outward (from the body) normal to the surface S that bounds the foreign body. Evidently the quadrupole field interior

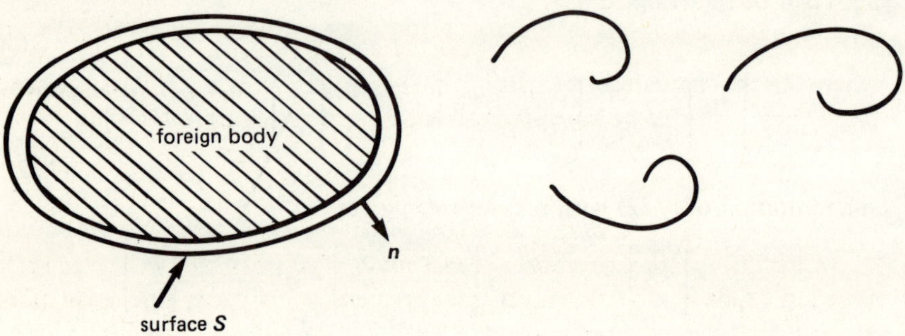

Fig. 7.8 — A stationary control surface enclosing a foreign body.

to the body is equivalent to the sum of a dipole and a monopole. Both these sources are concentrated on the bounding surface S and the dipole strength density is the stress applied to that surface while the monopole strength per unit surface area is the mass flux crossing the surface.

This result allows Lighthill's acoustic analogy to be applied over a limited volume, V, bounded by a stationary surface S, to give the exact expression of the sound field as

$$\rho'(\mathbf{x}, t) = \frac{1}{4\pi c^2} \frac{\partial^2}{\partial x_i \, \partial x_j} \int_V \left[\frac{T_{ij}}{r} \right] \mathrm{d}V - \frac{1}{4\pi c^2} \frac{\partial}{\partial x_j} \int_S n_i \left[\frac{\rho v_i v_j + p_{ij}}{r} \right] \mathrm{d}S + \frac{1}{4\pi c^2} \frac{\partial}{\partial t} \int_S \left[\frac{\rho \mathbf{v} \cdot \mathbf{n}}{r} \right] \mathrm{d}S. \tag{7.35}$$

If the foreign bodies are motionless and rigid then $\mathbf{n} \cdot \mathbf{v} = 0$ on their surfaces in which case they contribute to the sound in inviscid fluid a term

$$-\frac{1}{4\pi c^2} \frac{\partial}{\partial x_i} \int_S n_i \left[\frac{p'}{r} \right] \mathrm{d}S.$$

The retarded time variation is negligible for a sufficiently compact body in which case this field is simply

$$-\frac{1}{4\pi c^2} \frac{\partial}{\partial x_i} \left[\frac{F_i}{r} \right] \tag{7.36}$$

where $F_i = \int_S n_i p' \, \mathrm{d}S$ is the force applied by the body to the fluid. We have seen already in (7.19) how the application of an external force is acoustically equivalent to a dipole and confirm here that all compact rigid bodies in turbulent flow generate sound by virtue of the unsteady forces which act on them. The magnitude of these forces is characteristically

$$F \sim \rho_0 U^2 D^2 \tag{7.37}$$

where D is the characteristic surface dimension and, on setting $\partial/\partial x_i$ equivalent to M/D as in (7.25), we see that the sound of these forces is such that

$$\rho' \sim \rho_0 M^3 \frac{D}{r} \quad \text{and} \quad \overline{\rho'^2} \sim \rho_0^2 M^6 \frac{D^2}{r^2}. \tag{7.38}$$

These body forces generate sound more effectively by a factor of M^{-2} than low Mach number free turbulence because they are only dipole and more efficient than the compact quadrupoles of free turbulence. The wind can be heard to howl, for example, by virtue of the fluctuating force it exerts on obstacles in its path.

Whenever the foreign bodies dilate in volume, they are more efficient still, being monopole. The retarded time variation is again negligible when the bodies are compact and the mean density is uniform so that the third term in equation (7.35) which is then the dominant one becomes

$$\frac{1}{4\pi c^2}\frac{\partial}{\partial t}\left[\frac{Q}{r}\right], \tag{7.39}$$

where $Q/\rho_0 = \int_S v_n \, dS$ is the rate of volume flux out of the surface S into the fluid. We have already seen in (7.19) how volume creation induces monopole source terms and here we see that effect. The pulsating volume of an underwater bubble constitutes a monopole that generates sound more effectively than both the body forces on foreign bodies and the turbulence of surrounding flow. The same type of dimensional analysis will show that the sound field is then such that

$$\rho' \sim \rho_0 M^2 \frac{D}{r} \quad \text{and} \quad \overline{\rho'^2} \sim \rho_0^2 M^4 \frac{D^2}{r^2} \tag{7.40}$$

in these cases where M and D are respectively the Mach number of the surface motion and its linear scale.

It is a fact that if external means of injecting volume unsteadily into the fluid can be found, then that process will be an efficient monopole source of sound. The garden hose containing trapped air bubbles 'chuffs' as the bubbles escape through the nozzle. That process is acoustically very effective. Domestic hot water supplies often make noises for similar reasons.

Similarly aerodynamic loads on foreign bodies are acoustically equivalent to dipole sources of sound, much less efficient than monopoles when the sources are compact but noisier than their surrounding turbulent flow.

Turbulence in contact with foreign bodies makes sound very much more effectively than turbulence in free space. This can easily be demonstrated by blowing onto the edge of a piece of paper. To explain this effect easily, we have to resort to a powerful theorem that simplifies the analysis. That theorem will be described in the next chapter and will allow us to examine in a straightforward manner some more complicated aerodynamic sources which are much noisier and therefore, obviously, of greater practical importance than their homogeneous counterparts.

EXERCISES FOR CHAPTER 7

1. A source radiating a symmetric sound field $(p_0/r) \sin \omega(t - r/c)$ is placed a distance l from an 'anti-source' which generates a pressure $-(p_0/r') \sin \omega(t - r'/c)$ at a distance r' away. By calculating the mean squared pressure far from these sources, show that the acoustic power radiated by this double source combination is

$$\frac{4\pi p_0^2}{\rho_0 c}\left[1 - \frac{c}{\omega l}\sin\left(\frac{\omega l}{c}\right)\right].$$

Compare this with the power than would be radiated by each source in isolation especially in the limits of very small and very large values of l. Why is the presence of the neighbouring source able to prevent the radiation of acoustic power and is the question of 'where has that power gone to?' a sensible one?

2. A pipe of length l has one closed and one open end. Heat is added at a rate $h_0\, e^{i\omega t}$ uniformly across the cross-section of the pipe at a distance x_0 from the closed end. Calculate the sound pressure within the pipe, neglecting any energy loss from the open end.

3. A drummer strikes a drum of square cross-section with sides $0 \leqslant x_1 \leqslant 0.25$ m, $0 \leqslant x_2 \leqslant 0.25$ m. The skin vibrates with a frequency of 40 Hz in such a mode that the normal surface displacement is given by

$$\xi(x_1, x_2, t) = 0.01\, e^{i\omega t} \sin\left(\frac{\pi x_1}{0.25}\right) \sin\left(\frac{\pi x_2}{0.25}\right).$$

Calculate the radiated sound power.

4. A loudspeaker cone is mounted in a hole of 10 cm diameter in an otherwise closed cabinet. Calculate the sound power radiated by a 1 mm axial movement of this cone at a frequency of 100 Hz.

5. (i) Show that the pulsating compact sphere investigated in Chapter 2 is acoustically equivalent to a point monopole of strength $\rho_0 \ddot{V}(t)$, where $V(t)$ is the instantaneous volume of the sphere and the dot denotes differentiation. (ii) Show that the oscillating compact sphere in Chapter 2 is acoustically equivalent to a point dipole with strength $\mathbf{F}(t) + \rho_0 V\dot{\mathbf{U}}(t)$, where $\mathbf{F}(t)$ is the instantaneous force exerted on the fluid by the sphere, $\dot{\mathbf{U}}(t)$ is the sphere's acceleration, and V is the volume of the sphere.

6. The wind blowing past a long wire stretched from $(0, 0, -l)$ to $(0, 0, l)$ exerts a force in the x_1-direction on the wire of strength $L\, e^{i\omega t}$ per unit length of wire. Show that the far-field pressure distribution is

$$p'(\mathbf{x}, t) = -\frac{x_1 i\omega L\, e^{i\omega(t - r/c)}}{r^2} \frac{2\sin(\omega x_3 l/rc)}{\omega x_3/rc}$$

where $r = (x_1^2 + x_2^2 + x_3^2)^{\frac{1}{2}}$. Comment on its limiting form when $\omega l/c$ becomes small.

7. Two jet engines have the same exit temperature and produce the same thrust, but one has twice the diameter of the other. Use Lighthill's eighth power law to estimate which engine is the loudest and by how much.

Chapter 8

The Reciprocal Theorem and Sound Generated near Surfaces of Discontinuity

8.1 RECIPROCITY OF SOURCE AND FIELD

Acoustics problems, and indeed all wave problems, are usually greatly simplified when a harmonic time dependence is considered. There is no loss of generality in this simplification because (cf. section 1.10) Fourier's theorem allows all linear fields to be regarded as a superposition of harmonic waves. We consider again a wave field with harmonic time dependence at angular frequency ω. Since the operator $\partial/\partial t$ is then equal to the algebraic multiplier $i\omega$, the wave operator simplifies

$$\left(\frac{\partial^2 \varphi}{\partial t^2} - c^2\nabla^2\varphi\right) = (-\omega^2\varphi - c^2\nabla^2\varphi). \tag{8.1}$$

We divide this by c^2 and write ω/c as the parameter k, the 'wave number of the sound field; $k = 2\pi/\lambda$, λ being the length of plane sound waves at radian frequency ω.

Time harmonic sound waves conform with the equation

$$k^2\varphi + \nabla^2\varphi = 0. \tag{8.2}$$

This is known as the *Helmholtz equation.* If there is a source, then the right-hand side is not zero. We consider for example a time harmonic field $\varphi_1 \, e^{i\omega t}$ that is generated by a point monopole at $\mathbf{x}_0$ and a point quadrupole at $\mathbf{x}_1$. We therefore define the amplitude of the field, φ_1, to be a solution of the particular inhomogeneous form of the Helmholtz equation

$$k^2\varphi_1 + \nabla^2\varphi_1 = Q\,\delta(\mathbf{x} - \mathbf{x}_0) + T_{ij}\frac{\partial^2\,\delta(\mathbf{x} - \mathbf{x}_1)}{\partial x_i\,\partial x_j}. \tag{8.3}$$

Q is the amplitude of the time harmonic monopole strength and T_{ij} is similarly the amplitude of the time harmonic quadrupole strength tensor.

Now consider φ_2, the amplitude of a second field that is generated by a unit strength point monopole of frequency ω at $\mathbf{x}_2$

$$k^2\varphi_2 + \nabla^2\varphi_2 = \delta(\mathbf{x} - \mathbf{x}_2). \tag{8.4}$$

These equations define the two different fields $\varphi_1\, e^{i\omega t}$ and $\varphi_2\, e^{i\omega t}$ in a space of volume V that is bounded by a surface S. That surface may be made up of several parts, but we can consider, in particular, that it is in two distinct sections. One is a surface at infinity, on which the waves are outgoing with φ proportional to $\partial\varphi/\partial n$. That is because the distant wave is a function of $(r - ct)$, and being harmonic in time, it is also harmonic in r. The normal (radial) derivative is then equivalent to a multiplier and φ must be proportional to $\partial\varphi/\partial n$ or $\partial\varphi/\partial r$. The other part of S is a surface S_1 on which boundary conditions are to be specified.

Now we multiply equation (8.3) by φ_2 and subtract from it the product φ_1 times equation (8.4) to obtain,

$$\varphi_2\nabla^2\varphi_1 - \varphi_1\nabla^2\varphi_2 = \varphi_2 Q\,\delta(\mathbf{x}_0 - \mathbf{x}_0) + \varphi_2 T_{ij}\frac{\partial^2\,\delta(\mathbf{x} - \mathbf{x}_1)}{\partial x_i\,\partial x_j} - \varphi_1\,\delta(\mathbf{x} - \mathbf{x}_2) \tag{8.5}$$

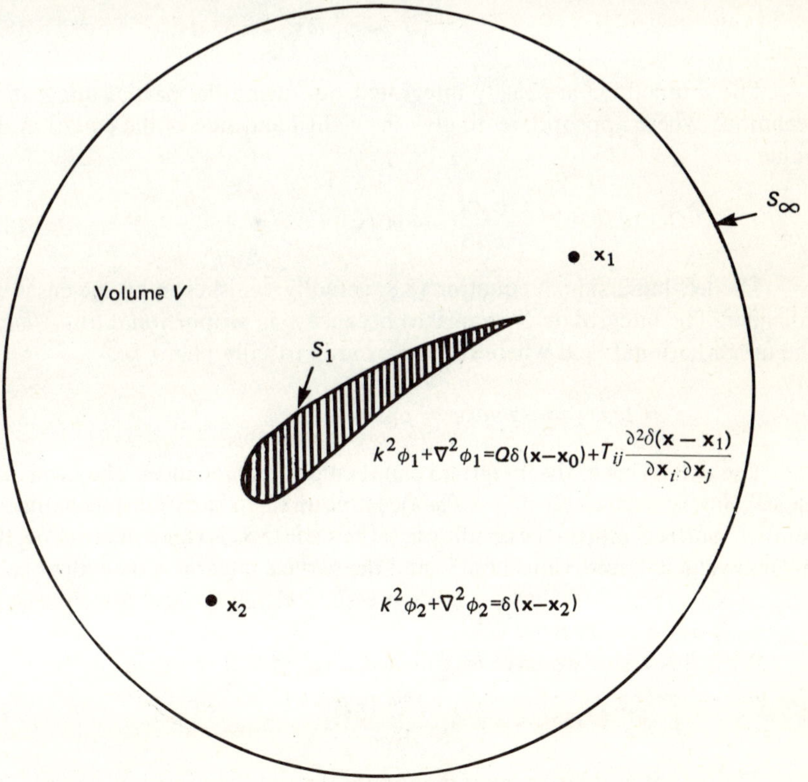

Fig. 8.1 — The reciprocal theorem for a volume V bounded by surfaces S_1 and S_∞ states that $\phi_1(x_2) = Q\phi_2(x_0) + T_{ij}\dfrac{\partial^2\phi_2(x_1)}{\partial x_i \partial x_j}$.

or, equivalently,

$$\frac{\partial}{\partial x_i}\left\{\varphi_2\frac{\partial\varphi_1}{\partial x_i}-\varphi_1\frac{\partial\varphi_2}{\partial x_i}\right\}$$

$$=\varphi_2 Q\,\delta(\mathbf{x}-\mathbf{x}_0)+T_{ij}\varphi_2\frac{\partial^2\,\delta(\mathbf{x}-\mathbf{x}_1)}{\partial x_i\,\partial x_j}-\varphi_1\,\delta(\mathbf{x}-\mathbf{x}_2).\quad(8.6)$$

This equation is now integrated over the volume V that is bounded by the surface S_1 and S_∞, and the divergence theorem used to transform the left -hand side into a surface integral.

$$\int_V \frac{\partial v_i}{\partial x_i}\,\mathrm{d}V \equiv \int_S v_n\,\mathrm{d}S \quad(8.7)$$

$$\int_{S_1+S_\infty}\left\{\varphi_2\frac{\partial\varphi_1}{\partial n}-\varphi_1\frac{\partial\varphi_2}{\partial n}\right\}\mathrm{d}S=\int_V\left\{\varphi_2 Q\,\delta(\mathbf{x}-\mathbf{x}_0)+T_{ij}\varphi_2\frac{\partial^2\,\delta(\mathbf{x}-\mathbf{x}_1)}{\partial x_i\,\partial x_j}\right.$$

$$\left.-\varphi_1\,\delta(\mathbf{x}-\mathbf{x}_2)\right\}\mathrm{d}V. \quad(8.8)$$

The δ functions are easily integrated out, using the partial integration technique where appropriate, to give the right-hand side of the equation the value

$$Q\varphi_2(\mathbf{x}_0)+T_{ij}\frac{\partial^2\varphi_2(\mathbf{x}_1)}{\partial x_i\,\partial x_j}-\varphi_1(\mathbf{x}_2). \quad(8.9)$$

The left-hand side of equation (8.8) actually vanishes in all the cases we consider. The integral on S_∞ vanishes because φ is proportional to $\partial\varphi/\partial n$ in the far radiational field where the waves are virtually plane, i.e.

$$p'=\rho_0 c v_n=-i\omega\rho_0\varphi=\rho_0 c\frac{\partial\varphi}{\partial n}. \quad(8.10)$$

The two terms in the integrand can then be seen to cancel. They will also cancel on the surface S_1 if $\varphi \sim \partial\varphi/\partial n$ is required to satisfy a mechanically imposed surface impedance condition. If the surface S_1 is rigid, then $\partial\varphi/\partial n$, the normal velocity, must vanish on S_1 and the surface integral is then identically zero. It is also zero if the surface S_1 is a pressure release surface, as is the case at the free surface of water.

With this result we have established a reciprocal theorem:

$$\varphi_1(\mathbf{x}_2)=Q\varphi_2(\mathbf{x}_0)+T_{ij}\frac{\partial^2\varphi_2(\mathbf{x}_1)}{\partial x_i\,\partial x_j}. \quad(8.11)$$

The field at $\mathbf{x}_2$ due to a point monopole at $\mathbf{x}_0$ and a point quadrupole at $\mathbf{x}_1$ is precisely the same as the field at $\mathbf{x}_0$ plus the double divergence of the field at $\mathbf{x}_1$ due to a monopole at $\mathbf{x}_2$.

In the particular case where T_{ij} is zero, the reciprocal theorem states that the field at $\mathbf{x}_2$ due to a source at $\mathbf{x}_0$ is identical to the field at $\mathbf{x}_0$ due to a source at $\mathbf{x}_2$. But, as is the case in most aero-acoustic applications, Q is zero, in which case the theorem states that the field at $\mathbf{x}_2$ due to a quadrupole at $\mathbf{x}_1$ is identical to the double gradient of the field at $\mathbf{x}_1$ due to a source at $\mathbf{x}_2$. It is often very much simpler to evaluate the field near a wave scattering body due to a distant source than it is to evaluate directly the distant field due to a complex source near the body. In those cases the reciprocal theorem is a powerful tool of analysis. We given some examples below.

8.2 SOUND SOURCES NEAR A PLANE SURFACE OF DISCONTINUITY

The rigid plane surface is a reflector of sound

The simplest of all boundary effects concern the plane surface. Consider the source distribution depicted in Figure 8.2. The distribution is arbitrary apart from the constraint that it is symmetric about the surface S that lies in the plane $x_3 = 0$, i.e. the source $Q(x_1, x_2, x_3)$ is such that it is equal to $Q(x_1, x_2, -x_3)$.

$$Q(x_1, x_2, x_3) \equiv Q(x_1, x_2, -x_3). \qquad (8.12)$$

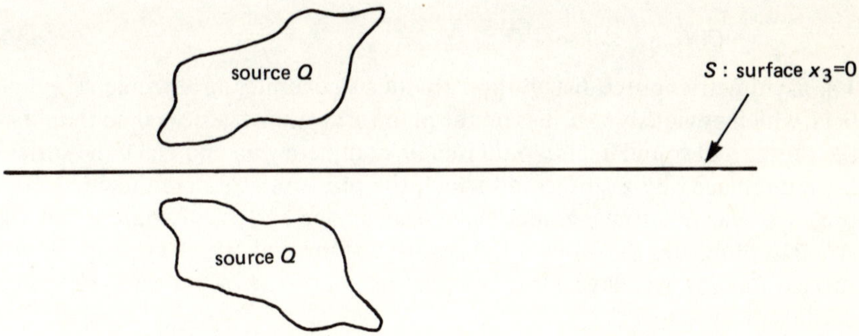

Fig. 8.2 — The source is symmetrical about $x_3 = 0$ and so is the sound field.

The symmetric source distribution obviously generates a sound field with the same symmetry about the plane $x_3 = 0$; if the potential in this sound field is φ then

$$\varphi(x_1, x_2, x_3) = \varphi(x_1, x_2, -x_3) \qquad (8.13)$$

and

$$\left.\frac{\partial\varphi}{\partial n}\right|_{x_3=0} = \frac{\partial\varphi}{\partial x_3}(x_1, x_2, x_3 = 0) - \frac{\partial\varphi}{\partial x_3}(x_1, x_2, -x_3 = 0) = 0; \quad (8.14)$$

the normal velocity evidently vanishes on S. It therefore follows that S could be replaced by a rigid surface without changing the source or sound fields. A rigid surface is identically equivalent to an image system of sources.

The pressure release surface is a 'negative' reflector of sound

A similar geometry to that already considered is depicted in Figure 8.3. This time the source is asymmetric about the surface S, i.e.

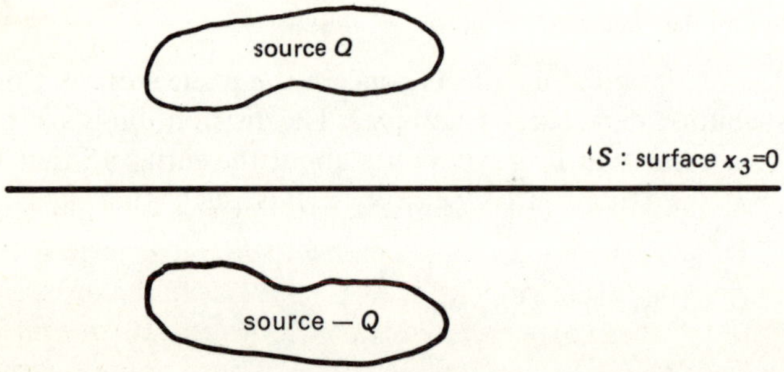

Fig. 8.3 — The source is antisymmetrical about $x_3 = 0$ and so is the sound field.

$$Q(x_1, x_2, x_3) = -Q(x_1, x_2, -x_3). \quad (8.15)$$

The asymmetric source distribution obviously generates an asymmetric sound field, which obviously vanishes on the plane of asymmetry $x_3 = 0$, so that both the source and sound fields would remain completely unchanged if the surface S were replaced by a surface on which the pressure was maintained zero. A pressure release surface is identical to an image system of sources but the strength of the image source is the negative of the real one. A pressure release surface has, as we have already seen in Chapter 4, a pressure reflection coefficient of -1.

Other plane surfaces are 'asymptotic' reflectors

We have seen in Chapter 4 that for plane waves incident on a plane surface of discontinuity, the reflection coefficient is, apart from the two cases considered above, a function of the incidence angle between the wave and surface. For sources near a surface, the waves are not plane so that in effect many angles of incidence are involved and the simple reflection property is lost. But there is nonetheless a simple distant reflection property and the plane wave reflection coefficient is relevant in the asymptotic sense that it describes the very distant sound generated by sources near a plane of discontinuity.

The distant field radiated from a monopole near a plane boundary is simple because, by the reciprocal relation, the problem corresponds to the evaluation of the local field due to a distant reciprocal source. The incident waves in the reciprocal problem are therefore virtually plane waves. We have already seen how plane waves are merely reflected at a plane homogeneous boundary. The reflection coefficient R, which is in general complex and dependent on both the angle of incidence and frequency, has modulus less than, or equal to, unity. The reciprocal local field is thus the superposition of incident and reflected waves with the reflection coefficient being that for *plane* acoustic waves at the angle determined by the position of the distant source. Thus, even though in the actual problem the local field is very far from plane, the distant field it generates is that of two sources, the real source and an image source at the position of specular reflection in the boundary, the strength of the image source being the reflection coefficient $\boldsymbol{R}$ times that of the real source. The reflection coefficient is that for plane acoustic waves at the angle of incidence equal to that of the distant rays.

When the boundary is plane but has inhomogeneous surface properties (stiffening ribs or perforations for example), then the reciprocal theorem is usually still valid but the interaction of plane waves with the surface produces, in addition to the specularly reflected waves, scattered waves centred on the surface inhomogeneities. These give the field much more character. The reflection coefficient then has no meaning because of the many waves that are generated or scattered from the single incident wave.

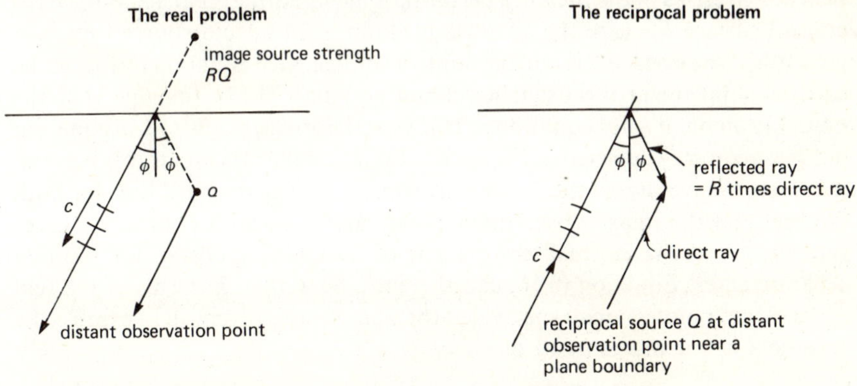

Fig. 8.4 — The distant field radiated from a monopole can be found by using reciprocity.

8.3 KIRCHHOFF'S THEOREM FOR PLANE SURFACES

Kirchhoff's theorem concerns a source free linear acoustic medium bounded by surfaces from which the sound emerges. It relates the sound to surface conditions in a way that can be derived from the expressions we have produced in Chapter 7. For example, provided the observation point **x** is never in the volume *V*, we have shown (equation 7.34) that

$$\frac{\partial^2}{\partial x_i\,\partial x_j}\int_V\left[\frac{T_{ij}}{r}\right]d^3\mathbf{y}$$

$$= -\frac{\partial}{\partial x_j}\int_S n_i\left[\frac{\rho v_i v_j + p_{ij}}{r}\right]d^2\mathbf{y} + \frac{\partial}{\partial t}\int_S\left[\frac{\rho\mathbf{v}\cdot\mathbf{n}}{r}\right]d^2\mathbf{y},$$

so that under the linear conditions to which Kirchhoff's theorem relates, $T_{ij} = \rho v_i v_j = 0$, $p_{ij} = p'\,\delta_{ij}$ and $\rho v_i = \rho_0 v_i$,

$$0 = -\frac{\partial}{\partial x_i}\int_S n_i\left[\frac{p'}{r}\right]d^2\mathbf{y} + \rho_0\frac{\partial}{\partial t}\int_S\left[\frac{\mathbf{v}\cdot\mathbf{n}}{r}\right]d^2\mathbf{y}, \qquad (8.16)$$

provided **x** and **y** are *never* coincident.

Under the same linearity constraints, Lighthill's acoustic analogy applied over a bounded region (equation 7.35) gives the sound field as,

$$4\pi c^2\rho'(\mathbf{x}, t) = -\frac{\partial}{\partial x_i}\int_S n_i\left[\frac{p'}{r}\right]d^2\mathbf{y} + \rho_0\frac{\partial}{\partial t}\int_S\left[\frac{\mathbf{v}\cdot\mathbf{n}}{r}\right]d^2\mathbf{y}. \qquad (8.17)$$

In both equations (8.16) and (8.17) the square brackets imply the retarded value of time $t - |\mathbf{x} - \mathbf{y}|/c$ and the unit vector **n** (components n_i) lies in the direction of the normal to ***S*** at **y** and leads into that region of space in which the observation point **x** lies.

Equation (8.17) is Kirchhoff's theorem and gives the value of the sound field at **x** in terms of the pressure perturbation and normal velocity at the fixed control surface *S* where the acoustic medium is linear and source free.

When the surface ***S*** is a plane control surface, then equation (8.16) can be used to supplement the usual Kirchhoff equation (8.17). Imagine that the observation point **x**′ for equation (8.16) lies at the 'image point' of **x** in equation (8.17), a geometry depicted in Figure 8.5. Equation (8.16) is applicable because the point **x**′ never lies in the real acoustic medium. The retarded time for both the real and the image observation points are identical for surface sources though n_i and the 'normal' component of $\partial/\partial x_i$ are opposites for the two separate cases. Equation (8.16) can therefore be written in terms of the real observation point geometry provided the sign of $n_i v_i$ is reversed, to express the zero field at the image point of **x** as:

$$0 = \frac{\partial}{\partial x_i}\int_S n_i\left[\frac{p'}{r}\right]d^2\mathbf{y} + \rho_0\frac{\partial}{\partial t}\int_S\left[\frac{\mathbf{v}\cdot\mathbf{n}}{r}\right]d^2\mathbf{y}. \qquad (8.18)$$

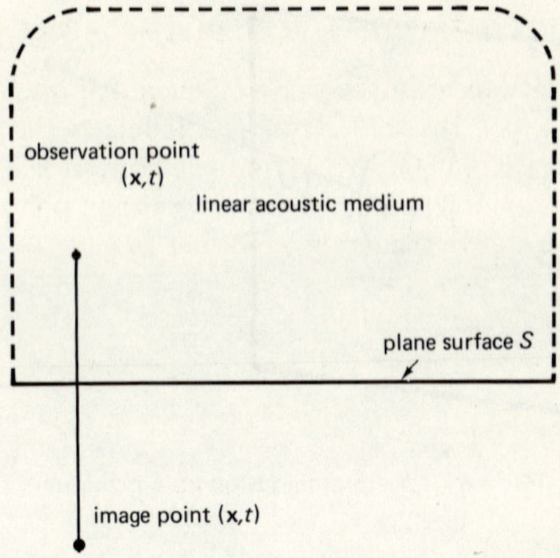

Fig. 8.5 — Equation (8.17) is used at $(\mathbf{x}, t)$ and equation (8.16) at $(\mathbf{x}', t)$ to show that for an infinite plane surface S

$$\rho'(x, t) = \frac{\rho_0}{2\pi c^2}\frac{\partial}{\partial t}\int_S\left[\frac{\mathbf{v}\cdot\mathbf{n}}{r}\right]\mathrm{d}S = \frac{-1}{2\pi c^2}\frac{\partial}{\partial x_i}\int_S n_i\left[\frac{p'}{r}\right]\mathrm{d}S.$$

The notation in this exact result is now precisely the same as in equation (8.17), so that by adding and subtracting the two equations we can express the sound field generated by boundary conditions on a plane surface in terms of either the surface pressure or the normal velocity. The two integrals in the more usual form of Kirchhoff's theorem (equation 8.17) are in fact equal for the plane control surface S.

$$\rho'(\mathbf{x}, t) = \frac{1}{2\pi c^2}\rho_0\frac{\partial}{\partial t}\int_S\left[\frac{\mathbf{v}\cdot\mathbf{n}}{r}\right]\mathrm{d}^2\mathbf{y} = \frac{-1}{2\pi c^2}\frac{\partial}{\partial x_i}\int_S n_i\left[\frac{p'}{r}\right]\mathrm{d}^2\mathbf{y}. \tag{8.19}$$

The sound of a baffled piston

As an example consider that a circular piston of radius a is mounted flush in an infinite hard wall and vibrates with velocity $U_0\,\mathrm{e}^{i\omega t}$. Determine the far-field radiated pressure.

The normal velocity is thus specified everywhere on a plane control surface so that equation (8.19) can be used to give the sound field as,

$$p'(\mathbf{x}, t) = c^2\rho'(\mathbf{x}, t) = \frac{\rho_0 i\omega U_0}{2\pi}\int_{S_0}\frac{\mathrm{e}^{i\omega(t-r/c)}}{r}\,\mathrm{d}S, \tag{8.20}$$

where $\mathbf{S}_0$ is the surface of the piston, r is the distance between the observer at $\mathbf{x}$ and each element of the source $\mathbf{y}$. If we take the surface to be on $x_1 = 0$ the

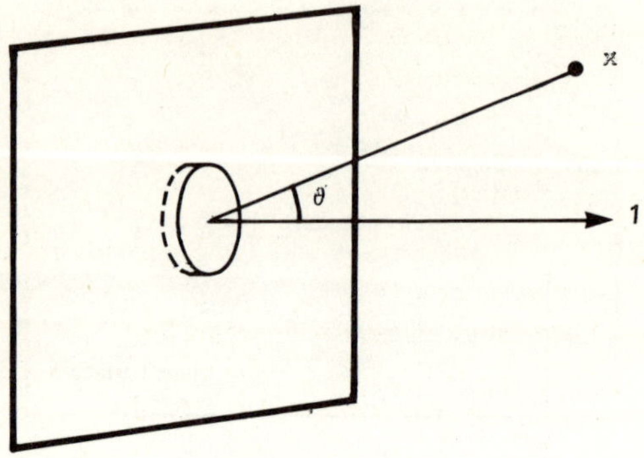

Fig. 8.6 — A vibrating piston on a plane wall.

source position **y** can be written as $(0, y_2, y_3)$ and the surface integral is then to be evaluated over the face of the piston: $y_2^2 + y_3^2 < a^2$.

It is convenient to express **y** in polar co-ordinates, and so we write

$$y_2 = \sigma \cos \mu, \quad y_3 = \sigma \sin \mu.$$

Then the integral in (8.20) becomes

$$\int_0^a \sigma \, \mathrm{d}\sigma \int_0^{2\pi} \frac{\mathrm{e}^{i\omega(t - r/c)}}{r} \, \mathrm{d}\mu.$$

We now write **x** in spherical polar co-ordinates

$$x_1 = R \cos \theta, \quad x_2 = R \sin \theta \cos \varphi, \quad x_3 = R \sin \theta \sin \varphi,$$

so that

$$\begin{aligned} r &= (x_1^2 + (x_2 - y_2)^2 + (x_3 - y_3)^2)^{\frac{1}{2}} \\ &= (R^2 - 2R\sigma \sin \theta \cos(\varphi - \mu) + \sigma^2)^{\frac{1}{2}}. \end{aligned}$$

In the far-field where $R \gg a$,

$$r \sim R - \sigma \sin \theta \cos(\varphi - \mu)$$

and

$$\frac{1}{r} \sim \frac{1}{R}.$$

The distant pressure disturbance can therefore be written as

$$p'(\mathbf{x}, t) = \frac{\rho_0 i\omega U_0}{2\pi R} \mathrm{e}^{i\omega(t - R/c)} \int_0^a \sigma \, \mathrm{d}\sigma \int_0^{2\pi} \mathrm{e}^{i\omega\sigma \sin \theta \cos(\phi - \mu)/c} \, \mathrm{d}\mu. \tag{8.21}$$

The μ-integral is a standard integral and can be evaluated in terms of the function $J_0(z)$ which is defined by

$$J_0(z) = \frac{1}{2\pi}\int_0^{2\pi} e^{iz\cos\theta}\,d\theta.$$

(J_0 is the Bessel function of zero order, and Tables giving its numerical values are published.)

Equation (8.21) can therefore be re-written as

$$p'(\mathbf{x}, t) = \frac{\rho_0 i\omega U_0}{R}\,e^{i\omega(t-R/c)}\int_0^a \sigma J_0\left(\frac{\omega\sigma\sin\theta}{c}\right)d\sigma.$$

The remaining integral can be expressed in terms of J_1, the Bessel function of order one, by using the result

$$zJ_1(z) = \int_0^z \zeta J_0(\zeta)\,d\zeta.$$

Hence

$$\int_0^a \sigma J_0\left(\frac{\omega\sigma\sin\theta}{c}\right)d\sigma = \frac{ca}{\omega\sin\theta}J_1\left(\frac{\omega a\sin\theta}{c}\right).$$

This leads to a final expression for the far-field pressure

$$p'(\mathbf{x}, t) = \frac{\rho_0 i\omega a^2 U_0}{2R}\,e^{i\omega(t-R/c)}\,\frac{2J_1\left(\dfrac{\omega a\sin\theta}{c}\right)}{\dfrac{\omega a\sin\theta}{c}}. \tag{8.22}$$

The values of the function $(2J_1(z)/z)$ can be computed from the Bessel function. They are listed in Tables and are plotted in Figure 8.7. We see that for small values of the Helmholtz number $\omega a/c$, $\dfrac{2J_1(\omega a\sin\theta/c)}{\omega a\sin\theta/c} \sim 1$ for all angles θ and so the pressure perturbation simplifies to

$$p'(\mathbf{x}, t) = \frac{1}{2}\frac{\rho_0}{R}\,i\omega a^2 U_0\,e^{i\omega(t-R/c)} = 2\,\frac{\rho_0[\dot{Q}]}{4\pi R} \tag{8.23}$$

where $\dot{Q}$ is the rate of change of volume displaced by the vibrating piston. In the compact limit the vibrating piston radiates sound omnidirectionally just like a point source but the strength of the field is twice that which the same volume displacement would generate in unbounded space, the factor 2 represents the constructive interference with the image source that is effectively induced by the rigid baffle surface. For larger values of $\omega a/c$ the acoustic pressure perturbation is a function of θ, the angle between the observer's position and an axis through the centre of the piston. On the axis,

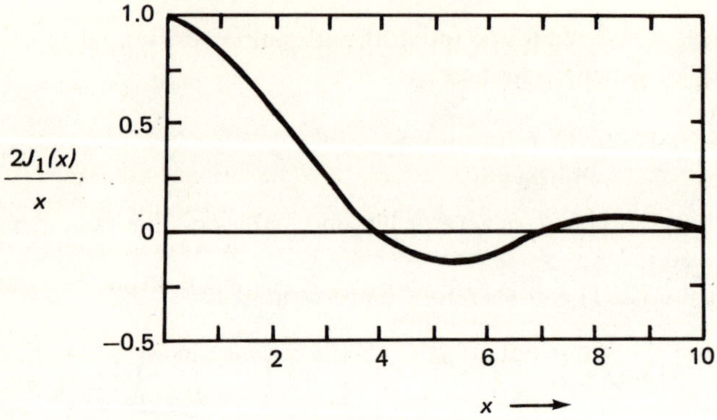

Fig. 8.7 — The variation of $2J_1(x)/x$ with x.

where $\theta = 0$, $\dfrac{2J_1(\omega a \sin\theta/c)}{\omega a \sin\theta/c}$ is again equal to 1. The pressure on the axis is therefore described by equation (8.23) and is identical to the pressure at the same radial distance from a compact piston. For sufficiently big $\omega a/c$ the pressure vanishes at an angle θ_1 such that $\sin\theta_1 = 3.83c/\omega a$. The region $|\theta| < \theta_1$ is called the major lobe for the acoustic pressure. Again if the piston is sufficiently non-compact the pressure perturbation has a second zero at an angle θ_2 such that $\sin\theta_2 = 7.02c/\omega a$. The secondary lobe is detected at angles between θ_1 and θ_2. From the shape of the $2J_1(z)/z$ graph we see that the maximum value of the intensity in the secondary lobe is smaller than in the major lobe. If the piston is grossly non-compact a large number of side lobes may be present, each with a smaller peak intensity than the preceding lobe.

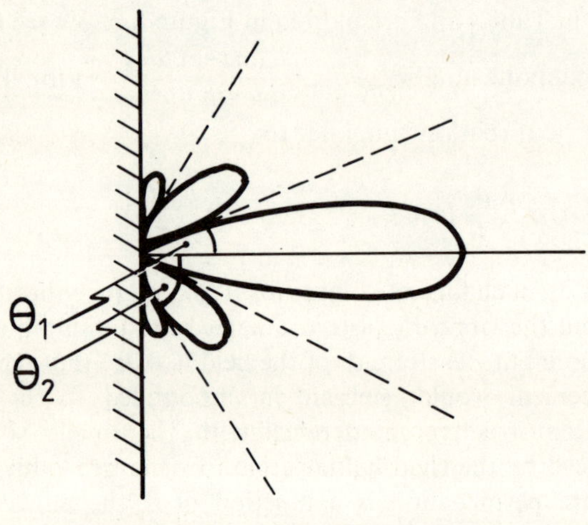

Fig. 8.8 — The radiation pattern produced by a non-compact vibrating piston.

8.4 SOUND SCATTERED BY A SPHERICAL BUBBLE IN TURBULENCE

The sound of turbulent water flows is usually heard because of entrained gas bubbles which resonate and radiate a musical sound even though the turbulent excitation is a broad band noise. The babbling brook is a natural example of this and the sound of domestic water flows is heard only because of entrained bubbles. Reciprocal relations allow the sound of a single spherical bubble to be described very simply.

Lighthill's analogy equates the sound of real turbulence to that induced by a quadrupole distribution in a perfect medium at rest, the bubble surface being one on which boundary conditions must be supplied to supplement the free quadrupole field. The boundary condition on a bubble surface is one in which φ is proportional to $\partial\varphi/\partial n$, the proportionality constant being determined from a consideration of the interior gas properties. Taking one frequency component at a time Lighthill's equation

$$\frac{\partial^2 \rho'}{\partial t^2} - c^2\nabla^2\rho' = \frac{\partial^2 T_{ij}}{\partial x_i\,\partial x_j} \tag{8.24}$$

reduces to

$$\frac{\omega^2}{c^2}\rho' + \nabla^2\rho' = \frac{-1}{c^2}\frac{\partial^2 T_{ij}}{\partial x_i\,\partial x_j} = k^2\rho' + \nabla^2\rho'. \tag{8.25}$$

By the reciprocal theorem, the sound field $\rho'\,e^{i\omega t}$ at $\mathbf{x}_0$ is exactly the same as $(-T_{ij}/c^2)(\partial^2\varphi/\partial x_i\,\partial x_j)\,e^{i\omega t}$, where φ is the amplitude of the hypothetical field at the real quadrupole location that would be generated by a point source at $\mathbf{x}_0$, i.e.

$$k^2\varphi + \nabla^2\varphi = \boldsymbol{\delta}(\mathbf{x} - \mathbf{x}_0). \tag{8.26}$$

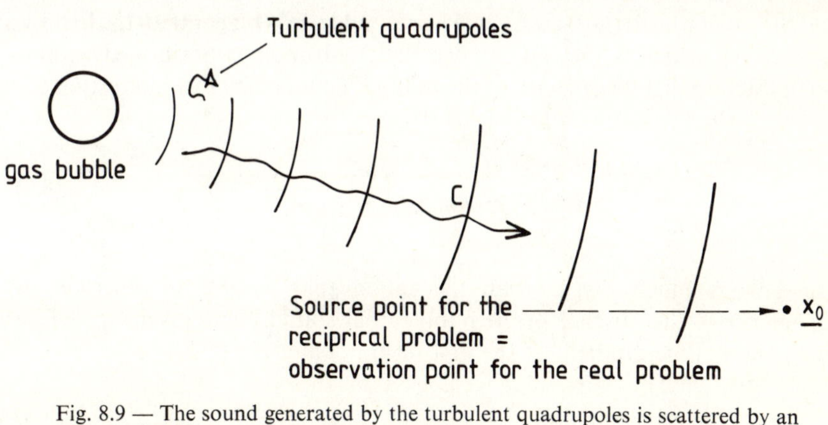

Fig. 8.9 — The sound generated by the turbulent quadrupoles is scattered by an adjacent bubble.

The bubble can be positioned, without loss of generality, to be near the co-ordinate origin so that the incident field in its vicinity due to the distant reciprocal source at $\mathbf{x}_0$ is

$$\varphi\, e^{i\omega t} = -\frac{1}{4\pi|\mathbf{x}_0|}\, e^{i\omega(t-|\mathbf{x}_0|/c)}. \tag{8.27}$$

Equation (6.23) describes the field scattered by a bubble radius a excited by a spatially uniform incident field. It shows that the scattered field at a distance r from the bubble centre is equal to

$$-\frac{a}{r}\, e^{-i\omega(r-a)/c}\left\{1-\frac{\omega_0^2}{\omega^2}\left(1+\frac{i\omega a}{c}\right)\right\}^{-1} \tag{8.28}$$

times the field incident on the bubble. Hence the reciprocal field in the vicinity of the bubble (i.e. the solution to equation (8.26) with the appropriate bubble boundary conditions) is approximately

$$\varphi\, e^{i\omega t} = \frac{-1}{4\pi|\mathbf{x}_0|}\left\{e^{i\omega(t-|\mathbf{x}_0-\mathbf{x}|/c)} - \frac{a}{r}\,\frac{e^{-i\omega(r-a)/c}\, e^{i\omega(t-|\mathbf{x}_0|/c)}}{\left(1-\dfrac{\omega_0^2}{\omega^2}(1+i\omega a/c)\right)}\right\}. \tag{8.29}$$

We have ignored the spatial variation of the incident field over the bubble surface in this expression.

The reciprocal theorem shows how the real scattered sound field is related to this reciprocal field:

$$\rho'\, e^{i\omega t} = \frac{-T_{ij}}{c^2}\,\frac{\partial^2\varphi}{\partial x_i\,\partial x_j}\, e^{i\omega t} \quad \text{where } r^2 = x_i x_i. \tag{8.30}$$

The gradient of φ arises from two parts, one due to the phase variation and the other due to the factor r^{-1}. The second term is bigger than the first by a factor $c/\omega r$, which is very large when the quadrupole is positioned within a small fraction of a wavelength of the bubble. That is the most interesting case and then the far sound field is given by:

$$\rho'\, e^{i\omega t} = \frac{-e^{i\omega(t-|\mathbf{x}_0|/c)}}{4\pi|\mathbf{x}_0|c^2}\,\frac{T_{ij}}{a^2}\,\frac{\partial^2}{\partial x_i\,\partial x_j}\left(\frac{a^3}{r}\right)\frac{1}{\left\{1-\dfrac{\omega_0^2}{\omega^2}(1+i\omega a/c)\right\}}. \tag{8.31}$$

This is much bigger than the sound that would be generated by turbulence in the absence of the bubble, that sound being given by $-T_{ij}/c^2$ times the double gradient of the first term in equation (8.29) i.e.

$$\rho'\, e^{i\omega t}\big|_{\text{no bubble}} = -\frac{e^{i\omega(t-|\mathbf{x}_0|/c)}}{4\pi|\mathbf{x}_0|c^2}\,\frac{\omega^2}{c^2}\, T_{ij}\hat{x}_{0i}\hat{x}_{0j}. \tag{8.32}$$

$\hat{x}_{0i}$ (and $\hat{x}_{0j}$) being the unit vector in the direction of the distant observation point $\mathbf{x}_0$.

Since $(\partial^2/\partial x_i\,\partial x_j)(a^3/r)$ is a term of order one, the magnitude of the bubble scattered field is bigger than the direct field of a quadrupole by a factor of order

$$\left(\frac{c}{\omega a}\right)^2\left(1-\frac{\omega_0^2}{\omega^2}(1+i\omega a/c)\right)^{-1}; \tag{8.33}$$

this is a very big number and explains why bubbles are so very important in the acoustics of unsteady water flows. The intensity is related to the square of the field amplitude, so that the bubble increases the sound output of a nearby quadrupole by a factor of order,

$$\left(\frac{c}{\omega a}\right)^4\left|1-\frac{\omega_0^2}{\omega^2}(1+i\omega a/c)\right|^{-2}$$

i.e.

$$\left|\left(\frac{\omega a}{c}\right)^2-\left(\frac{\omega_0 a}{c}\right)^2\left(1+\frac{i\omega a}{c}\right)\right|^{-2}. \tag{8.34}$$

Equation (6.22) gives the typical Helmholtz number at bubble resonance a value

$$\frac{\omega_0 a}{c}=0.014$$

at which condition (8.34) represents an amplification factor of $(c/\omega_0 a)^6$ i.e. 10^{11} or 110 decibels.

Even well below resonance there is still a very large amplifying effect of bubbles in turbulence because then the factor (8.34) is effectively $(c/\omega_0 a)^4$ which is about 70 dB.

8.5 THE SCATTERING OF AERODYNAMIC SOUND BY A SHARP EDGE

The problem we examine now is typical of those in which sound is generated by turbulent boundary layer flow leaving the sharp trailing edge of an aerofoil. Lighthill's equation shows that the sound at one frequency ω is governed by the quadrupole driven Helmholtz equation

$$k^2\rho'+\nabla^2\rho'=\frac{-1}{c^2}\frac{\partial^2 T_{ij}}{\partial x_i\,\partial x_j}. \tag{8.35}$$

The factor $e^{i\omega t}$ again being understood.

We wish to determine how much sound will be generated by an element of turbulence near the edge of a scattering surface. In fact we wish to examine the field of one turbulent eddy, of volume l^3, positioned a distance r_0 from the edge. By the reciprocal theorem this amounts to the same problem as determining the structure of the sound field that would be generated near the edge by a distant acoustic source. That is a much easier problem because, near the edge the sound satisfies the 'near-a-singularity' form of the wave equation i.e. it is simply a solution of Laplace's equation (cf. Chapter 2). Now the problem is reduced to that of determining the structure of the singularity near an edge in potential flow. That can be done by conformal mapping. The full

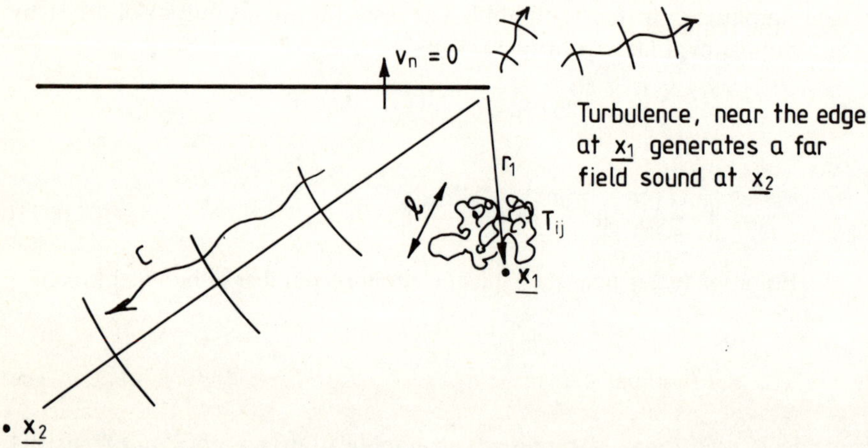

Fig. 8.10 — The sound generated by the turbulent quadrupoles is scattered by the sharp edge.

real axis is mapped onto the top and bottom sides of the negative half of the real axis by the transformation $Z^{\frac{1}{2}}$. The real part then gives the potential near the edge to be proportional to

$$\varphi \sim r_1^{\frac{1}{2}} \sin \tfrac{1}{2}\theta_1. \tag{8.36}$$

It is easily checked that $r_1^{\frac{1}{2}} \sin \frac{1}{2}\theta_1$ is the required solution of Laplace's equation. In two dimensions,

$$\nabla^2 \varphi = \frac{1}{r_1} \frac{\partial}{\partial r_1}\left(r_1 \frac{\partial \varphi}{\partial r_1}\right) + \frac{1}{r_1^2} \frac{\partial^2 \varphi}{\partial \theta_1^2}, \tag{8.37}$$

so that

$$\begin{aligned} \nabla^2(r_1^{\frac{1}{2}} \sin \tfrac{1}{2}\theta_1) &= \frac{1}{r_1} \frac{\partial}{\partial r_1} (\tfrac{1}{2} r_1^{\frac{1}{2}} \sin \tfrac{1}{2}\theta_1) - \frac{1}{4r_1^2} r_1^{\frac{1}{2}} \sin \tfrac{1}{2}\theta_1 \\ &= \tfrac{1}{4} r_1^{-3/2} \sin \tfrac{1}{2}\theta_1 - \tfrac{1}{4} r_1^{-3/2} \sin \tfrac{1}{2}\theta_1 \\ &= 0. \end{aligned} \tag{8.38}$$

Furthermore on the scattering surface,

$$\begin{aligned}\frac{\partial\varphi}{\partial n} &= \frac{1}{r_1}\frac{\partial\varphi}{\partial\theta_1}\\ &= \frac{1}{r_1}\frac{\partial}{\partial\theta_1}(r_1^{\frac{1}{2}}\sin\tfrac{1}{2}\theta_1)\\ &= \frac{1}{r_1^{\frac{1}{2}}}\tfrac{1}{2}\cos\tfrac{1}{2}\theta_1 = 0 \quad \text{when } \theta_1 = \pi;\end{aligned} \tag{8.39}$$

this is the required boundary condition of zero normal velocity.

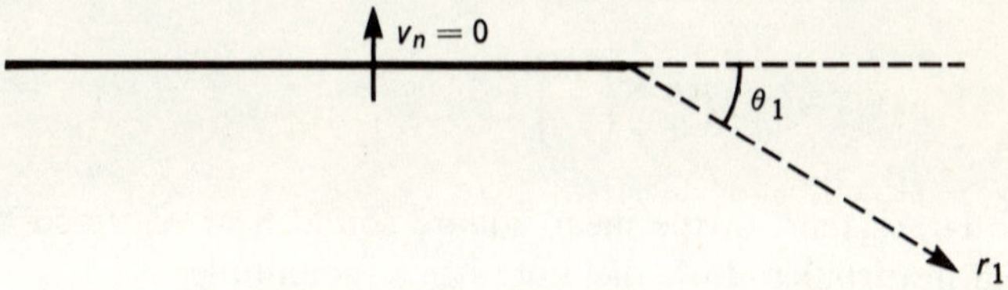

Fig. 8.11 $\phi \sim r_1^{\frac{1}{2}} \sin \frac{1}{2}\theta_1$ satisfies Laplace's equation and the boundary condition.

The density and the pressure perturbation are proportional to the potential in this linear flow so that equation (8.36) gives also the structure of the density field near the edge.

We will use the facts that the wavelength, $2\pi/k$, of the incident sound is the only length scale in the reciprocal edge problem, and that the sound has travelled from the distant source of unit strength to arrive near the edge with an amplitude proportional to r^{-1}. We can then determine the structure of the formula describing the edge scattered field by dimensional analysis alone. Since k^{-1} represents the only available scale for length in the local problem near the edge irradiated by the nearly plane waves of the distant source of the reciprocal field,

$$\varphi(\mathbf{x}_1) \sim (kr_1)^{\frac{1}{2}} \sin\tfrac{1}{2}\theta_1 \frac{1}{r} \tag{8.40}$$

is the form of the local solution to $k^2\varphi + \nabla^2\varphi = \delta(\mathbf{x} - \mathbf{x}_1)$, and

$$\frac{\partial^2\varphi}{\partial x_i\,\partial x_j}(\mathbf{x}_1) \sim k^{\frac{1}{2}} r_1^{-3/2} \sin\tfrac{1}{2}\theta_1 \frac{\hat{r}_{1i}\hat{r}_{1j}}{r} + \text{other gradient terms}; \tag{8.41}$$

the other gradient terms being similar. $\hat{r}_{1i}$ is the direction cosine $\partial r_1/\partial x_{1i}$.

The reciprocal relation, equation (8.11) with ρ' replacing φ_1, with Q set equal to zero and the relevant T_{ij} for equation (8.3) determined from Lighthill's equation (8.35), gives the amplitude of the density perturbation scattered from a quadrupole positioned at distance r_1 from the edge to be:

$$\rho' = \frac{-T_{ij}}{c^2}\frac{\partial^2\varphi}{\partial x_i\,\partial x_j} \tag{8.42}$$

$$\sim \frac{-T_{ij}}{c^2} k^{\frac{1}{2}} r_1^{-3/2} \sin\tfrac{1}{2}\theta_1 \frac{\hat{r}_{1i}\hat{r}_{1j}}{r}. \tag{8.43}$$

It then follows that because $k = Ml^{-1}$ in aerodynamically generated sound, the compactness ratio of an aerodynamic source being simply the flow Mach number, and because the total strength of one eddy $\int \hat{r}_{1i}\hat{r}_{1j}T_{ij}\,\mathrm{d}V$, is typically $\rho_0 U^2 l^3$,

$$\rho'(\mathbf{x}_2) \sim \rho_0 M^{5/2}\frac{l}{r}\left(\frac{l}{r_1}\right)^{3/2}\sin\tfrac{1}{2}\theta_1. \tag{8.44}$$

The largest part of the mean square sound field scattered by an edge immersed in turbulent flow therefore has a magnitude

$$\overline{\rho'^2} \sim \rho_0^2 M^5\left(\frac{l}{r_1}\right)^3\frac{l^2}{r^2}\sin^2\tfrac{1}{2}\theta_1. \tag{8.45}$$

This is more than the free space sound of the same eddy at low Mach number by a factor $M^{-3}l^3/r_1^3$. Scattering surfaces dominate the source structure of low Mach number flows. The noise scattered in this way from boundary layer turbulence by the trailing edges of lifting surfaces is currently though to be of the same order of magnitude as the engine noise of large jet transport aircraft as they approach the airfield to land.

Small compact surfaces also scatter the energy of turbulent eddies to make sound more effectively than could the eddy in free space. But the compact surface is not as effective as the scattering edge of the large rigid surface just considered. Small rigid bodies exert a force on the flow and forces are equivalent to acoustic dipoles. The magnitude of the force, which is the dipole strength, is $\rho_0 U^2 l^2$, and cf. equation (7.38) the dipole radiates a sound field proportional to $(k/c^2 r)$ times the dipole strength. By setting $kl = M$, as we have already seen, the sound of compact surfaces of scale l in turbulent flow with a characteristic velocity level U has magnitude

$$\overline{\rho'^2} \sim \rho_0^2 M^6\frac{l^2}{r^2}; \tag{8.46}$$

this is more effective than free turbulence at low Mach number by the factor M^{-2}.

EXERCISES FOR CHAPTER 8

1. A very small spherically symmetric acoustic transmitter is operated at a radian frequency ω at depth h below the free surface of the ocean. Show that the radiated sound power P is given by

$$P = \frac{(m\omega)^2 \rho_0}{4\pi c}\left[1 - \frac{c}{2h\omega}\sin\left(\frac{2\omega h}{c}\right)\right]$$

where m^2 is the mean square rate of volume outflow from the source and ρ_0 and c are the density and sound speed of the water.

To what depth should such a transmitter be lowered in order to achieve 50% of its deep water radiation efficiency at a frequency of 100 Hz.

2. Due to a misbalance in its components a compact machine exerts a fluctuating force $\mathbf{F}\,e^{i\omega t}$ on the surrounding air. If the sound is not reflected from any boundaries determine the rms pressure that would be radiated to a point $\mathbf{x}$ in the far-field.

This machine is placed on a hard floor and the centre of action of the force can be considered to be a distance $L(\ll \omega/c)$ above the floor. Using the method of images, or otherwise, determine the rms pressure radiated to a point $\mathbf{x}$ in the far-field if (i) $\mathbf{F}$ is in a direction parallel to the floor (ii) $\mathbf{F}$ is perpendicular to the floor.

3. A dynamic speaker cone of diameter 0.3 m is mounted on an infinitely large wall. Assuming that this cone may be regarded as a rigid circular piston of equal radius determine the velocity amplitude of the cone if the axial SPL 10 m away from the cone in air is 100 dB at a frequency of 2 kHz. Show that the major lobe is confined within the region $\theta < 44°$, where θ is the angle between the observer's position and the normal through the centre of the cone. What is the SPL on the wall at a distance of 10 m from the centre of the cone?

Note that $J_1(5.54) = -0.339$.

4. A square piston with sides of length a is mounted flush in an infinite hard wall and vibrates with a velocity $U_0\,e^{i\omega t}$. Determine the far-field radiated pressure. Comment on the form of the radiation pattern in the limits (i) $\omega a/c \ll 1$ and (ii) $\omega a/c \gg 1$.

5. An underwater transducer displaces an unsteady volume Q at a frequency ω. Its sound travels to a fish whose neutral buoyancy is established with a 'swim bladder' which may be assumed spherical, acoustically compact and filled with gas in an isothermal state at mean pressure p_0. Show that the sound pressure scattered back to this fish-seeking sonar arrives at an amplitude,

$$\frac{\rho_0 b\omega^2|Q|}{4\pi s^2|1 - 3p_0/\rho_0\omega^2 b^2|}$$

where b is the radius of the swim bladder, ρ_0 the density of water and s is the range of the fish from the sonar.

6. Use the reciprocal theorem to determine the distant sound field radiated by a quadrupole of strength T_{ij} positioned near a plane surface of infinite extent with impedance Z.

Hence deduce that an adjacent, infinite, plane, sound absorbing surface does not substantially increase the sound generated by turbulence.

Chapter 9

The Sound Field of Moving Sources

In many problems of sound generation the source is in motion relative to the observer. It might be, for example, an aircraft flying overhead or an express train rushing through a station. We saw in Chapter 8 that positioning a source near a boundary can drastically alter the radiated sound field. Source motion also has a considerable effect on the generation of sound. We begin our discussion of moving sources by considering the sound field due to a hypothetical moving point source and will postpone for the moment any discussion of how the source strength can be specified in a practical situation.

9.1 MOVING POINT SOURCES

In Chapter 7 equation (7.8) we showed that the pressure field generated by a source of strength $q(\mathbf{x}, t)$ is given by

$$p'(\mathbf{x}, t) = \int \frac{q(\mathbf{y}, \tau)}{4\pi|\mathbf{x} - \mathbf{y}|} \, \mathrm{d}^3\mathbf{y}, \tag{9.1}$$

where τ, the retarded time, varies over the source and is equal to $t - |\mathbf{x} - \mathbf{y}|/c$. Equation (9.1) is therefore entirely equivalent to

$$p'(\mathbf{x}, t) = \int \frac{q(\mathbf{y}, \tau)}{4\pi|\mathbf{x} - \mathbf{y}|} \, \delta(t - \tau - |\mathbf{x} - \mathbf{y}|/c) \, \mathrm{d}^3\mathbf{y} \, \mathrm{d}\tau \tag{9.2}$$

and this form is more amenable to manipulation to display the effects of source motion. Let us suppose that the source is concentrated at the single moving point $\mathbf{x} = \mathbf{x}_s(t)$, so that $q(\mathbf{x}, t)$ may be written as

$$q(\mathbf{x}, t) = Q(t) \, \delta(\mathbf{x} - \mathbf{x}_s(t)). \tag{9.3}$$

For such a source the sound field described by equation (9.2) is

$$p'(\mathbf{x}, t) = \int \frac{Q(\tau) \, \delta(\mathbf{y} - \mathbf{x}_s(\tau)) \, \delta(t - \tau - |\mathbf{x} - \mathbf{y}|/c)}{4\pi|\mathbf{x} - \mathbf{y}|} \, \mathrm{d}^3\mathbf{y} \, \mathrm{d}\tau. \tag{9.4}$$

The δ-function can be used to evaluate the $\mathbf{y}$-integrals immediately to obtain

$$p'(\mathbf{x}, t) = \int_{-\infty}^{\infty} \frac{Q(\tau)\,\delta(t - \tau - |\mathbf{x} - \mathbf{x}_s(\tau)|/c)}{4\pi|\mathbf{x} - \mathbf{x}_s(\tau)|}\,\mathrm{d}\tau. \tag{9.5}$$

Now δ-functions have the property that for any functions f, g

$$\int_{-\infty}^{\infty} f(\tau)\,\delta(g(\tau))\,\mathrm{d}\tau = \left[\frac{f(\tau)}{|\mathrm{d}g/\mathrm{d}\tau|}\right]_{\tau=\tau^*}, \tag{9.6}$$

where τ^* is a zero of g, i.e. $g(\tau^*) = 0$, and the right-hand side of (9.6) is to be summed over all zeros of g (if there are more than one). This relationship may be proved in a straightforward way by changing the integration variable in $\int f(\tau)\,\delta(g(\tau))\,\mathrm{d}\tau$ from τ to $g(\tau)$. We can use this property to evaluate the τ-integral in equation (9.5), where $g(\tau) = t - \tau - |\mathbf{x} - \mathbf{x}_s(\tau)|/c$, and

$$\frac{\mathrm{d}g}{\mathrm{d}\tau} = -1 + \frac{(x_i - x_{si})}{|\mathbf{x} - \mathbf{x}_s|c}\frac{\mathrm{d}x_{si}}{\mathrm{d}\tau}.$$

$\mathrm{d}x_{si}/\mathrm{d}\tau$ is simply the velocity at which the source position travels through the acoustic medium, $c\mathbf{M}(\tau)$, say, and $(x_i - x_{si})/|\mathbf{x} - \mathbf{x}_s|\,\mathrm{d}x_{si}/\mathrm{d}\tau$ is the component of this source velocity in the direction of the observer. We will denote it by cM_r. Then $|\mathrm{d}g/\mathrm{d}\tau| = |1 - M_r|$ and equation (9.5) yields

$$p'(\mathbf{x}, t) = \frac{Q(\tau^*)}{4\pi r|1 - M_r|}, \tag{9.7}$$

where $r = |\mathbf{x} - \mathbf{x}_s(\tau^*)|$ and τ^* satisfies

$$c(t - \tau^*) = |\mathbf{x} - \mathbf{x}_s(\tau^*)|. \tag{9.8}$$

If there is more than one solution to this equation the right-hand side of (9.7) is to be summed over all such τ^*. The sound heard by the observer at $\mathbf{x}$ and time t is emitted by the source at $\mathbf{x}_s(\tau^*)$ at time τ^*. If this point source were at rest it would emit an omnidirectional sound field, the factor $(1 - M_r)^{-1}$ in (9.7) means that when the source is in motion the pressure perturbation in the forward arc is more intense than that in the rear arc. Equation (9.7) also shows that a source in unsteady motion radiates an unsteady sound field even if $Q(\tau)$ is constant. We demonstrated in Chapter 7 that no sound is generated by the steady addition of heat or by the application of a steady force at a fixed point. If however the point of application accelerates, sound is generated.

We now restrict our attention to the case where the source moves with a constant velocity. Without loss of generality we will assume this uniform velocity to be in the 1-direction, and that at zero time the source is at the origin. Then $M_r = M\cos\theta$ where θ is the angle between the observer's position and the 1-axis measured from the source position at emission time, and equation (9.8) simplifies to

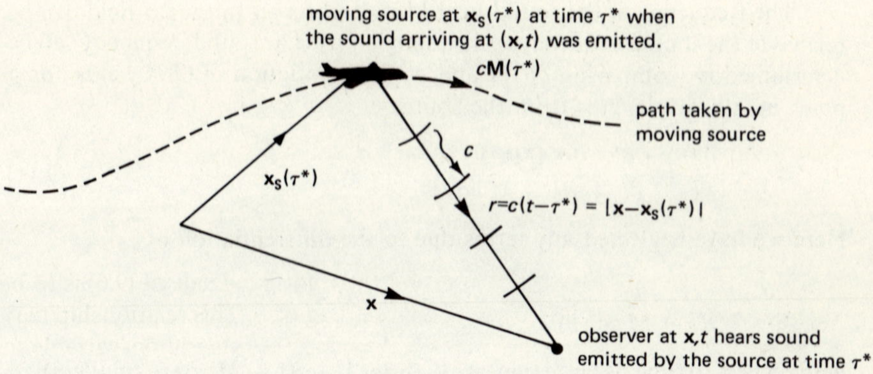

Fig. 9.1 — Emission from a moving source.

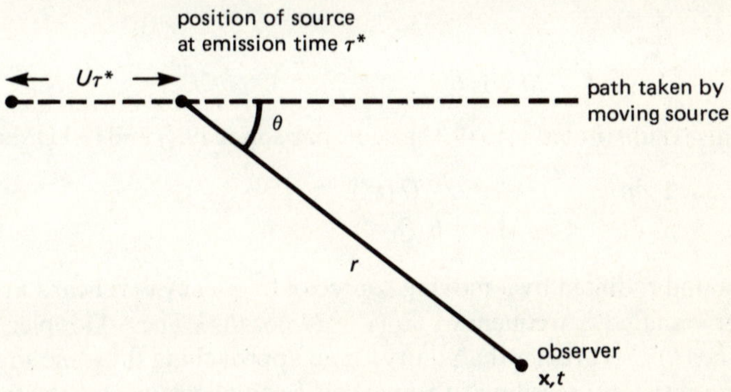

Fig. 9.2 — Emission from a source moving with a constant velocity.

$$c(t - \tau^*) = \{(x_1 - U\tau^*)^2 + x_2^2 + x_3^2\}^{\frac{1}{2}}. \tag{9.9}$$

If the source is near the origin at emission time and the observer is far away, this equation for τ^* may be rewritten as

$$c(t - \tau^*) = |\mathbf{x}|\left(1 - \frac{x_1}{|\mathbf{x}|^2} U\tau^*\right)$$

which has the solution

$$\tau^* = \frac{t - |\mathbf{x}|/c}{1 - M\cos\theta} \quad \text{for } |\mathbf{x}| \gg |U\tau^*|, \tag{9.10}$$

where for this distant observer $\cos\theta = x_1/|\mathbf{x}|\{1 + 0(U\tau^*/|\mathbf{x}|)\}$.

The frequency of the sound heard by an observer in the far-field can be related to the source frequency, ω, in a simple way. The sound frequency can be determined by comparing $\partial p'/\partial t$ and p'. Differentiation of (9.7) yields, for a point many wavelengths from the source,

$$\frac{\partial p'}{\partial t} = \frac{\partial \tau^*}{\partial t} \frac{Q'(\tau^*)}{4\pi r|1 - M \cos \theta|}. \tag{9.11}$$

Here we have neglected any terms due to the differentiation of

$$\frac{1}{r(1 - M \cos \theta)}$$

with respect to time because they are of order $U/\omega r(1 - M \cos \theta)$ smaller than the term retained in (9.11). From differentiating (9.9) we see that

$$1 - \frac{\partial \tau^*}{\partial t} = -M \cos \theta \frac{\partial \tau^*}{\partial t},$$

i.e.

$$\frac{\partial \tau^*}{\partial t} = \frac{1}{1 - M \cos \theta}.$$

When this is substituted into (9.11) a comparison of (9.7) and (9.11) shows that

$$\frac{1}{p'} \frac{\partial p'}{\partial t} = \frac{1}{1 - M \cos \theta} \frac{Q'(\tau^*)}{Q(\tau^*)} \tag{9.12}$$

i.e. the sound radiated by a moving source of frequency ω is heard at $\mathbf{x}$ at the 'Doppler shifted' frequency $\omega/(1 - \mathbf{M} \cos \theta)$. The Doppler factor $(1 - M \cos \theta)^{-1}$ is greater than unity for an approaching subsonic source and less than unity for a receding subsonic source and accounts for the frequency shifts commonly observed when noisy vehicles pass by.

Should the source be approaching at exactly twice the speed of sound, the sound would be heard in perfect time and pitch but backwards. This interpretation of the meaning of the Doppler factor being exactly -1 was made by Lord Rayleigh who imagined a violinist playing his instrument at Mach 2!

We see from (9.12) that if a uniformly moving source were non-evolving, i.e. *if* the only frequencies it contained were at zero frequency, then it can only make sound at non-zero frequency at the conditions $M \cos \theta = 1$. Thus all non-evolving sources in uniform subsonic motion are silent. Supersonic non-evolving sources are heard as sound only in the directions where $M \cos \theta = 1$. These are the 'Mach wave' directions along which weak ballistic shock waves travel.

Equation (9.9) is an implicit relation for the emission time τ^* of sound heard at $\mathbf{x}$ at time t:

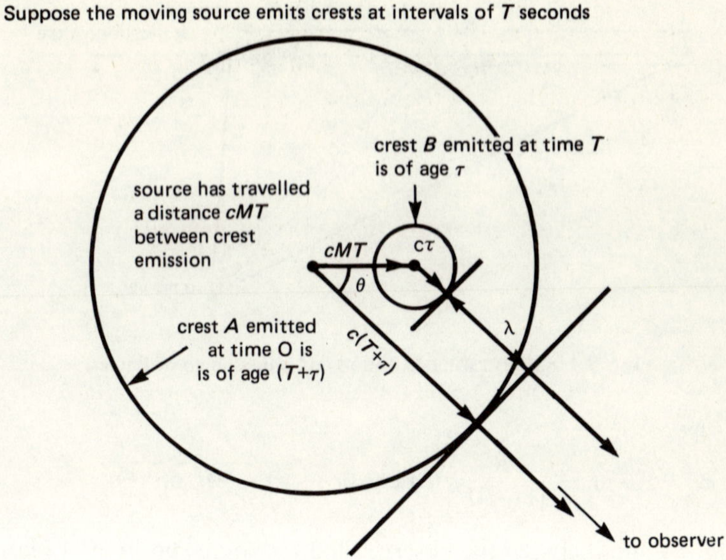

Fig. 9.3 — The Doppler factor for a moving source
Crest A is travelling in the direction of the distant observer and leads crest B by a distance λ
therefore $c(T + \tau) = \lambda + c\tau + cMT\cos\theta$
$\lambda = cT(1 - M\cos\theta)$.
i.e. the wavelength is contracted by the Doppler factor.

$$c(t - \tau^*) = \{(x_1 - U\tau^*)^2 + x_2^2 + x_3^2\}^{\frac{1}{2}}.$$

This equation can be solved to obtain τ^* explicitly for any observer's position. Squaring (9.9) leads to a quadratic equation for τ^* which has solutions

$$\tau^* = \frac{ct - Mx_1 \pm \sqrt{(x_1 - Ut)^2 + (1 - M^2)(x_2^2 + x_3^2)}}{c(1 - M^2)}. \tag{9.13}$$

Spurious roots may have been introduced by squaring and only solutions in (9.13) which have τ^* real and $\tau^* < t$ are solutions of the original equation (9.9). When the source velocity is subsonic only the solution in (9.13) with the lower choice of sign satisfies these conditions and there is only one value of τ^* satisfying (9.9) for each $(\mathbf{x}, t)$.

The radical in (9.13) appears to have a very complicated form, but it simplifies greatly if we introduce variables R, Θ based on the position of the source at time t. With $R = \{(x_1 - Ut)^2 + x_2^2 + x_3^2\}^{\frac{1}{2}}$ and $\cos\Theta = (x_1 - Ut)/R$, the solution for τ^* can be rewritten as

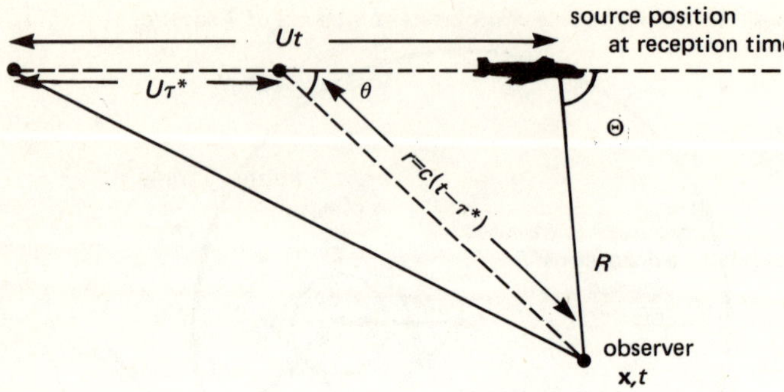

Fig. 9.4 — Illustrating reception and emission co-ordinates.

$$\tau^* = t - \frac{R}{c(1-M^2)}(M\cos\Theta + \sqrt{1-M^2\sin^2\Theta}).$$

R is the distance between the observer and the source position at reception time t, and Θ is the angle between the observer's position and the 1-axis measured from the source position at reception time. (R, Θ) are called *reception time* co-ordinates because they are measured from the source's position at reception time, while (r, θ) which are measured from the position of the source at the time the sound was emitted, are known as *emission time* co-ordinates. Equation (9.7) gave the pressure perturbation in 'emission' co-ordinates;

$$p'(\mathbf{x}, t) = \frac{Q(\tau^*)}{4\pi r|1 - M\cos\theta|}.$$

Rewriting this in terms of 'reception' co-ordinates we obtain

$$r|1 - M_r| = |c(t-\tau^*) - M(x_1 - U\tau^*)| = R\sqrt{1-M^2\sin^2\Theta}$$

and so

$$p'(\mathbf{x}, t) = \frac{1}{4\pi R\sqrt{1-M^2\sin^2\Theta}} \times Q\left(t - \frac{R}{c(1-M^2)}(M\cos\Theta + \sqrt{1-M^2\sin^2\Theta})\right). \tag{9.14}$$

Reception co-ordinates are just co-ordinates in a reference frame that moves with the source. They are therefore the 'natural' co-ordinates in a wind-tunnel geometry in which the source is at rest and the fluid is convected past it with a velocity $-\mathbf{U}$, i.e. R is then the radial distance from the observer to the fixed source, and Θ the angle between the observer's position and the 1-direction measured from the fixed source position in the wind tunnel.

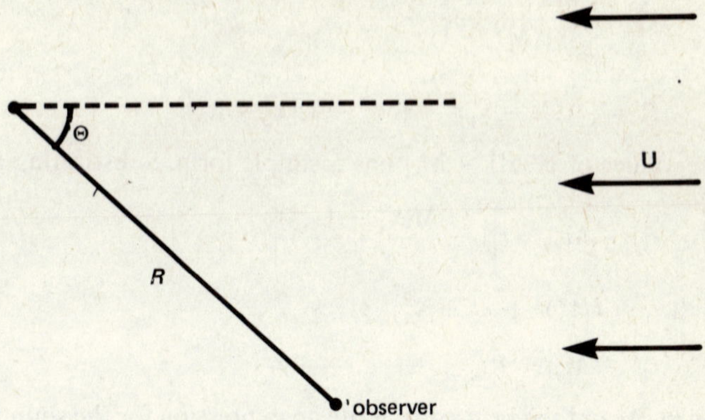

Fig. 9.5 — A fixed source at rest in a wind tunnel.

When the source is moving supersonically the pressure perturbation does not have quite such a simple form. From an inspection of (9.14) or alternatively by returning to the expression for the emission time given in equation (9.13) we see that when M is greater than unity sound is only heard within the region

$$\Theta < \sin^{-1}(1/M),$$

i.e. when

$$(M^2 - 1)^{\frac{1}{2}}(x_2^2 + x_3^2)^{\frac{1}{2}} < Ut - x_1;$$

that is for positions lying within the Mach cone. For positions within this cone

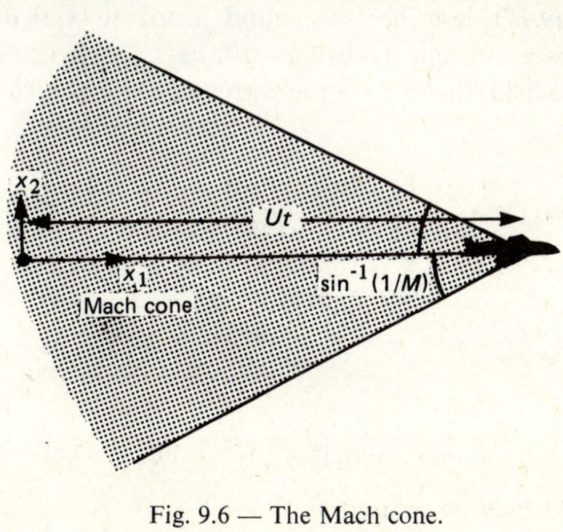

Fig. 9.6 — The Mach cone.

both values of τ^* in (9.13) are real and less than t. The sound emitted at two distinct times is heard simultaneous at $\mathbf{x}$. These two emission times are

$$\tau^*_{1,2} = \frac{Mx_1 - ct \pm \overline{R}}{c(M^2 - 1)}, \tag{9.15}$$

where

$$\overline{R} = \sqrt{(Ut - x_1)^2 - (M^2 - 1)(x_2^2 + x_3^2)}.$$

For these values of τ^*, $r|1 - M_r|$ has a simple form. Substituting for τ^* in

$$r|1 - M_r| = r\left|1 - \frac{M(x_1 - U\tau^*)}{r}\right|$$

with $r = \{(x_1 - U\tau^*)^2 + x_2^2 + x_3^2\}^{\frac{1}{2}}$ gives

$$r|1 - M_r| = \overline{R}. \tag{9.16}$$

Equation (9.7) can then be used to write an expression for the sound heard by an observer at a point $(\mathbf{x}, t)$ within the Mach cone. Since there are now two values of τ^* we sum the contributions to the sound field from both of them, according to the prescription in equation (9.6).

$$p'(\mathbf{x}, t) = \frac{1}{4\pi\overline{R}}\left\{Q\left(\frac{Mx_1 - ct + \overline{R}}{c(M^2 - 1)}\right) + Q\left(\frac{Mx_1 - ct - \overline{R}}{c(M^2 - 1)}\right)\right\}. \tag{9.17}$$

The two components of the sound heard at $(\mathbf{x}, t)$ have different frequencies and phases and there is some interference between these two sound elements. For a point on the Mach cone $\overline{R}$ vanishes and (9.17) shows that the pressure perturbation produced by a point source is singular. We will now go on to consider a source distribution of finite size to show that finite source dimension controls this singularity.

Equation (9.17) describes the sound heard at $(\mathbf{x}, t)$ due to a moving source that passes through the origin at time $\tau = 0$. It can be used to write down the sound field due to a source with velocity $\mathbf{U}$ which is initially at the point $(\eta_1, 0, 0)$:

$$p'(\mathbf{x}, t) = \frac{1}{4\pi\overline{R}}\left\{Q\left(\frac{M(x_1 - \eta_1) - ct + \overline{R}}{c(M^2 - 1)}\right) + Q\left(\frac{M(x_1 - \eta_1) - ct - \overline{R}}{c(M^2 - 1)}\right)\right\} \tag{9.18}$$

where now

$$\overline{R} = \sqrt{(Ut - x_1 + \eta_1)^2 - (M^2 - 1)(x_2^2 + x_3^2)}. \tag{9.19}$$

$\overline{R}$ factorises and may be written as

$$\bar{R} = \sqrt{(Ut - x_1 + \eta_1 + \sigma)(Ut - x_1 + \eta_1 - \sigma)}, \tag{9.20}$$

with $\sigma = \sqrt{(M^2 - 1)(x_2^2 + x_3^2)}$.

A moving source with length l in the 1-direction can be considered as a superposition of moving point sources, and integration of (9.18) shows that the sound field generated by such a source distribution is given by

$$p'(\mathbf{x}, t) = \frac{1}{4\pi} \int_{\eta_1 = 0}^{l} \frac{1}{\bar{R}} \left\{ Q\left(\eta_1, \frac{M(x_1 - \eta_1) - ct + \bar{R}}{c(M^2 - 1)}\right) + Q\left(\eta_1, \frac{M(x_1 - \eta_1) - ct - \bar{R}}{c(M^2 - 1)}\right) \right\} d\eta_1. \tag{9.21}$$

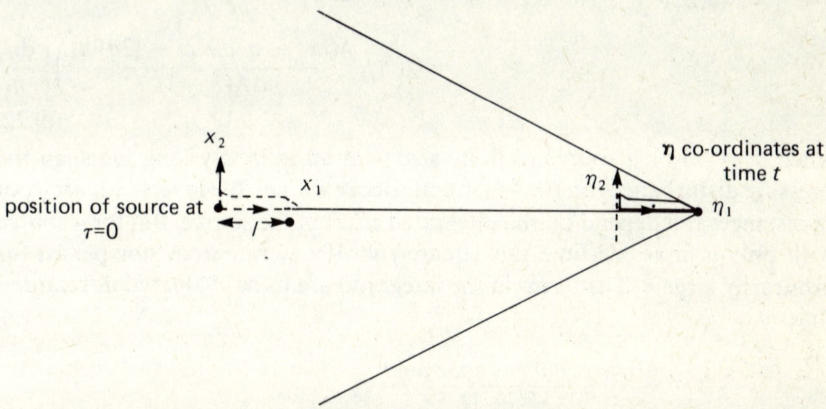

Fig. 9.7 — A supersonically moving source of length l.

If the source has a finite length in the 1-direction the sound heard is that due to an integral of the source elements $Q(\eta_1, \tau^*)$ over η_1, where the retarded time τ^* is a function of η_1. Now

$$c(t - \tau^*) = |\mathbf{x} - \boldsymbol{\eta} - \mathbf{U}\tau^*|$$

and so

$$\frac{d\tau^*}{d\eta_1} = \frac{x_1 - \eta_1 - U\tau^*}{cr(1 - M_r)}.$$

For directions in which the effective wavelength is much longer than the body dimension, $c(1 - M_r) \gg \omega l$, the effect of the variation of retarded time as η_1 varies over the range $(0, l)$ is negligible, and (9.21) shows that the observer at $(\mathbf{x}, t)$ hears sound emitted from the supersonically moving source at two distinct times. However for an observer on the Mach cone, $M_r \approx 1$, τ^* varies rapidly with η_1 and as η_1 just varies over the range $(0, l)$ the retarded time can vary over the entire history of the source. A distant observer hears the accumulated sound emitted by the source during its entire lifetime. To demonstrate this effect explcitly we consider a point on the Mach cone at which $Ut - x_1 = \sigma$, $\bar{R} = \eta_1^{\frac{1}{2}}(2\sigma + \eta_1)^{\frac{1}{2}}$. $\bar{R}$ therefore vanishes and the integrand in (9.21) is singular at $\eta_1 = 0$. This suggests that nearly all the important contributions to the integral come from near this point. We therefore approximate $\bar{R}$ by

$$\bar{R} \approx \eta_1^{\frac{1}{2}}(2\sigma)^{\frac{1}{2}} \quad \text{for } \sigma \gg l.$$

Hence, for a point on the Mach cone away from the apex,

$$p'(\mathbf{x}, t) = \frac{1}{4\pi(2\sigma)^{\frac{1}{2}}} \int_{\eta_1=0}^{l} \left\{ Q\left(\eta_1, \frac{M(x_1 - \eta_1) - ct + (2\sigma)^{\frac{1}{2}}\eta_1^{\frac{1}{2}}}{c(M^2 - 1)}\right) + Q\left(\eta_1, \frac{M(x_1 - \eta_1) - ct - (2\sigma)^{\frac{1}{2}}\eta_1^{\frac{1}{2}}}{c(M^2 - 1)}\right)\right\} \frac{d\eta_1}{\eta_1^{\frac{1}{2}}} \tag{9.22}$$

when $\sigma \gg l$. For a source of finite size with an infinitely long life-span the pressure disturbances on the Mach cone decay with $\sigma^{\frac{1}{2}}$, the inverse square-root of distance, and depend on the integrated effect of the source. But for a source with only a finite life-time this square-root decay rate does not persist for arbitrarily large σ. The terms in the integrand are to be evaluated at retarded times

$$\tau = \frac{M(x_1 - \eta_1) - ct \pm \eta_1^{\frac{1}{2}}(2\sigma)^{\frac{1}{2}}}{c(M^2 - 1)}.$$

If the source only emits sound for a finite time $0 < \tau < T$ say, then for an observer in the very distant far-field, $\sigma \gg c^2T^2/l$, the integrand is zero over most of the range of integration in (9.22), and the result of the integration in (9.22) is therefore small. Since the finite time duration of the source is then the crucial thing, it is appropriate to express the integrals in (9.22) in terms of an integral over retarded time, τ:

$$\frac{\partial \tau}{\partial \eta_1} = \frac{-M \pm \frac{1}{2}\eta_1^{-\frac{1}{2}}(2\sigma)^{\frac{1}{2}}}{c(M^2 - 1)}.$$

Equation (9.22) then shows that the Mach wave sound vanishes unless $Mx_1 = ct$. Now $Mx_1 = ct$ and $Ut - x_1 = \sigma$ combine to give $|\mathbf{x}| = ct$, and for such a position (9.22) gives the field on the distant Mach cone, a field that is usually called the *Mach wave* sound, as

$$p'(\mathbf{x}, t) \underset{|\mathbf{x}| \to \infty}{=} \frac{c(M^2 - 1)}{4\pi\sigma} \int_0^T Q(\eta_1(\tau), \tau)\, d\tau = \frac{U}{4\pi|\mathbf{x}|} \int_0^T Q(\eta_1(\tau), \tau)\, d\tau. \tag{9.23}$$

The Mach wave sound heard in the very far-field therefore decays inversely with distance. All the sound ever released during the entire life history of the source is heard by the distant observer in one big bang! $\eta_1(\tau) \sim 0$ and (9.23) is effectively an integral over $Q(\eta_1, \tau)$ at constant η_1. Now $\int Q(0, \tau)\,d\tau = \hat{Q}(0)$, where $\hat{Q}(\omega)$ is the Fourier transform of $Q(0, \tau)$, $\hat{Q}(\omega) = \int Q(0, \tau)\,e^{-i\omega\tau}\,d\tau$. It is only those elements of the moving source at zero frequency that generate sound in the distant Mach wave.

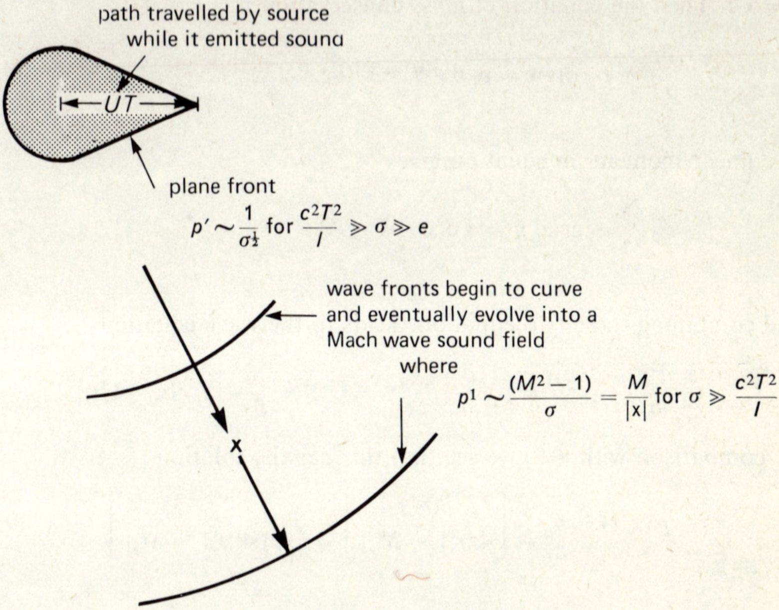

Fig. 9.8 — The very distant Mach cone.

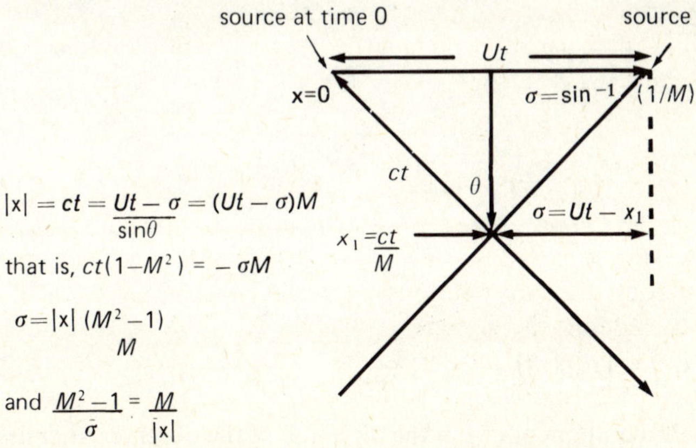

Fig. 9.9 — The Mach wave geometry.

So far we have described the sound due to a hypothetical moving source Q without addressing the question of how such a source could be realised in practice. We saw in Chapter 7 that the linear creation of mass and externally applied forces gave rise to simple source fields. We will now investigate the form of the sound field generated when fluid of density ρ_0 is injected at a rate $\rho_0\dot{\beta}(t)$ and a force $\mathbf{f}(t)$ (including any force due to the unsteady creation of momentum associated with $\dot{\beta}$) is applied at the uniformly moving point $\mathbf{x} = \mathbf{U}t$. Then the equation of mass conservation is

$$\frac{\partial \rho}{\partial t} + \rho_0 \operatorname{div} \mathbf{v} = \rho_0 \dot{\beta}\, \delta(\mathbf{x} - \mathbf{U}t). \tag{9.24}$$

The linear momentum equation gives

$$\rho_0 \frac{\partial \mathbf{v}}{\partial t} + \operatorname{grad} p' = \mathbf{f}\, \delta(\mathbf{x} - \mathbf{U}t) \tag{9.25}$$

and combining these two equations leads to the wave equation

$$\frac{1}{c^2}\frac{\partial^2 p'}{\partial t^2} - \nabla^2 p' = \rho_0 \frac{\partial}{\partial t}\{\dot{\beta}\, \delta(\mathbf{x} - \mathbf{U}t)\} - \frac{\partial}{\partial x_i}\{f_i\, \delta(\mathbf{x} - \mathbf{U}t)\}.$$

By comparison with (9.7) we see that this has the solution

$$p'(\mathbf{x}, t) = \rho_0 \frac{\partial}{\partial t}\left[\frac{\dot{\beta}(\tau^*)}{4\pi r|1 - M_r|}\right] - \frac{\partial}{\partial x_i}\left[\frac{f_i(\tau^*)}{4\pi r|1 - M_r|}\right]. \tag{9.26}$$

Many wavelengths from the source, the largest terms in (9.26), arise from the differentiation of τ^*, the emission time. Now differentiation of (9.9) gives

$$\frac{\partial \tau^*}{\partial t} = \frac{1}{1 - M_r}, \tag{9.27}$$

and

$$\frac{\partial \tau^*}{\partial x_i} = -\frac{(x_i - U_i\tau^*)}{cr(1 - M_r)}. \tag{9.28}$$

Hence we can write

$$p'(\mathbf{x}, t) = [\rho_0\ddot{\beta}(\tau^*) + \dot{f}_r(\tau^*)]\,\frac{1}{4\pi r(1 - M_r)|1 - M_r|} \tag{9.29}$$

where $\dot{f}_r(\tau^*)$ is the component of $\dot{\mathbf{f}}$ in the direction of the observer at emission time τ^*. Equation (9.29) shows that the effect of source motion is to modify the sound radiated by an injection of mass and by an external force by the factor $(1 - M_r)^{-2}$. In fact the situation is far more complicated than that, because most practical means of producing an apparent unsteady mass injection at a moving point also produce a momentum change and hence a fluctuating force. In general then $\beta(\tau)$ and $\mathbf{f}(\tau)$ are not independent and we need a far more careful analysis to investigate the sound produced by real moving sources. Now we saw in Chapter 7 that the Lighthill theory provides an exact way of describing sound generation, and that if there are any foreign bodies in the flow the Lighthill quadrupole sources within a body can be replaced by a distribution of monopoles and dipoles on a stationary control surface enclosing the body. If the foreign bodies are in motion, the most convenient control surface is one which moves with the body, because it is on this surface that we are most likely to be able to specify the flow conditions. The strength of the source terms on such a control surface is investigated next.

9.2 THE SOUND OF MOVING FOREIGN BODIES

Many problems of sound generation involve solid bodies in motion. These moving obstacles might be, for example, the rotating blades on the fan of an aircraft gas turbine, or on a ship's propeller. The Lighthill theory provides a formally exact way of describing sound generation in terms of quadrupole sources. Since we are unlikely to know the strength of these sources within the bodies, we surround them by a control surface S, and only apply Lighthill's equation in the fluid exterior to this control surface. Let us suppose that this surface moves with a velocity $\mathbf{u}(\mathbf{x}, t)$.

We now introduce a function $f(\mathbf{x}, t)$ where f is negative within the control surface and positive within the surrounding fluid. Such a function can always be defined provided the surface is sufficiently smooth. On the control surface

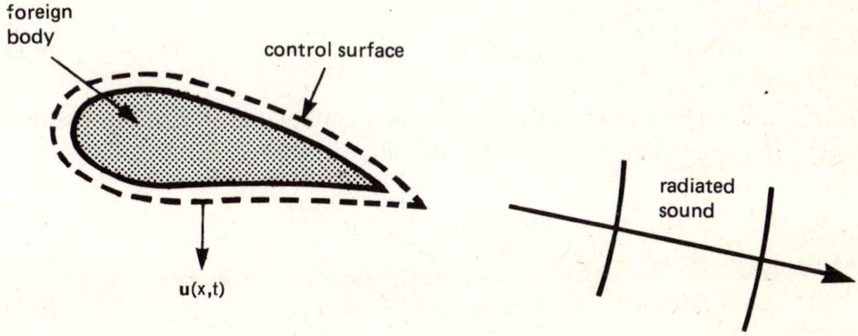

Fig. 9.10 — A foreign body enclosed by a control surface which moves with velocity $\boldsymbol{u}(x, t)$.

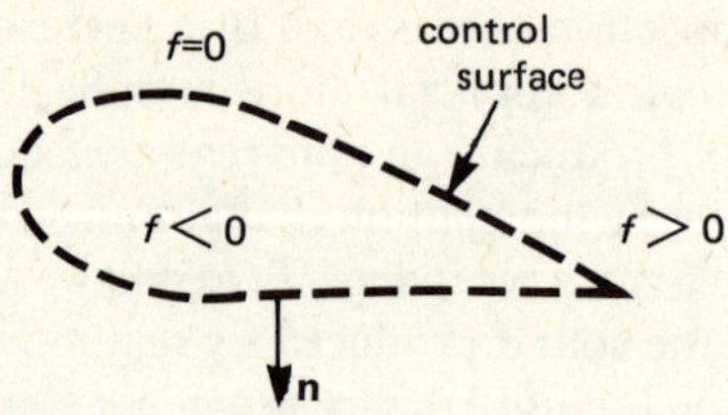

Fig. 9.11 — Definition of the function $f(x, t)$.

$f = 0$ at all times, and therefore grad f is in the direction of the outward normal:

$$\text{grad}\, f = \mathbf{n}|\text{grad}\, f|, \tag{9.30}$$

and

$$\frac{\partial f}{\partial t} + u_i \frac{\partial f}{\partial x_i} = 0 \quad \text{for points on the control surface.} \tag{9.31}$$

The δ-function $\delta(f)$ is only non-zero on the control surface and it is a property of the δ-function that for any function g

$$\int_\infty g(\mathbf{x})\, \delta(f)\, \mathrm{d}^3\mathbf{x} = \int_S \frac{g(\mathbf{x})}{|\text{grad}\, f|}\, \mathrm{d}S. \tag{9.32}$$

Let us define a Heaviside function $H(x)$, by

$$H(x) = \begin{cases} 1 & \text{for } x > 0 \\ 0 & \text{for } x < 0 \end{cases}.$$

Then $H(f)$ vanishes within the control surface and is equal to unity in the region exterior to the surface. By multiplying the exact equation of mass conservation (7.20) by $H(f)$ we obtain

$$H(f)\left\{\frac{\partial \rho'}{\partial t} + \frac{\partial}{\partial x_i}(\rho v_i)\right\} = 0, \tag{9.33}$$

an equation which is trivially satisfied at points within the control surface where $H(f)$ vanishes. (9.33) can be rearranged to give

$$\frac{\partial}{\partial t}(H\rho') + \frac{\partial}{\partial x_i}(H\rho v_i) = \rho' \frac{\partial H}{\partial t} + \rho v_i \frac{\partial H}{\partial x_i}$$

$$= \{\rho(v_i - u_i) + \rho_0 u_i\} \frac{\partial f}{\partial x_i}\, \delta(f). \tag{9.34}$$

The last step follows from replacing ρ' by $\rho - \rho_0$ and from (9.31) which shows that

$$\frac{\partial H}{\partial t} = -u_i \frac{\partial H}{\partial x_i} = -u_i \frac{\partial f}{\partial x_i} \delta(f).$$

We see from equation (9.34) that there is an apparent source due to the motion of the control surface and to any fluid crossing this control surface.

The non-linear momentum equation (7.21) may be manipulated in a similar way, so that

$$\frac{\partial}{\partial t}(\rho v_i) + \frac{\partial}{\partial x_j}(p_{ij} + \rho v_i v_j) = 0$$

becomes

$$\frac{\partial}{\partial t}(H\rho v_i) + \frac{\partial}{\partial x_j}(H(p_{ij} + \rho v_i v_j)) = \{\rho v_i(v_j - u_j) + p_{ij}\} \frac{\partial f}{\partial x_j} \delta(f). \quad (9.35)$$

The stress exerted across the control surface and the momentum flux through it are sources of momentum for the exterior fluid. Equations (9.34) and (9.35) give the wave equation.

$$\frac{\partial^2}{\partial t^2}(H\rho') - c^2\nabla^2(H\rho') = \frac{\partial^2(HT_{ij})}{\partial x_i \partial x_j} - \frac{\partial}{\partial x_i}\left(\{\rho v_i(v_j - u_j) + p_{ij}\} \frac{\partial f}{\partial x_j} \delta(f)\right)$$
$$+ \frac{\partial}{\partial t}\left(\{\rho(v_i - u_i) + \rho_0 u_i\} \frac{\partial f}{\partial x_i} \delta(f)\right) \quad (9.36)$$

where T_{ij} is Lighthill's quadrupole source

$$T_{ij} = \rho v_i v_j + p_{ij} - c^2(\rho - \rho_0)\,\delta_{ij}.$$

By comparison with (9.2) we see that the solution to (9.36) is

$$4\pi c^2 H\rho'(\mathbf{x}, t) = \frac{\partial^2}{\partial x_i \partial x_j} \int \frac{HT_{ij}}{|\mathbf{x} - \mathbf{y}|} \delta(t - \tau - |\mathbf{x} - \mathbf{y}|/c)\, d^3\mathbf{y}\, d\tau$$
$$- \frac{\partial}{\partial x_i} \int \frac{\{\rho v_i(v_j - u_j) + p_{ij}\}}{|\mathbf{x} - \mathbf{y}|} \frac{\partial f}{\partial y_j} \delta(f)\,\delta(t - \tau - |\mathbf{x} - \mathbf{y}|/c)\, d^3\mathbf{y}\, d\tau$$
$$+ \frac{\partial}{\partial t} \int \frac{\{\rho(v_i - u_i) + \rho_0 u_i\}}{|\mathbf{x} - \mathbf{y}|} \frac{\partial f}{\partial y_i} \delta(f)\,\delta(t - \tau - |\mathbf{x} - \mathbf{y}|/c)\, d^3\mathbf{y}\, d\tau.$$

(9.37)

If the control surface is moving, the source strengths described by these integrands will vary as a function of τ at fixed $\mathbf{y}$ simply due to source motion even if the source is non-evolving. The structure of the source is therefore more conveniently displayed by using a moving co-ordinate system $\boldsymbol{\eta}$, chosen so that the sources are at rest in th $\boldsymbol{\eta}$-frame If the source convection velocity is $\mathbf{U}(\boldsymbol{\eta}, \tau)$, we will define the $\boldsymbol{\eta}$ co-ordinate system by

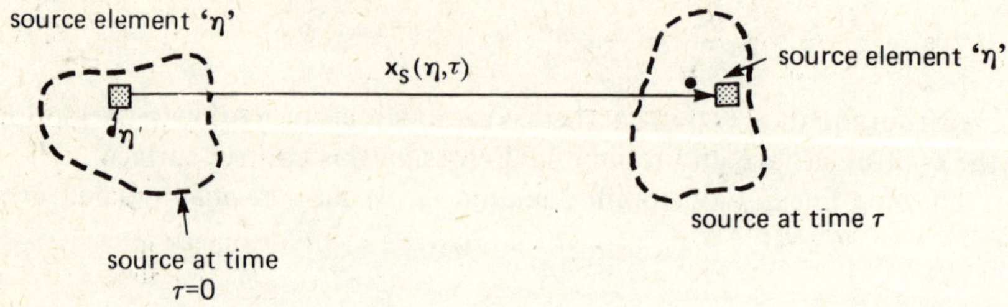

Fig. 9.12 — The moving co-ordinate system $\boldsymbol{\eta}$.

$$\mathbf{y} = \boldsymbol{\eta} + \mathbf{x}_s(\boldsymbol{\eta}, \tau),$$

where the $\mathbf{y}$ and $\boldsymbol{\eta}$ co-ordinates coincide at time $\tau = 0$ and $\mathbf{x}_s(\boldsymbol{\eta}, \tau) = \int_0^\tau \mathbf{U}(\boldsymbol{\eta}, \tau')\,\mathrm{d}\tau'$, is the distance moved by the source element $\boldsymbol{\eta}$ in the time interval $(0, \tau)$. Rewriting $\mathbf{y}$ in terms of $\boldsymbol{\eta}$ in (9.37) we obtain the equivalent form

$$\begin{aligned}
4\pi c^2 H\rho'(\mathbf{x}, t) = {} & \frac{\partial^2}{\partial x_i\,\partial x_j}\int \frac{HT_{ij}}{|\mathbf{x} - \boldsymbol{\eta} - \mathbf{x}_s|}\,\delta(t - \tau - |\mathbf{x} - \boldsymbol{\eta} - \mathbf{x}_s|/c)J\,\mathrm{d}^3\boldsymbol{\eta}\,\mathrm{d}t \\
& - \frac{\partial}{\partial x_i}\int \frac{\{\rho v_i(v_j - u_j) + p_{ij}\}}{|\mathbf{x} - \boldsymbol{\eta} - \mathbf{x}_s|}\,\frac{\partial f}{\partial y_j} \\
& \times \delta(f)\,\delta(t - \tau - |\mathbf{x} - \boldsymbol{\eta} - \mathbf{x}_s|/c)J\,\mathrm{d}^3\boldsymbol{\eta}\,\mathrm{d}\tau \\
& + \frac{\partial}{\partial t}\int \frac{\{\rho(v_i - u_i) + \rho_0 u_i\}}{|\mathbf{x} - \boldsymbol{\eta} - \mathbf{x}_s|}\,\frac{\partial f}{\partial y_i} \\
& \times \delta(f)\,\delta(t - \tau - |\mathbf{x} - \boldsymbol{\eta} - \mathbf{x}_s|/c)J\,\mathrm{d}^3\boldsymbol{\eta}\,\mathrm{d}\tau
\end{aligned}$$

where $\mathrm{d}^3\mathbf{y} = J\,\mathrm{d}^3\boldsymbol{\eta}$. J is the ratio of volume elements in the $\mathbf{y}$ and $\boldsymbol{\eta}$ spaces. We can again use the δ-function to evaluate the τ-integral just as in section 9.1 to obtain

$$\begin{aligned}
4\pi c^2 H\rho'(\mathbf{x}, t) = {} & \frac{\partial^2}{\partial x_i\,\partial x_j}\int \frac{HT_{ij}J}{r|1 - M_r|}\,\mathrm{d}^3\boldsymbol{\eta} \\
& - \frac{\partial}{\partial x_i}\int \frac{\{\rho v_i(v_j - u_j) + p_{ij}\}}{r|1 - M_r|}\,\frac{\partial f}{\partial y_j}\,\delta(f)J\,\mathrm{d}^3\boldsymbol{\eta} \\
& + \frac{\partial}{\partial t}\int \frac{\{\rho(v_i - u_i) + \rho_0 u_i\}}{r|1 - M_r|}\,\frac{\partial f}{\partial y_i}\,\delta(f)J\,\mathrm{d}^3\boldsymbol{\eta}.
\end{aligned} \tag{9.38}$$

The integrands are to be evaluated at a retarded time τ^* where τ^*, defined by

$$c(t - \tau^*(\boldsymbol{\eta})) = |\mathbf{x} - \boldsymbol{\eta} - \mathbf{x}_s(\boldsymbol{\eta}, \tau^*)|, \tag{9.39}$$

varies over the source region.

$$r = |\mathbf{x} - \boldsymbol{\eta} - \mathbf{x}_s(\boldsymbol{\eta}, \tau^*)| \quad \text{and} \quad M_r = \frac{U_i}{cr}(x_i - \eta_i - x_{si}(\boldsymbol{\eta}, \tau^*)). \quad (9.40)$$

$\delta(f)$ is only non-zero on the control surface and so the last two integrals in (9.38) will reduce to surface integrals. Now from (9.32)

$$\int g(\boldsymbol{\eta})\, \delta(f)\, \mathrm{d}^3\boldsymbol{\eta} = \int_S \frac{g(\boldsymbol{\eta})}{|\mathrm{grad}_{\boldsymbol{\eta}}\, f|}\, \mathrm{d}S, \quad \text{for any function } g,$$

and from (9.30)

$$\frac{\partial f}{\partial y_i} = n_i |\mathrm{grad}_{\mathbf{y}}\, f| \quad \text{on the surface}$$

By using both these relationships we can rewrite (9.38) as

$$\begin{aligned} 4\pi c^2 H\rho'(\mathbf{x}, t) = {} & \frac{\partial^2}{\partial x_i\, \partial x_j} \int_V \frac{J T_{ij}}{r|1 - M_r|}\, \mathrm{d}^3\boldsymbol{\eta} \\ & - \frac{\partial}{\partial x_i} \int_S \frac{\{\rho v_i(v_j - u_j) + p_{ij}\}}{r|1 - M_r|}\, n_j K\, \mathrm{d}S(\boldsymbol{\eta}) \\ & + \frac{\partial}{\partial t} \int_S \frac{\{\rho(v_i - u_i) + \rho_0 u_i\}}{r|1 - M_r|}\, n_i K\, \mathrm{d}S(\boldsymbol{\eta}), \end{aligned} \quad (9.41)$$

where $K = J|\mathrm{grad}_{\mathbf{y}}\, f|/|\mathrm{grad}_{\boldsymbol{\eta}}\, f|$. A natural choice of a convected co-ordinate system is one which moves with the control surface so that $\mathbf{u}(\mathbf{x}, t) = \mathbf{U}(\mathbf{x}, t)$ for points on the surface, and then the control surface is fixed in $\boldsymbol{\eta}$-space. In that case $|\mathrm{grad}_{\boldsymbol{\eta}}\, f|$ is a purely geometrical factor, and K turns out to be the ratio of the area elements of the surface S in the $\mathbf{y}$ and $\boldsymbol{\eta}$ spaces. If all the source elements move with the same linear or angular velocity, there is no change in the area or volume occupied by elements in $\boldsymbol{\eta}$-space and $J = K = 1$.

Equation (9.41) relates the sound heard at $(\mathbf{x}, t)$ to a volume distribution of quadrupoles, with additional monopole and dipole sources over a moving surface. If we consider a surface at rest, (9.41) reduces to (7.35), the expression we derived for a stationary control surface. If, on the other hand, the surface is impenetrable, the normal surface velocity must be equal to that of the flow

$$\mathbf{u}\cdot\mathbf{n} = \mathbf{v}\cdot\mathbf{n}$$

and (9.41) simplifies to

$$\begin{aligned} 4\pi c^2 H\rho'(\mathbf{x}, t)\big| = {} & \frac{\partial^2}{\partial x_i\, \partial x_j} \int_V \frac{J T_{ij}}{r|1 - M_r|}\, \mathrm{d}^3\boldsymbol{\eta} - \frac{\partial}{\partial x_i} \int_S \frac{p_{ij} n_j K}{r|1 - M_r|}\, \mathrm{d}S(\boldsymbol{\eta}) \\ & + \frac{\partial}{\partial t} \int_S \frac{\rho_0 \mathbf{v}\cdot\mathbf{n} K}{r|1 - M_r|}\, \mathrm{d}S(\boldsymbol{\eta}). \end{aligned} \quad (9.42)$$

We see that the sound is generated by Lighthill's quadrupoles in the fluid, the stress exerted across the control surface and by any local volume displacement by the bounding surface. The appearance of ρ_0 rather than ρ in the monopole source strength demonstrates that it is indeed the volume pulsation rather than mass fluctuation per se that generates sound. The description of the sound field given by equation (9.42) is used extensively to determine the noise of helicopter and fan blades. However application of this formula is far from straightforward because at high Mach numbers all three terms on the right-hand side can represent significant sources. In fact although (9.41) and (9.42) give formally exact expressions for the radiated sound field in terms of 'sources' it is often not easy to evaluate these source terms. To illustrate the difficulty we will investigate in some detail the sound radiated by one of the most elementary moving sources—a pulsating sphere.

9.3 A COMPACT PULSATING SPHERE MOVING AT LOW MACH NUMBER

A pulsating sphere is just about the simplest source that we can imagine. We will denote the radius of the sphere at time t by $A(t)$ and write the constant velocity of its centre by $(U, 0, 0)$. The Mach number $M(= U/c)$ is assumed small enough that M^2 is negligible in comparison with unity and we consider a pulsation that is both linear and compact. A compact pulsation is one in which $\omega a/c \ll 1$, where a is a typical size of the body, $a = A(0)$ say, ω is an average frequency of pulsation, $\omega = 0(A'/A)$ and the dash denotes differentiation. Since the pulsation is linear $A - a$ and derivatives of A are small and their products may be neglected. We will take moving co-ordinates $\boldsymbol{\eta}$ that are convected with a uniform velocity $\mathbf{U}$, and consider a spherical control surface of fixed radius b also moving with velocity $\mathbf{U}$, where b is to be just large enough to enclose the sphere at all times. Then equation (9.41) shows that the sound field is given by

$$4\pi c^2 H \rho'(\mathbf{x}, t) = \frac{\partial^2}{\partial x_i\, \partial x_j} \int_V \frac{T_{ij}}{r|1 - M_r|}\, \mathrm{d}^3\boldsymbol{\eta} - \frac{\partial}{\partial x_i} \int_S \frac{\{\rho v_i(v_j - u_j) + p_{ij}\}}{r|1 - M_r|}\, n_j\, \mathrm{s}S(\boldsymbol{\eta}) + \frac{\partial}{\partial t} \int_S \frac{\{\rho(v_i - u_i) + \rho_0 u_i\}}{r|1 - M_r|}\, n_i\, \mathrm{d}S(\boldsymbol{\eta}). \tag{9.43}$$

The terms in the integrand are to be evaluated at a retarded time τ^* which satisfies

$$c(t - \tau^*(\boldsymbol{\eta})) = |\mathbf{x} - \boldsymbol{\eta} - \mathbf{U}\tau^*|. \tag{9.44}$$

τ^* varies over the source region and for an observer in the far-field, $|\mathbf{x}| \gg |\mathbf{U}\tau^*|$,

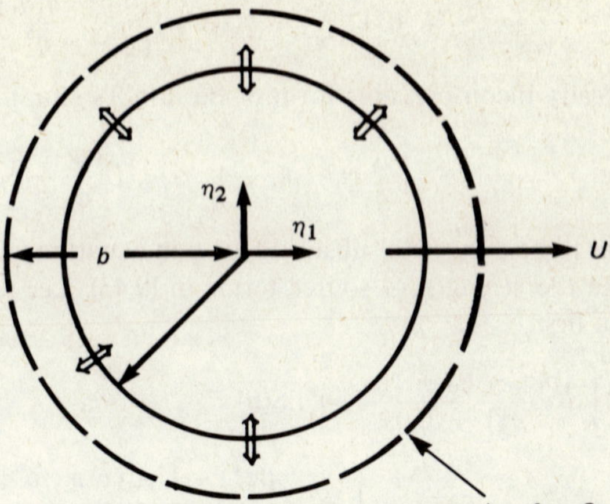

Fig. 9.13 — A moving pulsating sphere.

$$\tau^* = \left(t - \frac{|\mathbf{x}|}{c} + \frac{x_i n_i}{|\mathbf{x}|c}\right)\frac{1}{1 - M\cos\theta}. \qquad (9.45)$$

where $\cos\theta \approx x_1/|\mathbf{x}|$ for this distant observer (cf. with the form for τ^* in equation (9.10)).

Since the sphere is compact the surface terms in (9.43) are to be evaluated close to a singularity, where as we saw in section 2.1, the compressible equations reduce to the incompressible form. Therefore close to the body the velocity potential satisfies

$$\nabla^2\varphi = 0 \qquad (9.46)$$

and when M^2 is small the fluid is essentially incompressible there with $\rho = \rho_0$. On $\eta = |\boldsymbol{\eta}| = A$ the normal velocity of the sphere and fluid are equal, and hence at time τ

$$\mathbf{n}\cdot\text{grad}\,\varphi = \dot{A}(\tau) + \frac{U\eta_1}{\eta} \quad \text{on } \eta = A(\tau).$$

The solution of (9.46) that satisfies this boundary condition is

$$\varphi = -\frac{\dot{A}(\tau)a^2}{\eta} - \tfrac{1}{2}UA^3\frac{\eta_1}{\eta^3}. \qquad (9.47)$$

Both the fluid velocity and pressure perturbation can be calculated from this velocity potential.

$$\mathbf{v} = \frac{\partial \varphi}{\partial y_i} = \frac{\partial \varphi}{\partial \eta_i} = \dot{A}(\tau)a^2 \frac{\eta_i}{\eta^3} - \tfrac{1}{2}UA^3(\tau)\left\{\frac{\delta_{i1}}{\eta^3} - \frac{\eta_1\eta_i}{\eta^5}\right\}. \tag{9.48}$$

From the unsteady incompressible form of Bernoulli's equation

$$p' = -\rho_0 \frac{\partial \varphi}{\partial t}\bigg|_{\mathbf{y}} - \tfrac{1}{2}\rho v^2 = -\rho_0 \frac{\partial \varphi}{\partial t}\bigg|_{\boldsymbol{\eta}} + \rho_0 U \frac{\partial \varphi}{\partial \eta_1} - \tfrac{1}{2}\rho v^2. \tag{9.49}$$

All that remains to be done is to substitute for φ in equations (9.48) and (9.49) and to evaluate the strength of source terms in (9.43). Let us consider the monopole term first.

$$\frac{\partial}{\partial t}\int_S \frac{\{\rho(v_i - u_i) + \rho_0 u_i\}}{r(1 - M_r)} n_i \,\mathrm{d}S(\boldsymbol{\eta})$$
$$= \frac{\partial}{\partial t}\int_S \rho_0 \frac{\{\dot{A}(\tau^*) + UA^3(\tau^*)\eta_1/b^4\}}{r(1 - M_r)} \mathrm{d}S(\boldsymbol{\eta}). \tag{9.50}$$

By using (9.45) to expand the variation of retarded time over the body as a power series in $\boldsymbol{\eta}$ we see that

$$\dot{A}(\tau^*) + UA^3(\tau^*)\frac{\eta_1}{b^4} = \dot{A}(\tau_0^*) + UA^3(\tau_0^*)\frac{\eta_1}{b^4} + 3U\dot{A}(\tau_0^*)\frac{\eta_1\eta_i}{a^2}\frac{x_i}{|\mathbf{x}|c(1 - M\cos\theta)}$$

$$+ \text{ higher time derivatives and products of } \dot{A}(b - a)$$

where

$$\tau_0^* = \frac{t - |\mathbf{x}|/c}{1 - M\cos\theta} \tag{9.51}$$

is the retarded time appropriate for a source at the centre of the sphere $\boldsymbol{\eta} = \mathbf{0}$. Substituting into (9.50) and evaluating the integrals for large $|\mathbf{x}|$ gives

$$\frac{\partial}{\partial t}\int_S \frac{\{\rho(v_i - u_i) + \rho_0 u_i\}}{r(1 - M_r)} n_i \,\mathrm{d}S(\boldsymbol{\eta})$$
$$= \frac{\partial}{\partial t}\left\{\frac{4\pi a^2 \rho_0 \dot{A}(\tau_0^*)}{|\mathbf{x}|(1 - M\cos\theta)} + \frac{4\pi a^2\rho_0 M\cos\theta\,\dot{A}(\tau_0^*)}{|\mathbf{x}|}\right\}$$
$$= \frac{\partial}{\partial t}\left\{\frac{4\pi a^2\rho_0\dot{A}(\tau_0^*)}{|\mathbf{x}|(1 - M\cos\theta)^2}\right\} \quad \text{to order } M \tag{9.52}$$

where, since $|\mathbf{x}| \gg |\mathbf{U}\tau^*|$, we have written $(r|1 - M_r|)^{-1} = (|\mathbf{x}|(1 - M\cos\theta))^{-1}$, with $\cos\theta = x_1/|\mathbf{x}|$. From (9.51) $(\partial\tau_0^*/\partial t) = 1/(1 - M\cos\theta)$ and so the monopole source term gives a contribution $\dfrac{4\pi a^2\rho_0 A(\tau_0^*)}{r(1 - M\cos\theta)^3}$ to the radiated sound field.

When p' and $\mathbf{v}$ in (9.48) and (9.49) are substituted into the dipole source strength and the integrals are evaluated it turns out that the dipole contribution to the far-field is given by

$$-\frac{\partial}{\partial x_i}\int_S \frac{\{\rho v_i(v_j - u_j) + p\,\delta_{ij}\}}{r(1 - M_r)} n_j\,\mathrm{d}S = \frac{2\pi a^2\rho_0 M\cos\theta}{|\mathbf{x}|}\ddot{A}(\tau_0^*). \tag{9.53}$$

Also we note that any terms arising from the quadrupole are of order $\rho_0 a^2 M^2 \ddot{A}/|\mathbf{x}|$, and so by adding together (9.52) and (9.53) we obtain the final result

$$c^2\rho'(\mathbf{x}, t) = \frac{\rho_0 a^2 \ddot{A}(\tau_0^*)}{|\mathbf{x}|(1 - M\cos\theta)^{3\frac{1}{2}}}. \tag{9.54}$$

Motion amplifies the pressure perturbation by three and a half Doppler factors, which is far more complicated than the effect of motion on the point sources considered in section 9.1. This is because a moving pulsating body, producing a volume flux, necessarily has a momentum flux associated with it. Hence the sound field produced is that due to volume creation and a coupled fluctuating force acting at a moving point. Moreover the strength of the sound field generated by the force is only a Mach number smaller than the leading term and so affects the amplification factor.

EXERCISES FOR CHAPTER 9

1. Show that the distant sound field produced by a source of strength

$$q(\mathbf{x}, t) = \delta(x_1 - Ut)H(l^2 - x_2^2)\,\delta(x_3)Q\,\mathrm{e}^{i\omega t}$$

is

$$p'(\mathbf{x}, t) = \frac{c}{2\pi\omega x_2}\exp\left[\frac{i\omega(t - |\mathbf{x}|/c)}{1 - M\cos\theta}\right]\sin\left(\frac{\omega x_2 l}{|\mathbf{x}|c(1 - M\cos\theta)}\right)$$

when $|\mathbf{x}|$ is large in comparison with l and $|\mathbf{U}\tau^*|$ and where $\cos\theta = x_1/|\mathbf{x}|$. Comment on the form when the source is compact.

2. Show that the sound field due to an infinitely long line source of strength $Q\,\mathrm{e}^{i\omega t}$ lying parallel to the x_2-axis and moving in the 1-direction with a speed U is

$$p'(\mathbf{x}, t) = \frac{Qc}{2\pi}\int \mathrm{e}^{i\omega\tau}\frac{H(c(t-\tau) - [(x_1 - U\tau)^2 + x_3^2]^{\frac{1}{2}})}{\{c^2(t-\tau)^2 - [(x_1 - U\tau)^2 + x_3^2]\}^{\frac{1}{2}}}\,\mathrm{d}\tau.$$

Use the definition of the Hankel function,

$$H_0^{(2)}(x) = \frac{2i}{\pi}\int_1^\infty \frac{\mathrm{e}^{-ixt}}{(t^2 - 1)^{\frac{1}{2}}}\,\mathrm{d}t,$$

to express this as

$$p'(x, t) = \frac{-iQ}{4(1 - M^2)^{\frac{1}{2}}} \exp\left[\frac{i\omega}{1 - M^2}\left(t - \frac{Ux_1}{c^2}\right)\right]$$
$$\times H_0^{(2)}\left(\frac{\omega[(x_1 - Ut)^2 + (1 - M^2)x_3^2]^{\frac{1}{2}}}{c(1 - M^2)}\right).$$

3. Show that the distant pressure perturbation on $x_2 = 0$ produced by a source of constant strength moving round a circle of radius a (i.e. by a source of strength $q(\mathbf{x}, t) = m\,\delta(x_1)\,\delta(x_2 - a\cos\omega t)\,\delta(x_3 - a\sin\omega t)$) can be expressed as

$$p'(x_1, 0, x_3, t) = \frac{m}{4\pi\omega|\mathbf{x}|}\frac{\mathrm{d}\theta}{\mathrm{d}t}$$

where θ satisfies

$$\theta = \omega\left(t - \frac{|\mathbf{x}|}{c}\right) + \frac{\omega x_3}{c|\mathbf{x}|}\sin\theta$$

and $\mathrm{d}\theta/\mathrm{d}t$ is to be evaluated at retarded time θ/ω.

4. Show that the pressure perturbation generated by a linear fluid velocity v_3 normal to the plane surface $x_3 = 0$ can be expressed as

$$p'(\mathbf{x}, t) = \rho_0 \frac{\partial}{\partial t}\int_S \left[\frac{v_n}{2\pi|\mathbf{x} - \mathbf{y}|}\right]\mathrm{d}S.$$

A compact piston of radius a embedded in an infinite plane wall has a flow with a uniform velocity $\mathbf{U}$ over it, as shown in Figure 9.14. The displacement of the surface of the piston is $\xi = \xi_0\cos(\pi r/2a)\,\mathrm{e}^{i\omega t}$ where $r = (x_1^2 + x_2^2)^{\frac{1}{2}}$ and $a > r \geqslant 0$. Express this problem in terms of a co-ordinate system, $\mathbf{x}$, fixed in the fluid and then show that if M^2 is negligible in comparison with unity the distant pressure field is given by

$$p'(\mathbf{x}, t) = \frac{-(2\pi - 4)}{\pi^2}\frac{\rho_0\xi_0\omega^2a^2}{|\mathbf{x}|(1 + M\cos\theta)^3}\exp i\omega\left[\frac{t - |\mathbf{x}|/c}{1 + M\cos\theta}\right].$$

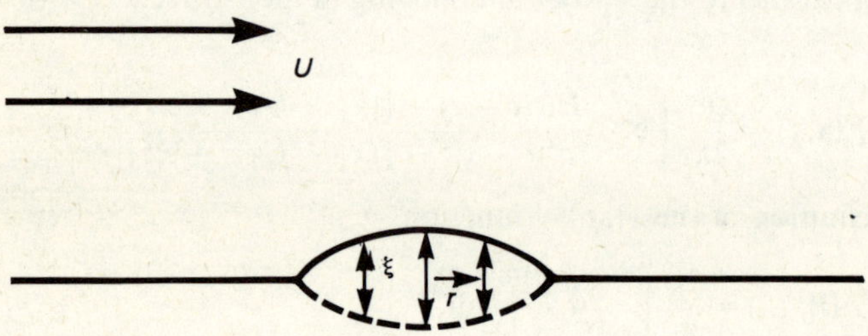

Fig. 9.14 — Flow over a vibrating piston.

Chapter 10

Fourier Synthesis, Spectral Analysis and Digital Techniques

10.1 FOURIER DECOMPOSITION OF A WAVE FIELD

We saw in Chapter 1 that Fourier's theorem states that a function $p(t)$ can be transformed into a function $\hat{p}(\omega)$ with no information being lost and that the inverse transform from $\hat{p}(\omega)$ to $p(t)$ is equally straightforward.

$$p(t) = \frac{1}{2\pi} \int_{-\infty}^{\infty} \hat{p}(\omega)\, e^{i\omega t}\, d\omega \tag{10.1}$$

$$\hat{p}(\omega) = \int_{-\infty}^{\infty} p(t)\, e^{-i\omega t}\, dt. \tag{10.2}$$

$p(t)$ and $\hat{p}(\omega)$ constitute a Fourier transform pair. The sign of ω and the scaling that places the 2π in this particular position is actually arbitrary but is convenient for what follows. We will regard $p(t)$ to be a continuous function of time, the output of an anemometer or microphone signal for example so that the variable ω then plays the role of angular frequency. The circumflex, ˆ, is used to denote transform variables that are functions of frequency only and not of time.

If $p(t)$ is a real function of time, the complex conjugate of equation (10.2) shows that

$$\hat{p}(-\omega) = \hat{p}^*(\omega) \tag{10.3}$$

where the star denotes the complex conjugate. Hence when $p(t)$ is real, all the information required to reconstruct $p(t)$ is contained in half the spectral decomposition, i.e. in $\hat{p}(\omega)$, $\omega \geqslant 0$.

Just as in equations (10.1) and (10.2) where a noise signal is analysed into its spectrum by Fourier decomposition so we could analyse a function of position $\mathbf{x}$, into a wave vector spectrum by repetition of Fourier's theorem. $a(\mathbf{x})$ can be written as

$$a(\mathbf{x}) = a(x_1, x_2, x_3) = \frac{1}{(2\pi)^3} \int_{\infty} \hat{a}(\mathbf{k})\, e^{i\mathbf{k}\cdot\mathbf{x}}\, d^3\mathbf{k},$$

where

$$\hat{a}(\mathbf{k}) = \int_{\infty} a(\mathbf{x})\, \mathrm{e}^{-i\mathbf{k}\cdot\mathbf{x}}\, \mathrm{d}^3\mathbf{x};$$

the integration ranges over the infinite three-dimensional space and **k** is called the wave vector.

Sound is a function of both position and time so that the four-dimensional Fourier transform is appropriate. That allows the pressure $p'(\mathbf{x}, t)$ to be regarded as a superposition of wavelets with wave vector **k** and frequency ω, the strength of each wavelet being the Fourier transform $\hat{p}(\mathbf{k}, \omega)$.

$$p'(\mathbf{x}, t) = \frac{1}{(2\pi)^4} \int_{-\infty}^{\infty} \int_{\infty} \hat{p}(\mathbf{k}, \omega)\, \mathrm{e}^{i\mathbf{k}\cdot\mathbf{x}}\, \mathrm{e}^{i\omega t}\, \mathrm{d}^3\mathbf{k}\, \mathrm{d}\omega$$

$$\hat{p}(\mathbf{k}, \omega) = \int_{-\infty}^{\infty} \int_{\infty} p'(\mathbf{x}, t)\, \mathrm{e}^{-i\mathbf{k}\cdot\mathbf{x}}\, \mathrm{e}^{-i\omega t}\, \mathrm{d}^3\mathbf{x}\, \mathrm{d}t. \quad (10.4)$$

The phase velocity $\mathbf{c_p}$, say, of a wavelet is anti-parallel to the wave vector and is defined as the ratio of the frequency ω to the magnitude of that wave vector.

$$\mathbf{c_p} = -\frac{\omega\mathbf{k}}{|\mathbf{k}|^2}$$

$$\omega = -\mathbf{c_p}\cdot\mathbf{k}.$$

If, for example, the pressure perturbation is caused by sound propagating as a plane wave in the direction of the unit vector **l**, then

$$p'(\mathbf{x}, t) = f(t - \mathbf{x}\cdot\mathbf{l}/c),$$

and

$$\hat{p}(\mathbf{k}, \omega) = \int_{-\infty}^{\infty} \int_{\infty} f(t - \mathbf{x}\cdot\mathbf{l}/c)\, \mathrm{e}^{-i\mathbf{k}\cdot\mathbf{x} - i\omega t}\, \mathrm{d}^3\mathbf{x}\, \mathrm{d}t$$

$$= \int_{-\infty}^{\infty} f(t - \mathbf{x}\cdot\mathbf{l}/c)\, \mathrm{e}^{-i\omega(t - \mathbf{x}\cdot\mathbf{l}/c)}\, \mathrm{d}(t - \mathbf{x}\cdot\mathbf{l}/c)$$

$$\times \int_{\infty} \mathrm{e}^{-i[\mathbf{k} + (\omega/c)\mathbf{l}]\cdot\mathbf{x}}\, \mathrm{d}^3\mathbf{x}$$

$$= (2\pi)^3 \hat{f}(\omega)\, \delta\left(\mathbf{k} + \frac{\omega}{c}\mathbf{l}\right). \quad (10.5)$$

$\delta(\mathbf{k}) = \delta(k_1)\,\delta(k_2)\,\delta(k_3)$ and we have applied the useful relationship given in equation (1.25):

$$\int_{-\infty}^{\infty} \mathrm{e}^{-ixk}\, \mathrm{d}x = 2\pi\, \delta(k).$$

Equation (10.5) shows that for this plane wave the wave vector spectrum of the sound at frequency ω is discrete being non-zero only at the wave vector $\mathbf{k} = -(\omega/c)\mathbf{l}$.

A sound field is generated by elements of the source with sonic phase velocity.

In order to demonstrate the versatility of Fourier transforms we will use them to simplify the relationship between the far-field pressure perturbation and the source. We say in Chapter 7, equation (7.8), that a sound field $p'(\mathbf{x}, t)$ is related to its source $q(\mathbf{x}, t)$ by

$$p'(\mathbf{x}, t) = \int_{\infty} \frac{q(\mathbf{y}, t - |\mathbf{x} - \mathbf{y}|/c)}{4\pi|\mathbf{x} - \mathbf{y}|} \, \mathrm{d}^3\mathbf{y}. \tag{10.6}$$

Let $\hat{q}(\mathbf{k}, \omega)$ be the Fourier transform of the source strength:

$$\hat{q}(\mathbf{k}, \omega) = \int_{-\infty}^{\infty} \int_{\infty} q(\mathbf{x}, t) \, \mathrm{e}^{-i\mathbf{k}\cdot\mathbf{x}} \, \mathrm{e}^{-i\omega t} \, \mathrm{d}^3\mathbf{x} \, \mathrm{d}t.$$

For an observer whose distance from the source is much larger than the source dimension, l,

$$\frac{1}{|\mathbf{x} - \mathbf{y}|} = \frac{1}{|\mathbf{x}|} + 0\left\{\frac{l}{|\mathbf{x}|^2}\right\},$$

and

$$|\mathbf{x} - \mathbf{y}| = |\mathbf{x}| - \frac{\mathbf{x}\cdot\mathbf{y}}{|\mathbf{x}|} + 0\left\{\frac{l^2}{|\mathbf{x}|}\right\}.$$

Hence, in the very far-field, equation (10.6) simplifies to

$$p'(\mathbf{x}, t) = \frac{1}{4\pi|\mathbf{x}|} \int_{\infty} q, \left(\mathbf{y}, t - \frac{|\mathbf{x}|}{c} + \frac{\mathbf{x}\cdot\mathbf{y}}{|\mathbf{x}|c}\right) \mathrm{d}^3\mathbf{y}.$$

This can be Fourier transformed in time to give

$$\hat{p}(\mathbf{x}, \omega) = \frac{1}{4\pi|\mathbf{x}|} \int_{\infty} \int_{-\infty}^{\infty} q\left(\mathbf{y}, t - \frac{|\mathbf{x}|}{c} + \frac{\mathbf{x}\cdot\mathbf{y}}{|\mathbf{x}|c}\right) \mathrm{e}^{-i\omega t} \, \mathrm{d}t \, \mathrm{d}^3\mathbf{y}.$$

After changing the time variable to $\tau = t - |\mathbf{x}|/c + \mathbf{x}\cdot\mathbf{y}/|\mathbf{x}|c$ we obtain

$$\begin{aligned} \hat{p}(\mathbf{x}, \omega) &= \frac{1}{4\pi|\mathbf{x}|} \int_{\infty} \int_{-\infty}^{\infty} q(\mathbf{y}, \tau) \, \mathrm{e}^{-i\omega[|\mathbf{x}|/c - \mathbf{x}\cdot\mathbf{y}/|\mathbf{x}|c]} \, \mathrm{d}^3\mathbf{y} \, \mathrm{d}\tau \\ &= \frac{\mathrm{e}^{-i\omega|\mathbf{x}|/c}}{4\pi|\mathbf{x}|} \int_{\infty} \hat{q}(\mathbf{y}, \omega) \, \mathrm{e}^{i(\omega\mathbf{x}/c|\mathbf{x}|)\cdot\mathbf{y}} \, \mathrm{d}^3\mathbf{y}, \end{aligned}$$

and so

$$\hat{p}(\mathbf{x}, \omega) \underset{|\mathbf{x}| \to \infty}{\sim} \frac{\mathrm{e}^{-i\omega|\mathbf{x}|/c}}{4\pi|\mathbf{x}|} \hat{q}\left(\mathbf{k} = -\frac{\omega}{c}\frac{\mathbf{x}}{|\mathbf{x}|}, \omega\right). \tag{10.7}$$

In Chapter 7 we demonstrated that a knowledge of the sound field is not sufficient to completely determine the source. We see in equation (10.7) which elements of the source are related directly to the distant sound field. It is apparent that the sound at frequency ω is only determined by those elements of the source that exactly match it in frequency, and have sonic phase speeds. In fact a comparison with (10.5) demonstrates that the distant sound field in a direction $\mathbf{x}$ from the source is influenced only by the spectral elements of the source that have the wavenumber-frequency relationship appropriate for a plane wave travelling in the $\mathbf{x}$-direction.

The sound of moving sources

Suppose now that the source distribution $q(\mathbf{x}, t)$ is specified with respect to a reference frame that is moving with respect to $\mathbf{x}$ at a constant velocity $\mathbf{U}$.

$$\mathbf{x} = \boldsymbol{\eta} + \mathbf{U}t$$

relates the moving frame co-ordinates $\boldsymbol{\eta}$ to the still atmosphere space variable $\mathbf{x}$. The moving source distribution is specified in terms of $(\boldsymbol{\eta}, t)$ as

$$q(\mathbf{x}, t) = q_m(\boldsymbol{\eta}, t) = q_m(\mathbf{x} - \mathbf{U}t, t).$$

The source spectrum, $\hat{q}(\mathbf{k}, \omega)$, is obtained by Fourier transform

$$\begin{aligned}\hat{q}(\mathbf{k}, \omega) &= \int_{-\infty}^{\infty}\int_{\infty} q_m(\mathbf{x} - \mathbf{U}t, t)\, \mathrm{e}^{-i\mathbf{k}\cdot\mathbf{x}}\, \mathrm{e}^{-i\omega t}\, \mathrm{d}^3\mathbf{x}\, \mathrm{d}t \\ &= \int_{-\infty}^{\infty}\int_{\infty} q_m(\mathbf{x} - \mathbf{U}t, t)\, \mathrm{e}^{-ik\cdot(\mathbf{x} - \mathbf{U}t)}\, \mathrm{e}^{-i(\omega + \mathbf{k}\cdot\mathbf{U})t}\, \mathrm{d}^3(\mathbf{x} - \mathbf{U}t)\, \mathrm{d}t.\end{aligned}$$

Hence

$$\hat{q}(\mathbf{k}, \omega) = \hat{q}_m(\mathbf{k}, \omega + \mathbf{U}\cdot\mathbf{k}).$$

The frequency spectrum in a moving frame is shifted from that in the fixed frame by the amount $\mathbf{U}\cdot\mathbf{k}$.

Equation (10.7) relates the spectrum of the radiated sound to the source spectrum. In terms of the spectrum of the source in moving co-ordinates it gives

$$\hat{p}(\mathbf{x}, \omega) = \frac{\mathrm{e}^{-i\omega|\mathbf{x}|/c}}{4\pi|\mathbf{x}|}\, \hat{q}_m\left(\mathbf{k} = -\frac{\mathbf{x}\omega}{|\mathbf{x}|c}, \omega\left(1 - \frac{\mathbf{U}\cdot\mathbf{x}}{c|\mathbf{x}|}\right)\right)$$

or equivalently

$$\hat{p}\left(\mathbf{x}, \frac{\omega}{1 - M_r}\right) = \frac{\mathrm{e}^{-i\omega|\mathbf{x}|/c(1 - M_r)}}{4\pi|\mathbf{x}|}\, \hat{q}_m\left(\mathbf{k} = -\frac{\omega\mathbf{x}}{c|\mathbf{x}|(1 - M_r)}, \omega\right),$$

where $M_r = \mathbf{U}\cdot\mathbf{x}/|\mathbf{x}|c$ is the 'Mach number' at which the source is approaching the observation point at $\mathbf{x}$. As we saw in Chapter 9 the sound radiated by a moving source of frequency ω is heard at $\mathbf{x}$ at the 'Doppler shifted' frequency $\omega/(1 - M_r)$.

10.2 THE STATISTICAL ANALYSIS OF CONTINUOUS RANDOM SIGNALS

We will begin our analysis of random signals by investigating the case where the signal, $p(t)$, is simply a function of time. $p(t)$ might be, for example, the signal from a microphone reading the fluctuating sound pressure at one position. We insist that $p(t)$ has zero mean.

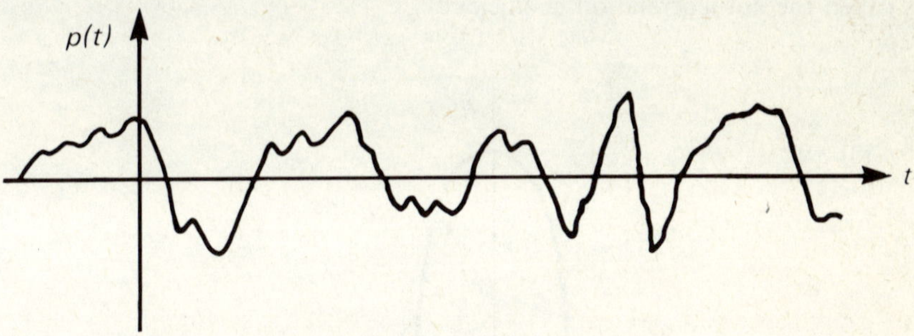

Fig. 10.1 — $p(t)$, a random signal.

Since $p(t)$ is a stochastic, or noise, signal, it can only be meaningfully described by its statistical properties, the mean square, or standard deviation, being the simplest. The mean square, or any other statistic of p, could be established by performing the experiment several times and measuring $p^2(t)$ in each 'realisation' of p and then taking the mean. In order for mean values to have a meaning, the field must be statistically stationary i.e. the mean value must not depend on the point around which the value was taken. So that $\overline{p^2(t)}$ is the same as $\overline{p^2(t + \tau)}$, where the over-bar signifies a mean value.

One of the most significant statistical measures of $p(t)$ is its *autocorrelation function*, $P(\tau)$, defined as

$$P(\tau) = \overline{p(t)p(t + \tau)}. \tag{10.8}$$

$p(t)$ is statistically independent of $p(t + \tau)$ for large enough τ and so then the mean value of their product will be zero. On the other hand when τ is zero, the autocorrelation function is simply the mean square signal level.

$$\begin{aligned} P(\tau) &\underset{\tau \to \infty}{\sim} 0 \\ P(0) &= \overline{p^2}. \end{aligned} \tag{10.9}$$

We can easily show that the autocorrelation function is an even function of τ.

$$P(\tau) = \overline{p(t)p(t+\tau)} = \overline{p(t'-\tau)p(t')} \quad \text{where } t' = t + \tau$$

$$= P(-\tau), \text{ because the average is independent of the common reference time } t'. \qquad ((10.10)$$

Often the correlation function is normalised to have a maximum value of unity; the normalised function

$$\frac{P(\tau)}{P(0)} = \frac{P(\tau)}{\overline{p^2}} \qquad (10.11)$$

is called the autocorrelation coefficient.

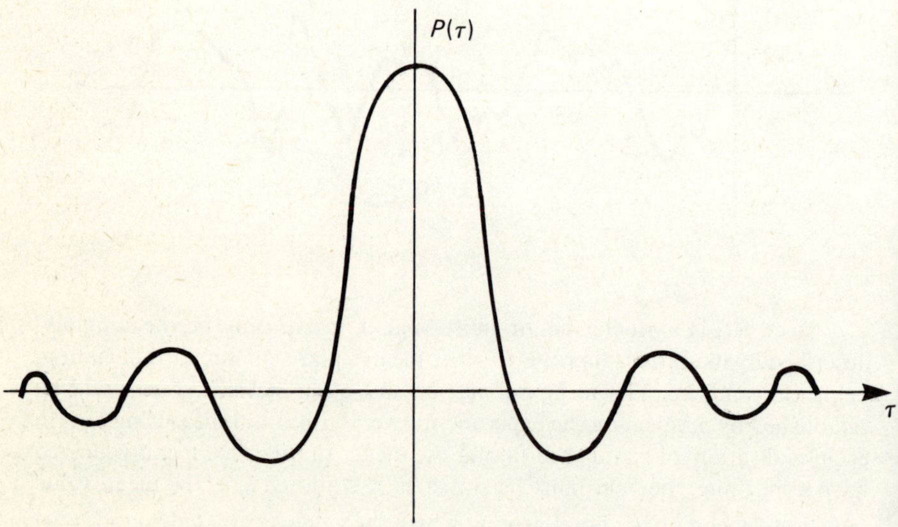

Fig. 10.2 — A typical autocorrelation function.

The Fourier transform of the autocorrelation function is called the *power spectral density*; we write it as $\hat{P}(\omega)$.

$$\hat{P}(\omega) = \int_{-\infty}^{\infty} P(\tau)\, e^{-i\omega\tau}\, d\tau \qquad (10.12)$$

$$= 2\int_{0}^{\infty} P(\tau) \cos \omega\tau\, d\tau \quad \text{since } P(\tau) \text{ is even.} \qquad (10.13)$$

We see that $\hat{P}(\omega)$ like $P(\tau)$ is a real and even function. The inverse of this transform states that

$$P(\tau) = \frac{1}{2\pi} \int_{-\infty}^{\infty} \hat{P}(\omega)\, e^{i\omega\tau}\, d\omega = \frac{1}{\pi} \int_{0}^{\infty} \hat{P}(\omega) \cos \omega\tau\, d\omega \tag{10.14}$$

and in particular, when $\tau = 0$,

$$P(0) = \overline{p^2} = \int_{0}^{\infty} \frac{\hat{P}(\omega)}{\pi}\, d\omega. \tag{10.15}$$

This identity allows $\overline{p^2}$ to be thought of as made up of various components at frequencies ω, the components when summed, constituting the whole. The quantity $\hat{P}(\omega)/\pi$ is the contribution from unit frequency interval and that is the reason for it being termed the spectral density. The power in a 'band' is the integral of the power spectral density over the frequency range in that band.

The spectrum is often characterised by its 'slope' in a given frequency interval. This is different for a constant bandwidth than it is for bandwidths based on the octave. If the power spectral density at frequency f_0 is locally proportional to f_0^n, then the octave based band levels will be proportional there to f_0^{n+1}, the band levels being an integral of the power spectral density over all frequencies in the band. The range of frequencies contained in these bands increase with frequency in proportion to the band centre frequency and, for this reason, the bandwidths are often referred to as 'proportional bandwidths'.

The autocorrelation function $P(\tau)$ is a perfectly normal integrable function. For random signals $P(\tau)$ tends to zero for large τ and so the power spectral density defined in equation (10.13) exists as a normal function. If however the function $p(t)$ is not stochastic, i.e. if $p(t)$ is statistically stationary and *deterministic*, $\hat{P}(\omega)$ will degenerate into a generalised function. As an example suppose that $p(t)$ is a harmonic signal with angular frequency α,

$$p(t) = \cos \alpha t = \tfrac{1}{2}(e^{i\alpha t} + e^{-i\alpha t}).$$

Then

$$P(\tau) = \overline{p(t)p(t+\tau)} = \tfrac{1}{2} \cos \alpha\tau.$$

The autocorrelation function of this harmonic, deterministic function, is also a harmonic function with the same frequency α. Its Fourier transform, the power spectral density, does not exist as an ordinary function, only as a generalised function because *all* the energy exists at the discrete frequency α so that the energy density, the power per unit frequency interval must there be singular:

$$\hat{P}(\omega) = \int_{-\infty}^{\infty} P(\tau)\, e^{-i\omega\tau}\, d\tau$$

$$= \frac{1}{4}\int_{-\infty}^{\infty} \{e^{i(\alpha-\omega)\tau} + e^{-i(\alpha+\omega)\tau}\}\, d\tau$$

$$= \frac{\pi}{2}\{\delta(\alpha-\omega) + \delta(\alpha+\omega)\}.$$

The power of the discrete frequency signal $p(t)$ is then equally held in concentrated elements at $\omega = \pm\alpha$.

The power spectral density is related to the Fourier transform of the signal

In general when $p(t)$ is a stochastic function, its Fourier transform similarly exists only as a generalised function but a generalised function that can be related to the power spectral density.

$$\hat{p}(\omega) = \int_{-\infty}^{\infty} p(t)\, e^{-i\omega t}\, dt$$

$$= \int_{-\infty}^{\infty} p(t+\tau)\, e^{-i\omega(t+\tau)}\, d\tau$$

so that

$$\overline{\hat{p}(\omega)\hat{p}(\alpha)} = \int_{-\infty}^{\infty}\int_{-\infty}^{\infty} \overline{p(t)p(t+\tau)}\, e^{-i\omega t}\, e^{-i\alpha(t+\tau)}\, dt\, d\tau.$$

The 'average' operation does not affect the exponential terms which are 'deterministic'. The average produces the autocorrelation function;

$$\overline{\hat{p}(\omega)\hat{p}(\alpha)} = \int_{-\infty}^{\infty} P(\tau) \int_{-\infty}^{\infty} e^{-i(\omega+\alpha)t}\, dt\, e^{-i\alpha\tau}\, d\tau$$

$$= 2\pi \int_{-\infty}^{\infty} P(\tau)\, e^{-i\alpha\tau}\, d\tau\, \delta(\omega+\alpha)$$

$$= 2\pi\hat{P}(\alpha)\, \delta(\omega+\alpha). \tag{10.16}$$

The mean product of the Fourier transforms of $p(t)$ thus constitutes a generalised function.

$$\overline{\hat{p}(\omega)\hat{p}(\alpha)} = 0 \quad \alpha \neq -\omega$$

$$\frac{1}{2\pi}\int_{-\infty}^{\infty} \overline{\hat{p}(\omega)\hat{p}(\alpha)}\, d\alpha = \hat{P}(\omega) = \hat{P}(-\omega). \tag{10.17}$$

Fourier elements of a stochastic function are statistically unrelated unless their frequencies *sum to zero.* When they do sum to zero, they are 'superrelated'. They then form a generalised function which has to be integrated into a normal function.

We have seen that the power spectral density is related to the Fourier transform of the signal in a simple way. This is often a clue to the simplest way of solving noise transmission problems, where the transmission coefficient is a function of frequency.

Example

A sound wave of level 100 dB is normally incident on an unsupported sheet of glass, 5 mm thick. The power spectral density of the incident sound is proportional to $(10^3 + \omega^2)^{-1}$, ω being the radian frequency in units of s^{-1}. Determine the frequency spectrum of the transmitted sound, and hence calculate the transmitted pressure level.

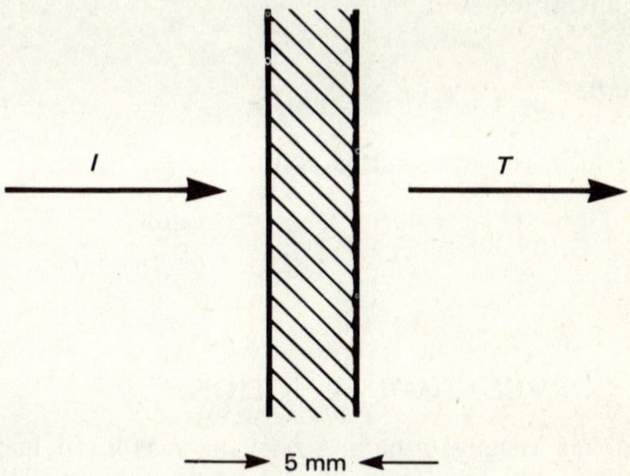

Fig. 10.3 — Sound transmission through a sheet of glass.

The power spectrum of the incident sound, $P_I(\omega) = A/(10^3 + \omega^2)$. The constant A can be calculated because

$$\overline{p^2} = p_{\text{ref}}^2 \times 10^{10} = \frac{A}{2\pi}\int_{-\infty}^{\infty} \frac{d\omega}{10^3 + \omega^2},$$

where $p_{\text{ref}} = 2 \times 10^{-5}$ N/m^2. Using the standard integral

$$\int_{-\infty}^{\infty} \frac{dx}{x^2 + a^2} = \frac{\pi}{a},$$

we find $A = 2 \times 10^{11.5} p_{\text{ref}}^2$.

We found in Chapter 4 (equation 4.7) that the transmission coefficient through a wall of mass m per unit is $\eta/(\eta + i\omega)$, where $\eta = 2\rho_0 c/m$. The Fourier transform of the transmitted sound is therefore related to the transform of the incident sound by

$$\hat{p}_T(\omega) = \frac{\eta}{\eta + i\omega}\,\hat{p}_I(\omega),$$

where for 5 mm thick glass $\eta = 65.28\ \text{s}^{-1}$.

The frequency spectrum of the transmitted sound, $\hat{P}_T(\omega)$, is, from (10.17),

$$\hat{P}_T(\omega) = \frac{1}{2\pi}\int \overline{\hat{p}_T(\omega)\hat{p}_T(\alpha)}\,\mathrm{d}\alpha$$

$$= \frac{1}{2\pi}\int \frac{\eta}{\eta + i\omega}\,\frac{\eta}{\eta + i\alpha}\,\overline{\hat{p}_I(\omega)\hat{p}_I(\alpha)}\,\mathrm{d}\alpha.$$

After using (10.16) we find a direct relationship between the spectra of the incident and transmitted sound:

$$\hat{P}_T(\omega) = \frac{\eta^2}{\eta^2 + \omega^2}\,\hat{P}_I(\omega).$$

The pressure level in the transmitted sound is

$$\frac{1}{2\pi}\int \hat{P}_T(\omega)\,\mathrm{d}\omega = \frac{A\eta^2}{2\pi}\int_{-\infty}^{\infty}\frac{\mathrm{d}\omega}{(10^3 + \omega^2)(\eta^2 + \omega^2)} = \underline{98.3\ \text{dB}}\,.$$

10.3 CROSS CORRELATION FUNCTIONS

Consider now that a signal p having zero mean is a stochastic statistically stationary function of two independent variables x and t. For definiteness $p(x, t)$ might be thought of as the noise pressure variation in a flow at downstream position x and time t, though the arguments and results have a much more general validity.

The *cross correlation* function p is defined as

$$P(\Delta, \tau) = \overline{p(x, t)p(x + \Delta, t + \tau)}. \tag{10.18}$$

This is a function only of the space separation Δ and time delay τ.

The mean square value of p is the value of the cross correlation function at zero argument,

$$P(0, 0) = \overline{p^2}. \tag{10.19}$$

The cross correlation function P has $\overline{p^2}$ as its maximum value. The '*space*' *correlation function*, $B(\Delta)$ say, is the value of the cross correlation function at zero time delay

$$B(\Delta) = \overline{p(x, t)p(x + \Delta, t)} = P(\Delta, 0). \tag{10.20}$$

The autocorrelation function of p, $P(\tau)$ say, is similarly the value of the cross correlation function at zero space separation.

$$P(\tau) = \overline{p(x, t)p(x, t + \tau)} = P(0, \tau). \tag{10.21}$$

$P(\Delta, \tau)$ displays a 'diagonal symmetry' in that it is symmetric along any straight line in the Δ, τ plane passing through the co-ordinate origin. This can be demonstrated by writing $x = x' - \Delta$ and $t = t' - \tau$ in the defining equation (10.18).

$$\begin{aligned} P(\Delta, \tau) &= \overline{p(x, t)p(x + \Delta, t + \tau)} \\ &= \overline{p(x' - \Delta, t' - \tau)p(x', t')} = P(-\Delta, -\tau). \end{aligned} \tag{10.22}$$

Finally, for this statistically stationary function the mean value $\overline{p(x, t)p(y, s)}$ is a function only of the space and time separations $x - y$ and $t - s$.

$$\overline{p(x, t)p(y, s)} = P(x - y, t - s) = P(y - x, s - t). \tag{10.23}$$

Typical forms of these correlation functions are presented graphically in Figure 10.4. The first is a set of curves of the cross correlation for various fixed values of time delay as a function of the space separation Δ. The second is an equivalent set of curves for fixed space separation as the time delay is varied. The third is a contour plot of iso-correlation curves as a function of both Δ and τ.

Even though p is a chaotically varying function, there are nevertheless various ways in which scales that characterise the signal can be defined. For example, $P/(\partial^2 P/\partial\tau^2)|_{\tau=0}$ defines the square of a characteristic time scale that is sometimes called the microscale following the terminology introduced by G. I. Taylor. Scales known as integral scales are particularly important. The integral time scale L_τ is defined from the autocorrelation function

$$L_\tau = \int_{-\infty}^{\infty} \frac{P(\tau)}{P(0)} \, d\tau \tag{10.24}$$

and the integral length scale L_x is similarly defined from the space correlation $B(\Delta)$

$$L_x = \int_{-\infty}^{\infty} \frac{B(\Delta)}{B(0)} \, d\Delta. \tag{10.25}$$

These scales will usually be proportional to the various dimensions of the system being examined. For example, again for definiteness only and without losing any generality, one might consider that p represents the random pressure fluctuation on the flat boundary surface of a flow. The boundary layer of that flow might be turbulent with a scale δ over which the mean velocity

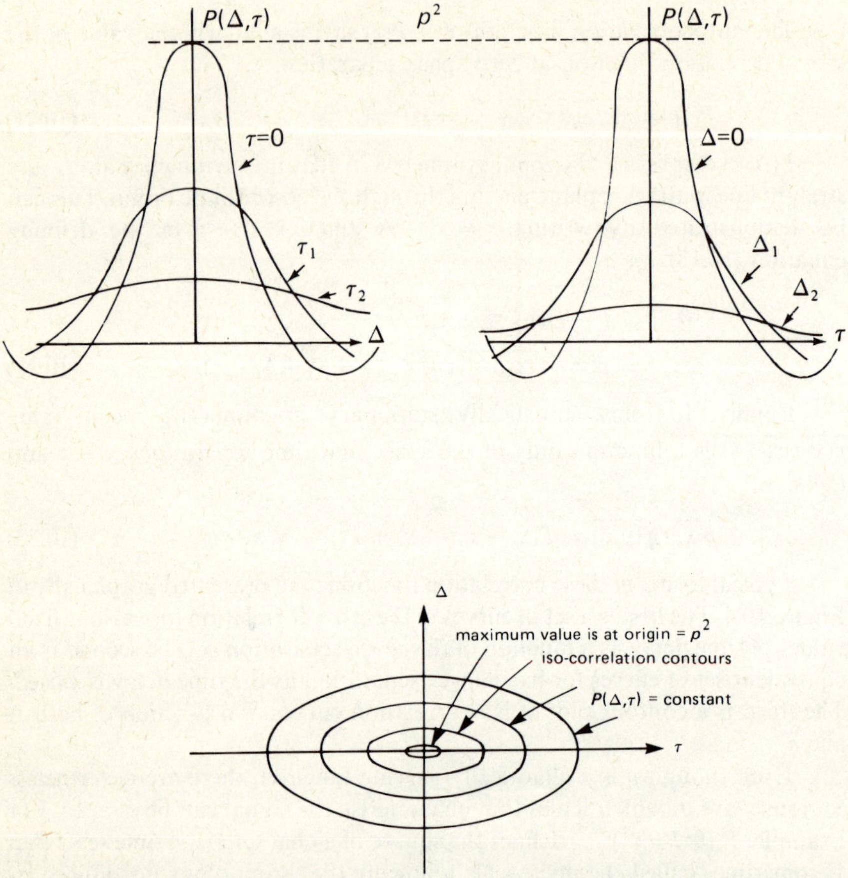

Fig. 10.4 — Sketches of a typical cross correlation function $P(\Delta, \tau)$.

rises continuously from zero at the surface to a maximum U, the free stream velocity. The flowing fluid might have constant density, ρ, in which case, by dimensional analysis, the magnitude of the pressure fluctuations will be proportional to ρU^2. The length scale δ and velocity scale U set a characteristic frequency scale U/δ; the pressure signal can often be normalised on these parameters. We can write

$$\overline{p^2} = (\rho U^2)^2 \overline{p_n^2}$$

where p_n is a pure number. Similarly

$$P(\Delta, \tau) = (\rho U^2)^2 P_n(\Delta/\delta, U\tau/\delta)$$

where P_n is a non-dimensional function of two independent non-dimensional variables; it has a maximum value of unity. If the characteristic scales have

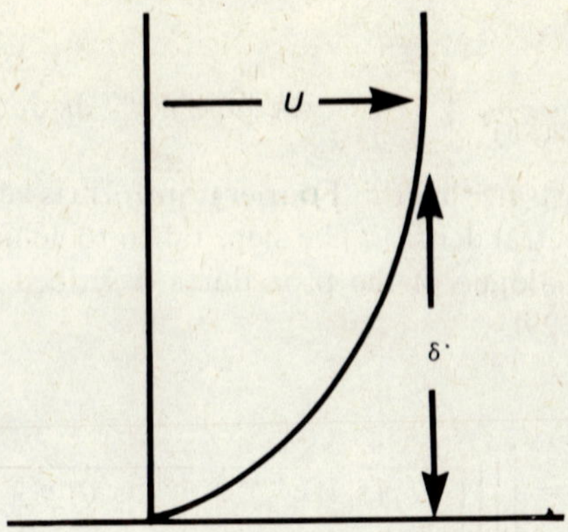

Fig. 10.5 — Length and velocity scales in a boundary layer.

been well chosen $P_n(\alpha, \beta)$ will be non-zero only for values of α and β in the range of order -1 to $+1$.

The Fourier transform of the cross correlation function is defined as the '*cross power spectral density*', $\hat{P}(k, \omega)$.

$$\hat{P}(k, \omega) = \int_{-\infty}^{\infty} \int_{-\infty}^{\infty} P(\Delta, \tau)\, e^{-ik\Delta}\, e^{-i\omega\tau}\, d\Delta\, d\tau \tag{10.26}$$

$$= \hat{P}(-k, -\omega) \quad \text{from (10.22)}$$

the inverse of which is

$$P(\Delta, \tau) = \frac{1}{(2\pi)^2} \int_{-\infty}^{\infty} \int_{-\infty}^{\infty} \hat{P}(k, \omega)\, e^{ik\Delta}\, e^{i\omega\tau}\, dk\, d\omega \tag{10.27}$$

and, as a special case when $\Delta = \tau = 0$,

$$P(0, 0) = \overline{p^2} = \frac{1}{(2\pi)^2} \int_{-\infty}^{\infty} \int_{-\infty}^{\infty} \hat{P}(k, \omega)\, dk\, d\omega. \tag{10.28}$$

It is this formula suggesting that the 'power', $\overline{p^2}$, is the integral of $\hat{P}(k, \omega)/(2\pi)^2$ over all wave number k and frequency ω that suggests $\hat{P}(k, \omega)/(2\pi)^2$ is the power per unit k and unit ω, and is the reason why $\hat{P}(k, \omega)/(2\pi)^2$ is called the cross power spectral density.

Fourier elements of p are almost always orthogonal

A two-dimensional Fourier transform pair can be defined by

$$\hat{p}(k, \omega) = \int_{-\infty}^{\infty} \int_{-\infty}^{\infty} p(x, t)\, e^{-ikx}\, e^{-i\omega t}\, dx\, dt$$

and

$$p(x, t) = \frac{1}{(2\pi)^2} \int_{-\infty}^{\infty} \int_{-\infty}^{\infty} \hat{p}(k, \omega)\, e^{ikx}\, e^{i\omega t}\, dk\, d\omega. \tag{10.29}$$

We will now demonstrate that this Fourier transform is very closely related to the cross power spectral density. The steps taken to achieve this are just the two-dimensional analogue of the procedures described in section 10.2. By definition, from (10.29),

$$\overline{\hat{p}(k, \omega)\hat{p}(k', \omega')}$$

$$= \iiiint_{-\infty}^{\infty} \overline{p(x, t)\, e^{-ikx}\, e^{-i\omega t}\, dx\, dt\, p(y, t')^{-ik'y}\, e^{-i\omega' t'}}\, dy\, dt'$$

The average affects only the non-deterministic elements of the integral which are, from equation (10.23),

$$\overline{p(x, t)p(y, t)} = P(y - x, t' - t).$$

Now change variables and let $y - x = \Delta$, $t' - t = \tau$, $dy = d\Delta$ and $dt' = d\tau$.

$$\overline{\hat{p}(k, \omega)p(k', \omega')} = \iint P(\Delta, \tau)\, e^{-ik'\Delta}\, e^{-i\omega'\tau}\, d\Delta\, d\tau \iint e^{-i(k+k')x}\, e^{-i(\omega+\omega')t}\, dx\, dt. \tag{10.30}$$

The $dx\, dt$ integral is over a deterministic function and gives

$$(2\pi)^2\, \delta(k + k')\, \delta(\omega + \omega').$$

The $d\Delta\, d\tau$ integral is the definition of the cross power spectral density $\hat{P}(k', \omega')$ cf. equation (10.26). Therefore equation (10.30) is

$$\begin{aligned} \overline{\hat{p}(k, \omega)\hat{p}(k', \omega')} &= (2\pi)^2 \hat{P}(k', \omega')\, \delta(k + k')\, \delta(\omega + \omega'), \\ &= (2\pi)^2 \hat{P}(-k, -\omega)\, \delta(k + k')\, \delta(\omega + \omega') \quad . \qquad (10.31) \\ &= (2\pi)^2 \hat{P}(k, \omega)\, \delta(k + k')\, \delta(\omega + \omega') \end{aligned}$$

$\hat{P}(k, \omega)$ is positive definite. We show this as follows. $\hat{p}(k, \omega)$ is the complex conjugate of $\hat{p}(-k, -\omega)$ (from equation 10.29) and

$$\begin{aligned} (2\pi)^2 \hat{P}(k, \omega) &= \int_{-\infty}^{\infty} \int_{-\infty}^{\infty} \overline{\hat{p}(k, \omega)\hat{p}(k', \omega')}\, dk'\, d\omega' \\ &= \int_{-\infty}^{\infty} \int_{-\infty}^{\infty} \overline{\hat{p}(k, \omega)\hat{p}^*(-k', -\omega')}\, dk'\, d\omega'. \end{aligned}$$

The integrand here is zero everywhere other than on the line $k' = -k$ and $\omega' = -\omega$ from (10.31). On that line the integrand is positive definite and real being $= \overline{|\hat{p}(k, \omega)|^2}$ (a highly singular, but positive definite function!) from which it follows that the integral also is a real positive definite function.

Equation (10.31) shows that unless $k + k' = 0$ and $\omega + \omega' = 0$, the Fourier elements are orthogonal. But that case is special and is the reason why the cross power spectral density can only be formed from the product of Fourier transforms by integration.

Equation (10.31) when integrated over all (k', ω') produces

$$(2\pi)^2 \hat{P}(-k, -\omega) = \int_{-\infty}^{\infty} \int_{-\infty}^{\infty} \overline{\hat{p}(k, \omega)\hat{p}(k', \omega')}\, \mathrm{d}k'\, \mathrm{d}\omega' \qquad (10.32)$$

$$= (2\pi)^2 \hat{P}(k, \omega).$$

Example

A plane incident sound wave makes an angle θ with a hard surface. What cross power spectral density would be obtained from measurements on the surface?

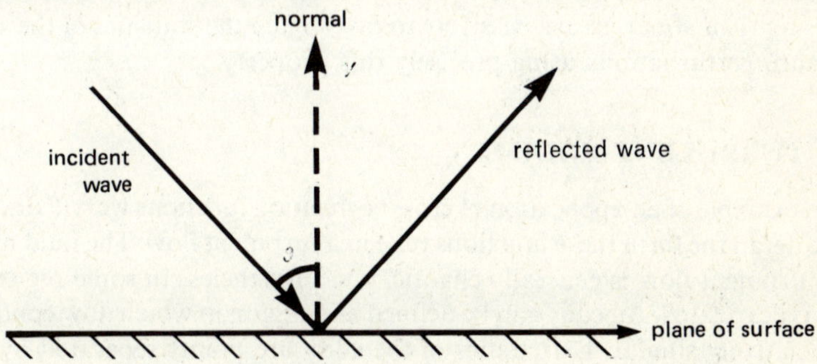

Fig. 10.6 — A plane wave is obliquely incident on a plane surface.

Let the pressure perturbations in the incident plane wave be $f(t - x \sin \theta/c + y \cos \theta/c)$. A wave of equal strength is reflected from the surafce and on $y = 0$ the pressure is given by

$$p(x, t) = 2f(t - x \sin \theta/c).$$

Hence

$$\hat{p}(k, \omega) = 2 \int_{-\infty}^{\infty} \int_{-\infty}^{\infty} f(t - x \sin \theta/c)\, \mathrm{e}^{-ikx - i\omega t}\, \mathrm{d}x\, \mathrm{d}t$$

$$= 2 \int_{-\infty}^{\infty} f(\tau)\, \mathrm{e}^{-i\omega\tau}\, \mathrm{d}\tau \int_{-\infty}^{\infty} \mathrm{e}^{-i(k + \omega \sin \theta/c)x}\, \mathrm{d}x$$

$$= 4\pi \hat{f}(\omega)\, \delta\left(k + \frac{\omega \sin \theta}{c}\right).$$

It then follows from (10.32) that

$$\hat{P}(k, \omega) = 4 \iint \overline{f(\omega)f(\omega')}\, \delta\left(k + \frac{\omega \sin\theta}{c}\right) \delta\left(k' + \frac{\omega' \sin\theta}{c}\right) \mathrm{d}k'\, \mathrm{d}\omega'$$

$$= 8\pi \iint \hat{F}(\omega)\, \delta(\omega + \omega')\, \delta\left(k + \frac{\omega \sin\theta}{c}\right) \delta\left(k' + \frac{\omega' \sin\theta}{c}\right) \mathrm{d}k'\, \mathrm{d}\omega'$$

,

where $\hat{F}(\omega)$ is the power spectral density of $f(t)$. The cross power spectral density is therefore equal to

$$\hat{P}(k, \omega) = 8\pi \hat{F}(\omega)\, \delta\left(k + \frac{\omega \sin\theta}{c}\right).$$

We see that the cross power spectral density has a strong peak at $\omega/k = -c/\sin\theta$. Detection of the phase velocities of any peaks in the surface pressure spectrum will therefore yield an estimate for the directions θ of any incoming sound waves. Submariners endeavour to obtain the bearings of distant vessels by using their sonar arrays passively to investigate the statistics of the surface pressure perturbations using precisely this property.

10.4 TURBULENT EDDIES

As an example of an application of cross correlation functions we will discuss in some detail the form these functions take in a turbulent flow. The fluid motion in a turbulent flow is generally chaotic, but nevertheless in some regions the flow is correlated. An eddy can be defined as a region in which flow conditions are relatively similar. Extremities of the eddy are points separated by more than the correlation scale. Within the eddy flow quantities are well correlated; they are uncorrelated over distances larger than the eddy dimension or, equivalently, the correlation scale. This usage of the term 'eddy' concerns a stochastic motion in which only statistical measures are significant. One eddy is not isolated or identifiable from another in any single realisation of the problem, nor indeed in any statistical sense. The eddy dimension indicates only the scale over which the stochastic variable is correlated. But eddying motions are nonetheless somewhat identifiable. Filaments of cigarette smoke are part of a chaotic turbulent flow, yet when viewed on a small enough scale the flow can be quite coherent and, in particular, eddies may be seen to move. Eddies are formed in non-uniformly moving flows and tend to move with the mean stream. Correlations are higher for points that have small separation from the eddy centre, so to speak. If the eddy is convected at constant speed U_c, and the space separation Δ is arranged so that at time delay τ, Δ is equal to $U_c\tau$, there will be no separation of the sample points relative to the eddy and the correlation will consequently be higher than at other values of the separation

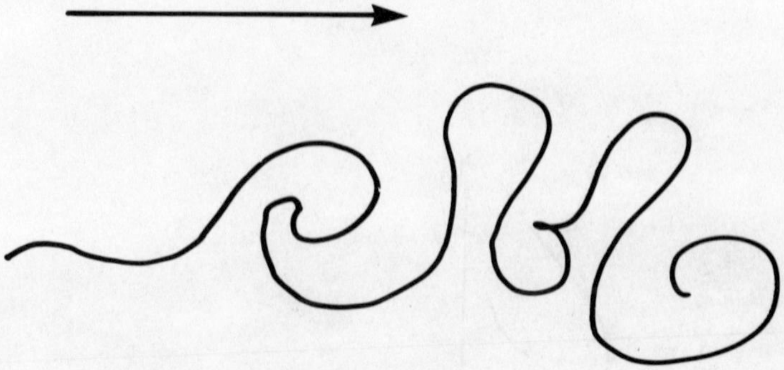

Fig. 10.7 — Eddies in a non-uniformly moving stream.

variable. Convected fields are recognisable by the tendency for maximum correlation to lie along a space time trajectory where Δ is proportional to τ.

When an eddy moves with speed U_c a co-ordinate $x - U_c t$ represents a definite point fixed in the reference frame moving with the eddy. An observer in that frame would measure the function $p(x + U_c t, t)$ rather than the $p(x, t)$ measured in the fixed reference frame. But the moving observer would also see a statistically stationary signal and one could form the cross correlation function of that signal, a function we denote by $P_m(\Delta, \tau)$ the suffix m implying a moving eddy-fixed observer.

$$P_m(\Delta, \tau) = \overline{p(x + U_c t, t)p(x + \Delta + U_c(t + \tau), t + \tau)}$$
$$= P(\Delta + U_c\tau, \tau). \tag{10.33}$$

The space correlation is uninfluenced by convection and is the same, for a given separation variable Δ, in both fixed and moving co-ordinate systems.

$$B(\Delta) = P(\Delta, 0) = P_m(\Delta, 0) = B_m(\Delta).$$

But the autocorrelation function is different.

$$P(\tau) = P(0, \tau) = P_m(-U_c\tau, \tau) \neq P_m(\tau)$$

$$P_m(\tau) = P_m(0, \tau) = P(U_c\tau, \tau).$$

The moving axis autocorrelation function $P_m(\tau)$ is the envelope of the maxima of $P(\Delta, \tau)$; the value of Δ which maximises the autocorrelation function is by definition the value $\Delta = U_c\tau$ for which there is no displacement relative to the eddy centre.

The cross correlation of a signal measured by a moving observer is simply related to that observed by a fixed observer. The relation amounts to nothing more than a skewing of the co-ordinate system provided the observer's speed, the convection speed, is uniform. The situation is illustrated in Figure 10.8. The eddy moves in the reference system for which its time scale is maximum.

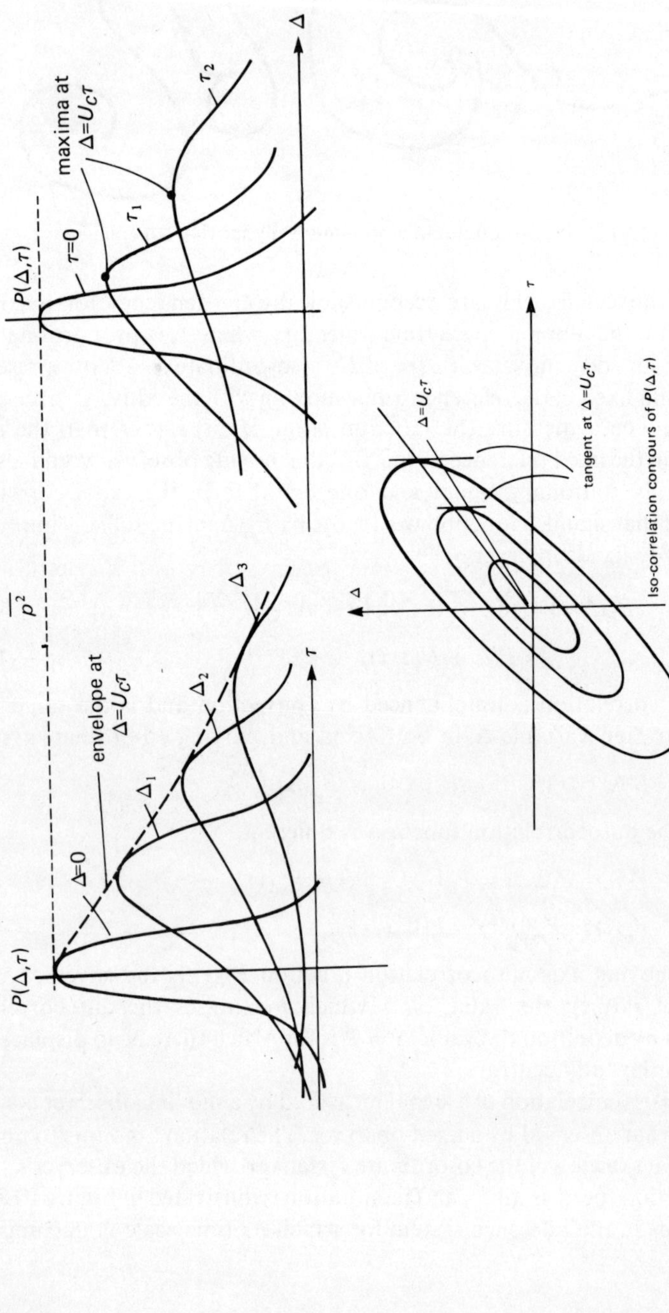

Fig. 10.8 — Typical cross correlations for eddies moving with a velocity U_c.

Cross spectra with eddy convection

The moving axis cross spectral density is the Fourier transform of the moving axis cross correlation function $P_m(\Delta, \tau)$.

$$\hat{P}_m(k, \omega) = \int_{-\infty}^{\infty} \int_{-\infty}^{\infty} P_m(\Delta, \tau)\, e^{-ik\Delta}\, e^{-i\omega\tau}\, d\Delta\, d\tau \tag{10.34}$$

or from (10.33)

$$\hat{P}_m(k, \omega) = \int_{-\infty}^{\infty} \int_{-\infty}^{\infty} P(\Delta + U_c\tau, \tau)\, e^{-ik\Delta}\, e^{-i\omega\tau}\, d\Delta\, d\tau.$$

Now change variables and let $\Delta + U_c\tau = \Delta'$ and $d\Delta = d\Delta'$.

$$\hat{P}_m(k, \omega) = \int_{-\infty}^{\infty} \int_{-\infty}^{\infty} P(\Delta', \tau)\, e^{-ik\Delta'}\, e^{-i(\omega - U_c k)\tau}\, d\Delta'\, d\tau \tag{10.35}$$

$$= \hat{P}(k, \omega - U_c k) \quad \text{from (10.26).}$$

Simply by a change of variable we find

$$\hat{P}(k, \omega) = \hat{P}_m(k, \omega + U_c k). \tag{10.36}$$

In the moving frame, the eddy has a maximum time scale—by definition. Therefore, in a moving frame the spectrum is concentrated at low frequencies—zero frequency in fact. The fixed axis cross power spectral density $\hat{P}(k, \omega)$ of a convected eddy pattern is therefore maximum at a value of ω which minimises the 'moving axis' frequency $\omega + U_c k$, cf. equation (10.36). The fixed axis spectrum is maximum when $\omega = -U_c k$.

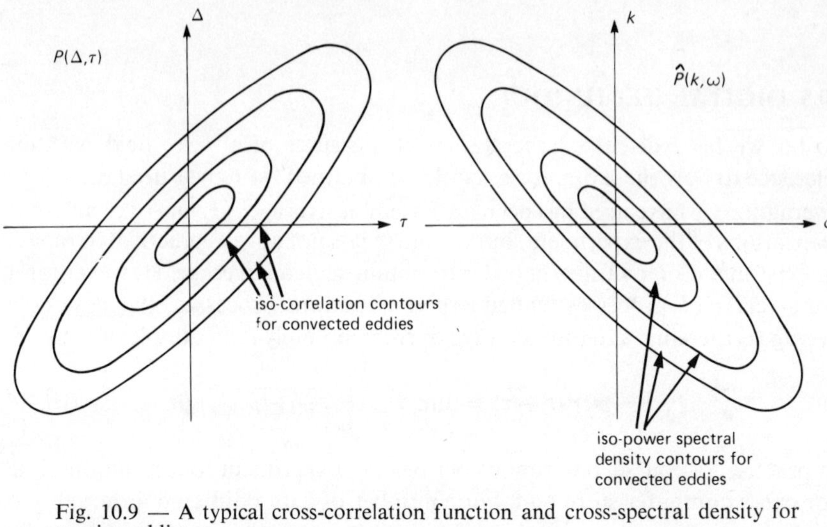

Fig. 10.9 — A typical cross-correlation function and cross-spectral density for moving eddies.

Taylor's hypothesis

The moving axis time scale is relatively large always—otherwise there would be little evidence of convective motion. As an approximation Taylor suggested that the moving axis time scale could be regarded as infinite, in which case

$$P_m(\Delta, \tau) = P_m(\Delta, 0) = B_m(\Delta) \tag{10.37}$$

$$P(\Delta, \tau) = P_m(\Delta - U_c\tau, \tau) = B_m(\Delta - U_c\tau). \tag{10.38}$$

Then

$$P(\Delta, 0) = B(\Delta) = B_m(\Delta),$$

and

$$P(0, \tau) = P(\tau) = B_m(-U_c\tau)$$

so that

$$P(\tau) = B(-U_c\tau) = B(U_c\tau)$$

i.e. Taylor's hypothesis is that the time variable is simply a rescaling of the space variable.

The moving axis spectrum is given by (10.35) and (10.38), and in the Taylor's hypothesis case is

$$\begin{aligned}\hat{P}_m(k, \omega) &= \int_{-\infty}^{\infty}\int_{-\infty}^{\infty} B_m(\Delta)\, \mathrm{e}^{-ik\Delta}\, \mathrm{e}^{-i\omega\tau}\, \mathrm{d}\Delta\, \mathrm{d}\tau \\ &= 2\pi\hat{B}(k)\,\delta(\omega),\end{aligned} \tag{10.39}$$

where $B(k)$ is the 'wavenumber' spectrum

$$\hat{B}(k) = \int_{-\infty}^{\infty} B(\Delta)\, \mathrm{e}^{-ik\Delta}\, \mathrm{d}\Delta.$$

10.5 DIGITAL TECHNIQUES

So far we have discussed the statistical measures of a wave field without reference to how these quantities could be obtained from measured data. The 'averaging' we have used has been an 'ensemble average', i.e. one over different realizations of the experiment, but of course it is not always practical to repeat an experiment many times in order to obtain such an average. However for a useful class of processes, called ergodic random processes, the ensemble average is the same as a time average over an infinitely long sample of data e.g.

$$P(\tau) = \overline{p(t)p(t+\tau)} = \lim_{T\to\infty} \frac{1}{2T}\int_{-T}^{T} \overline{p(t)p(t+\tau)}\, \mathrm{d}t. \tag{10.40}$$

In practise, of course, one cannot perform an experiment for an infinite time nor can a computer store an infinite number of data points, so although $p(t)$ might be measured as a continuous signal, only some of these values can be stored for subsequent digital analysis. It is customary to 'sample' and store the values of the signal $p(t)$ at equal time intervals. So let us suppose that $p(t)$ is measured for a finite time interval T, and sampled at N different times at time lags Δ i.e. at $t = n\Delta$, where $n = 0, 1, \ldots, N-1$ and $\Delta = T/N$. Provided $p(t)$ does not vary rapidly between two sampling points we can obtain an estimate of the power spectral density of p from this data.

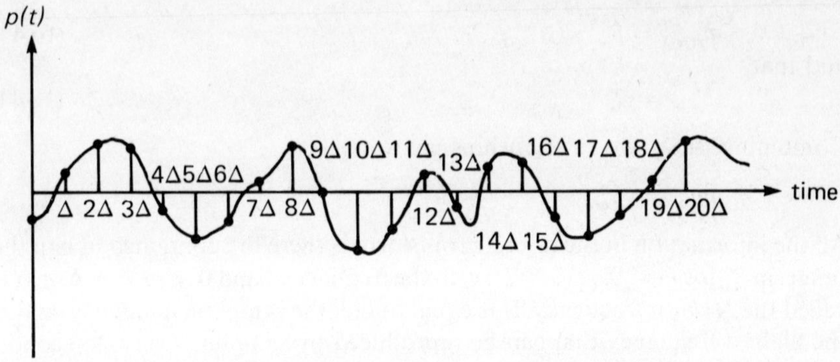

Fig. 10.10 — The continuous signal $p(t)$ is sampled at intervals of Δ seconds.

Although the power spectral density was defined as the Fourier transform of the autocorrelation, we saw in (10.16) that it could be related directly to the Fourier transform of the signal. Modern FFT (Fast Fourier Transform) techniques enable the transform of a signal to be evaluated quickly and accurately, and this is usually the most convenient way of determining its spectrum. To see how that can be achieved we begin by noting that a discrete Fourier transform of $p(n\Delta)$, $n = 0, 1, \ldots, N-1$, can be defined by

$$\hat{p}_m = \frac{1}{N} \sum_{n=0}^{N-1} p(n\Delta)\, e^{-2\pi imn/N}, \tag{10.41}$$

and that a knowledge of the $\hat{p}_m$'s is sufficient to reconstruct $p(n\Delta)$ from

$$p(n\Delta) = \sum_{m=0}^{N-1} \hat{p}_m\, e^{2\pi imn/N}. \tag{10.42}$$

This last relationship can be proved by substituting the expression for $\hat{p}_m$ in (10.41) into (10.42) to obtain

$$\sum_{m=0}^{N-1} \hat{p}_m\, e^{2\pi imn/N} = \frac{1}{N} \sum_{l=0}^{N-1} p(l\Delta) \sum_{m=0}^{N-1} e^{2\pi mi(n-l)/N}.$$

The double sum simplifies to $p(n\Delta)$ because

$$\sum_{m=0}^{N-1} e^{2\pi i m(n-l)/N} = 0 \quad \text{if } l \neq n$$

$$= N \quad \text{if } l = n.$$

Equation (10.42) shows that $\hat{p}_m$ can be considered as the 'amplitude' of the sampled disturbance with angular frequency $2\pi m/N\Delta$.

It is apparent from (10.41) that

$$\hat{p}_{m+N} = \hat{p}_m \tag{10.43}$$

and that

$$\hat{p}_{-m} = \hat{p}^*_m. \tag{10.44}$$

Combining these two relationships gives

$$\hat{p}_{N-m} = \hat{p}^*_m.$$

All the information in the discrete transform is therefore contained in half the range, in $\hat{p}_m$ for $m = 0, \ldots, N/2$, i.e. in the frequency band $0 < \omega < \pi/\Delta$. π/Δ is called the *Nyquist* frequency. It is equal to half the sampling frequency, and is the highest frequency that can be reproduced by sampling every Δ seconds.

The Fourier coefficient $\hat{p}_m$ could be evaluated by summing the series in (10.41) directly, but the calculation of the complete set $\hat{p}_m$, $m = 0, 1, \ldots, N/2$, would require about N^2 arithmetic operations. Fast Fourier transforms (FFTs) are based on a cunning regrouping of these summations which drastically reduces the number of steps required. Suppose that the sample length has been chosen so that N is an integer power of 2, $N = 2^l$ say. The terms to be evaluated are

$$\hat{p}_m = \frac{1}{N} \sum_{n=0}^{N-1} p(n\Delta)\, e^{-2\pi i m n/N} \quad 0 < m < N/2. \tag{10.45}$$

This sum can be split up by partitioning $p(n\Delta)$ into two shorter sequences, q_n and r_n say, where

$$q_n = p(2n\Delta), \quad r_n = p((2n+1)\Delta) \quad \text{for } 0 \leqslant n \leqslant N/2.$$

The transforms of q_n and r_n are given by

$$\hat{q}_m = \frac{1}{N/2} \sum_{n=0}^{N/2} q_n\, e^{-2\pi i m n/(N/2)}$$

$$\hat{r}_m = \frac{1}{N/2} \sum_{n=0}^{N/2} r_n\, e^{-2\pi i m n/(N/2)}$$

and it therefore follows that

$$\hat{p}_m = \tfrac{1}{2}(\hat{q}_m + e^{-2\pi i m/N}\hat{r}_m).$$

The sequences q_n and r_n can similarly each be subdivided into two sets each with $N/4$ elements. After $\log_2 N$ such partitionings N sets are obtained each with only one element. Now a sequence with one element has only one Fourier coefficient, and that is equal to the element itself. Working back from these N sets it is possible to evaluate $\hat{p}_m$ in $N \log_2 N$ arithmetic operations, a considerable saving on the N^2 operations required to sum (10.41) directly.

We have seen that if $p(t)$ is sampled at N data points, $t = n\Delta$; $n = 0, 1, \ldots, N-1$, it is possible to introduce a discrete Fourier transform $\hat{p}_m$ of $p(n\Delta)$. We will now investigate the form of the relationship between $\hat{p}_m$ and $\hat{p}(\omega)$, the Fourier transform of the continuous function $p(t)$. In equation (10.41) $\hat{p}_m$ is defined by

$$\hat{p}_m = \frac{1}{N} \sum_{n=0}^{N-1} p(n\Delta)\, \mathrm{e}^{-2\pi i m n/N}.$$

p(nΔ) in this equation can be rewritten as

$$p(n\Delta) = \frac{1}{2\pi} \int_{-\infty}^{\infty} \hat{p}(\omega)\, \mathrm{e}^{i\omega n\Delta}\, \mathrm{d}\omega,$$

and this leads to

$$\hat{p}_m = \frac{1}{2\pi N} \int_{-\infty}^{\infty} \hat{p}(\omega) \sum_{n=0}^{N-1} \mathrm{e}^{in(\omega\Delta - 2\pi m/N)}\, \mathrm{d}\omega.$$

The series in this expression is a GP and can be summed in a straightforward way to give

$$\hat{p}_m = \frac{1}{2\pi} \int_{-\infty}^{\infty} \hat{p}(\omega) \hat{f}\left(\omega - \frac{2\pi m}{T}\right) \mathrm{d}\omega \qquad (10.46)$$

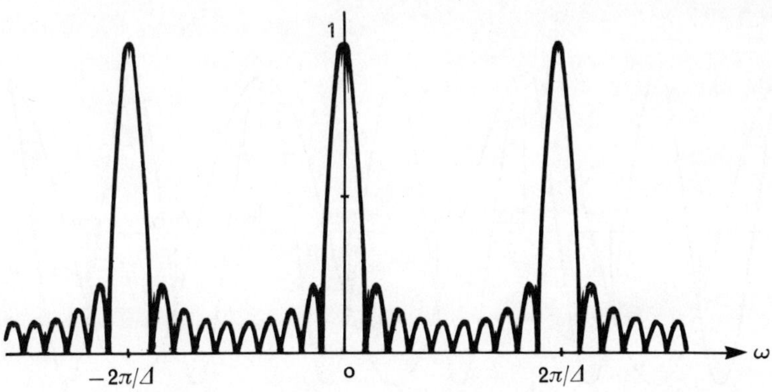

Fig. 10.11 — The variation of $|\hat{f}(\omega)|$ with frequency ω.

where

$$\hat{f}(\omega) = \frac{\sin\left(\dfrac{N\Delta\omega}{2}\right)}{N\sin\left(\dfrac{\Delta\omega}{2}\right)}\,e^{i(N-1)\Delta\omega/2}.$$

A sketch of $|\hat{f}(\omega)|$ is given in Figure 10.11. For large N, we note that $|\hat{f}(\omega)|$ is small except near $\sin(\Delta\omega/2) = 0$, and that for ω near $2\pi M/\Delta$, M integer, $|\hat{f}(\omega)|$ increases to a value of 1. Equation (10.46) therefore says that $\hat{p}_m$ is influenced by all pressure perturbations $\hat{p}(\omega)$ with $\omega \sim \pm(2\pi M/\Delta) + (2\pi m/T)$. $\hat{p}_m$ is not just related to $\hat{p}(2\pi m/T)$ as one would like, but contains terms due to all disturbances with frequencies $(2\pi M/\Delta) \pm (2\pi m/T)$. This contamination is called *aliasing*, it arises because it is impossible to distinguish between disturbances at a frequency ω and $2\pi M/\Delta \pm \omega$ by sampling at a rate Δ. Suppose, for example, that we sample the function $\cos[(2\pi M/\Delta) \pm \omega]t$ at $t = n\Delta$. Then since

$$\cos\left(\frac{2\pi M}{\Delta} \pm \omega\right)n\Delta = \cos(\omega n\Delta),$$

disturbances at these different frequencies are indistinguishable. Aliasing was sometimes observed in the motion of waggon wheels in old Westerns. There the frame speed of the film was too slow to reproduce the rapid motion of the wheels, the motion was aliased and the wheels appeared to be rotating at a much slower speed, or to be going backwards. Aliasing makes it impossible to interpret the information contained in $\hat{p}_m$, and the only way of obtaining useful data is to ensure that the sampling rate is sufficiently fast that the Nyquist frequency, π/Δ, is larger than the maximum frequency in the signal. Since in practice the signal is contaminated by noise, the highest frequency in the signal

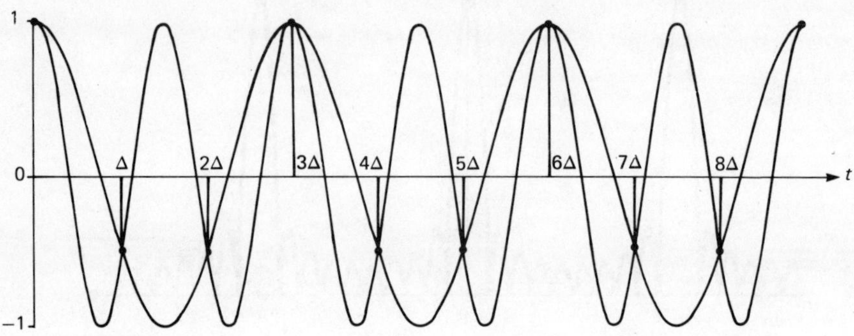

Fig. 10.12 — Cos $2\omega_0 t$ and cos $\omega_0 t$ are indistinguishable when sampled every $2\pi/3\omega_0$ seconds.

is not known. The usual procedure is to decide the highest frequency of interest and then to pass the signal through a low-pass filter which cuts off higher frequencies. The sampling rate is then chosen to ensure that the Nyquist frequency is at least as high as this cut-off frequency. This ensures that $\hat{p}(\omega)$ is zero at all values of $\omega = (2\pi M/\Delta) \pm (2\pi m/T)$, except when $M = 0$. The values of the discrete Fourier coefficients $\hat{p}$ are then only influenced by disturbances with frequencies nearly equal to $2\pi m/T$.

We can relate the power spectral density to the product of $\hat{p}_m$'s. From (10.46)

$$\hat{p}_m \hat{p}_m^* = \frac{1}{4\pi^2} \int \hat{p}(\omega)\hat{p}(-\alpha)\hat{f}\left(\omega - \frac{2\pi m}{T}\right)\hat{f}\left(-\alpha + \frac{2\pi m}{T}\right) \mathrm{d}\omega\, \mathrm{d}\alpha. \quad (10.47)$$

If we average this over several different time sequences of data we can use

$$\overline{\hat{p}(\omega)\hat{p}(-\alpha)} = 2\pi \hat{P}(\omega)\, \delta(\omega - \alpha),$$

to obtain

$$\overline{\hat{p}_m \hat{p}_m^*} = \frac{1}{2\pi} \int \hat{P}(\omega) \left| \hat{f}\left(\omega - \frac{2\pi m}{T}\right) \right|^2 \mathrm{d}\omega. \quad (10.48)$$

$|\hat{f}|^2$ is often called a window function. The measured data points enable us to evaluate the average of the true power spectrum over the frequency band where $\hat{f}$ is non-zero. The main peak of $\hat{f}(\omega)$ near the origin has a width of $4\pi/T$ and this is a measure of the frequency resolution that can be obtained with this window function.

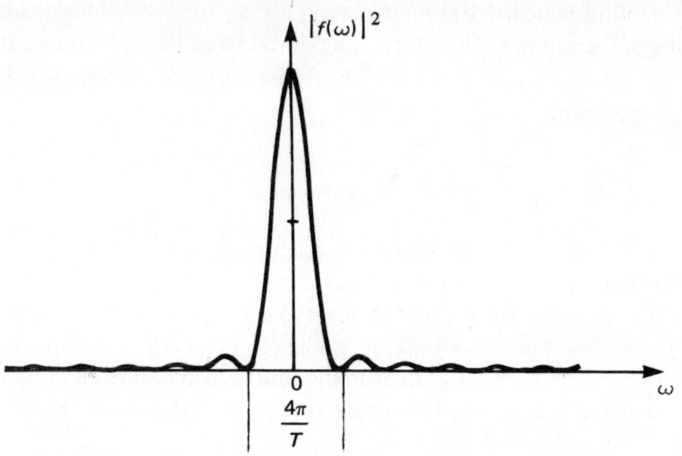

Fig. 10.13 — The variation of $|\hat{f}(\omega)|^2$ with ω for very large N.

EXERCISES FOR CHAPTER 10

1. An identical signal is fed into both channels of a perfect stereophonic hi-fidelity system and the sound produced is observed at a position directly in front of one speaker and at a distance d, equal to that separating the two speakers. Determine the frequency spectrum of the observed sound when the system is fed with white noise.

2. Noise with power spectral density $\hat{P}_I(\omega)$ propagates along a pipe of cross-sectional area A_1 and enters a single expansion chamber muffler of length l and cross-sectional area A_2. Determine the frequency spectrum of the transmitted sound.

3. A spherical gas bubble is trapped inside a region where fluid of density ρ_0 and sound speed c is in turbulent motion. The random pressure field within the turbulence changes slowly on the bubble scale so that the region surrounding the bubble is effectively all at pressure $p(t)$. The mean value of $p(t)$ is p_0. The bubble volume changes in response to this field.

By assuming that the motion induced by the bubble response is weak and irrotational and that the gas within the bubble is in an isothermal state, show that $A(\omega)$, the spectrum of the variation in bubble radius, is

$$A(\omega) = \frac{P(\omega)}{\left(\dfrac{3p_0}{a} - \rho_0 a\omega^2\right)^2 + 9p_0^2\omega^2/c^2},$$

where $P(\omega)$ is the spectrum of $p(t) - p_0$ and a is the mean radius of the bubble. Surface tension should be neglected.

4. A steady omnidirectional source of sound positioned 10 metres above a hard flat ground is heard 100 metres away by an observer whose ears are 2 metres above the ground. The sound that would be radiated by the source to a distance of 100 metres in free space has mean square amplitude $\overline{p^2}$ and a frequency spectrum

$$W(\omega) = \frac{2\pi\overline{p^2}}{10^3\sqrt{\pi}}\exp\left\{-\frac{\omega^2}{10^6}\right\}$$

per unit frequency, ω being the frequency in radians per second.

Determine, to an accuracy of 1 per cent, the mean square sound level heard by the observer. Show that the observed autocorrelation of the pressure is equal to $2P(\tau) + P(\tau - \tau_0) + P(\tau + \tau_0)$ where $P(\tau)$ is the autocorrelation of the pressure at a distance of 100 m from the source in free space, and τ_0 is a time interval. Hence calculate the frequency spectrum of the sound heard by the observer.

Note that

$$\int_{-\infty}^{\infty} e^{-\alpha^2 x^2} e^{ikx} \, dx = \frac{\sqrt{\pi}}{\alpha} \exp\left(-\frac{k^2}{4\alpha^2}\right).$$

5. Sketch the form of the cross power spectral densities and cross correlation functions of eddies conforming with Taylor's hypothesis.

6. Determine the forms of the cross correlation function and of the cross power spectral density of the function $v = \partial^2 p(x, t)/\partial x \, \partial t$ and express them as explicit functions of $p(x, t)$'s statistics.

If $\hat{P}(k, \omega)$ is Gaussian i.e.

$$\hat{P}(k, \omega) = A \exp\left[-\frac{k^2}{k_0^2} - \frac{\omega^2}{\omega_0^2}\right]$$

for some constants A, k_0, ω_0 sketch the contours of the cross power spectral density of the function v.

7. Show that for an ergodic random process $\overline{f(\partial f/\partial t)} = 0$ and $\overline{f(\partial^2 f/\partial t^2)} = -\overline{(\partial f/\partial t)^2}$. Hence show that the mean intensity associated with the three-dimensional sound pressure field

$$p'(\mathbf{x}, t) = \frac{\partial}{\partial x_1}\left\{\frac{\frac{\partial}{\partial t} f(r - ct)}{r}\right\},$$

where $r = |\mathbf{x}|$, is

$$\bar{\mathbf{I}} = \overline{\left(\frac{\partial^2 f}{\partial t^2}\right)^2} \frac{x_1^2 \mathbf{x}}{r^5 \rho_0 c^3}.$$

Check that $\bar{\mathbf{I}}$ is solenoidal.

8. The cross-correlation between two signals $p(t)$ and $q(t)$ is defined by $C_{pq}(\tau) = \overline{p(t)q(t + \tau)}$. Use the techniques of section 10.2 to show that the cross-spectral density function $\hat{C}_{pq}(\omega)$ $(= \int_{-\infty}^{\infty} C_{pq}(\tau) \, e^{-i\omega\tau} \, d\tau)$ is related to the Fourier transforms of p and q by

$$\overline{\hat{p}(\omega)\hat{q}(\omega')} = 2\pi \hat{C}_{pq}(\omega') \, \delta(\omega + \omega').$$

The coherence $\gamma(\omega)$ is defined by

$$\gamma^2 = \frac{|\hat{C}_{pq}(\omega)|^2}{\hat{P}(\omega)\hat{Q}(\omega)},$$

where $\hat{P}$ and $\hat{Q}$ are the power spectral densities of p and q respectively. Show that, if $\hat{q}(\omega) = \hat{s}(\omega) + \hat{n}(\omega)$, where the signal $\hat{s}(\omega) = T(\omega)\hat{p}(\omega)$ and $\hat{n}(\omega)$ is random noise, the signal-to-noise ratio is equal to $\gamma^2(1 - \gamma^2)^{-1}$.

Chapter 11

Flow Induced Vibration and Instability

11.1 THE FORCED HARMONIC OSCILLATOR

The equation

$$\frac{\partial^2 \varphi}{\partial t^2} + 2\beta \frac{\partial \varphi}{\partial t} + (k^2 + \beta^2)\varphi = f(t) \tag{11.1}$$

models several cases of the forced vibrations that are found in everyday life. It describes the behaviour of driven resonators which have many of the qualities found in the flow induced vibration of complex structures. The techniques

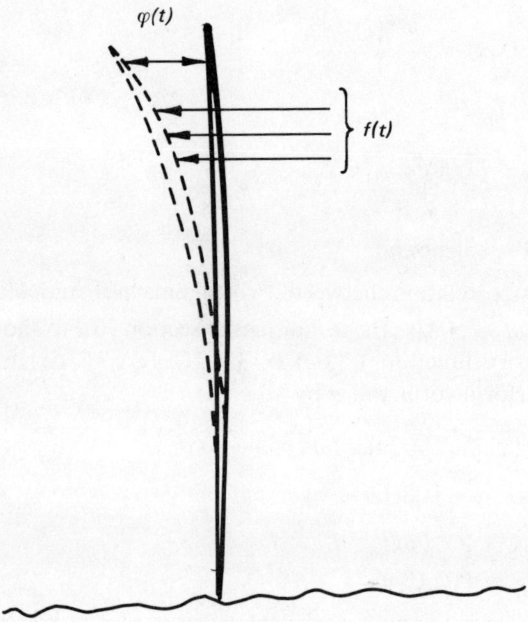

Fig. 11.1 — A mast will wave in an unsteady wind deflecting an amount $\phi(t)$ in response to an oscillating force $f(t)$.

used for their study are typical of those needed for the analysis of the simplest case which we now deal with as an introduction to the general subject.

For example a flexible panel in the wall of a wind tunnel containing turbulent flow might respond with displacement φ to a time varying force f per unit mass. Or the motion of a pendulum, or of a flexible mast, disturbed from equilibrium by an unsteady breeze might be studied in this model. β represents the damping in the oscillator and $(k^2 + \beta^2)$ the stiffness per unit mass expressed in such a way as to minimise the algebraic detail.

If there is no forcing then $\varphi(t)$ can have only the general form

$$\varphi(t) = \{A \cos kt + B \sin kt\}\, e^{-\beta t}, \tag{11.2}$$

A and B being constants determined by prescribed boundary conditions. Damping ensures the continual decay of the harmonic motion at angular frequency k, which persists indefinitely in a strictly harmonic manner only in the undamped oscillator i.e. when $\beta = 0$. Our interest here is in describing the response of the oscillator, either damped or undamped, when deriven by a randomly varying force $f(t)$. For definiteness we shall consider that the motion φ and the force f are zero for negative times but that $f(t)$ is statistically stationary for positive times: the problem is to describe the statistics of the resonator response induced by f. We proceed by Fourier transforming equation (11.1)

$$\begin{Bmatrix} \varphi(t) \\ f(t) \end{Bmatrix} = \frac{1}{2\pi} \int_{-\infty}^{\infty} \begin{Bmatrix} \hat{r}(\omega) \\ \hat{f}(\omega) \end{Bmatrix} e^{i\omega t}\, d\omega \tag{11.3}$$

so that (11.1) is reduced to the algebraic equation:

$$(k^2 + \beta^2 - \omega^2)\hat{\varphi} + 2i\omega\beta\hat{\varphi} = \hat{f} \tag{11.4}$$

which can be alternatively be written as

$$\hat{\varphi}(\omega) = \frac{-\hat{f}(\omega)}{(\omega + k - i\beta)(\omega - k - i\beta)}. \tag{11.5}$$

Thus having related the Fourier transforms of the force and response the power spectral densities follow immediately by forming the integral of the average product of $\hat{\varphi}(\omega)$ and $\hat{\varphi}(\alpha)$, as in equation (10.16), and then integrating over α.

$$\int_{-\infty}^{\infty} \overline{\hat{\varphi}(\omega)\hat{\varphi}(\alpha)\, d\alpha} = 2\pi \int_{-\infty}^{\infty} \hat{\Phi}(\omega)\, \delta(\omega + \alpha)\, d\alpha = 2\pi\hat{\Phi}(\omega) \tag{11.6}$$

$$\hat{\Phi}(\omega) = \frac{\hat{F}(\omega)}{(k^2 + \beta^2 - \omega^2)^2 + 4\omega^2\beta^2} \tag{11.7}$$

where $\hat{F}(\omega) = (1/2\pi) \int_{-\infty}^{\infty} \overline{\hat{f}(\omega)\hat{f}(\alpha)}\, d\alpha$ is the power spectral density of the

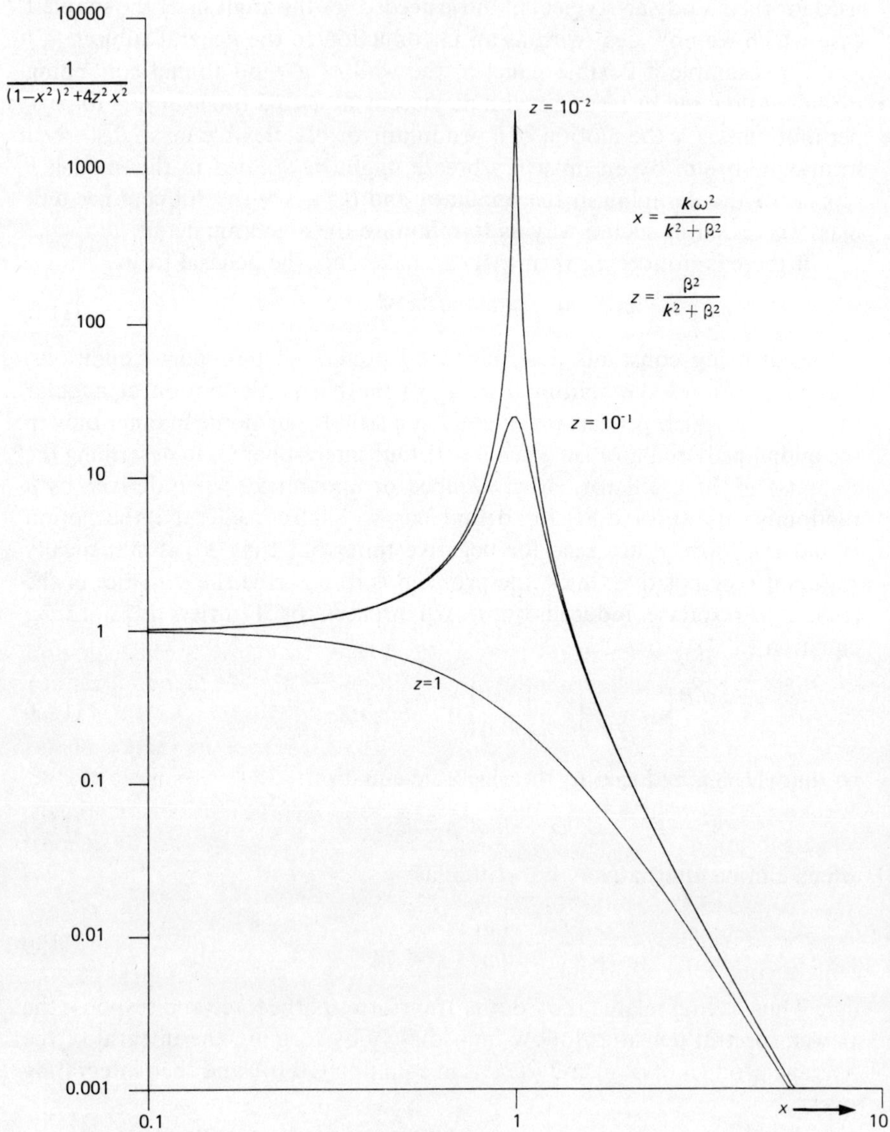

Fig. 11.2 — The response level of a harmonic oscillator as a function of frequency; that is Eq. (11.7).

statistically stationary force f that drives the oscillator to respond with power spectral density $\Phi(\omega)$. If the damping is very small, $\beta^2 \ll k^2$, the response is dominated by the peak at the resonance frequency $\omega = \pm k$, and it is in the vicinity of that peak that most of the vibrational energy lies. It is then appropriate to approximate equation (11.7) by setting $\hat{F}(\omega) = \hat{F}(k)$ and writing

$$\hat{\Phi}(\omega) = \hat{F}(k)\frac{1}{(k^2 - \omega^2)^2 + 4\omega^2\beta^2}. \tag{11.8}$$

The quality factor 'Q' of the oscillator is a measure of the sharpness of the resonance peak. The response spectrum is 3 dB down on its maximum level when the frequency is $\omega = k \pm \beta$. The difference, 2β, between these two frequencies is the bandwidth of the resonance and Q is defined as the ratio $k/2\beta$, i.e. it is the ratio of the resonance frequency to the bandwidth.

The mean square response is the integral of the power spectral density

$$\begin{aligned}\overline{\varphi^2} &= \frac{1}{2\pi}\int_{-\infty}^{\infty} \hat{\Phi}(\omega)\,d\omega = \frac{\hat{F}(k)}{\pi}\int_0^{\infty}\frac{d\omega}{(k^2 - \omega^2)^2 + 4\omega^2\beta^2} \\ &= \frac{\hat{F}(k)}{4k^2\beta} = \frac{Q\hat{F}(k)}{2k^3}.\end{aligned} \tag{11.9}$$

i.e. $\overline{\varphi^2} = \beta$ times the maximum value of $\hat{\Phi}(\omega)$ which is $\hat{F}(k)/4k^2\beta^2$.

This is the long term relation between force and response statistics, but in the initial value problem we are considering, there is a settling down period prior to the establishment of the statistically steady condition this equation describes. We are considering the initial value problem in order to illustrate the time taken to establish a given vibration level and also to have a technique for describing the response that occurs at the resonance frequency of the oscillator when the damping is zero. Equations (11.7) and (11.9) simply predict an infinite response for that condition whereas in fact the oscillator responds with a vibration of linearly increasing intensity as the force works to steadily increase the energy stored in the oscillation.

To describe the vibrational response while it is still growing as time increases it is clearly preferable to know $\varphi(t)$ rather than its Fourier transform constituents. That is obtained by Fourier transforming equation (11.5)

$$\begin{aligned}\varphi(t) &= \frac{-1}{2\pi}\int_{-\infty}^{\infty}\frac{\hat{f}(\omega)\,e^{i\omega t}\,d\omega}{(\omega + k - i\beta)(\omega - k - i\beta)} \\ &= \frac{-1}{2\pi}\int_{-\infty}^{\infty}\int_{-\infty}^{\infty}\frac{f(\tau)\,e^{-i\omega\tau}\,e^{i\omega t}\,d\omega\,d\tau}{(\omega + k - i\beta)(\omega - k - i\beta)}.\end{aligned} \tag{11.10}$$

The integral over ω can be written down immediately from reference to tables of Fourier transforms, or worked out by partial integration and Cauchy's theorem; its value is:

$$\frac{1}{2\pi}\int_{-\infty}^{\infty}\frac{e^{i\omega(t-\tau)}}{(\omega - k - i\beta)(\omega + k - i\beta)}\,d\omega = -\frac{e^{-(t-\tau)\beta}}{k}H(t-\tau)\sin k(t-\tau)$$

where H is the Heaviside function. Consequently

$$\varphi(t) = \int_{-\infty}^{t} f(\tau)\frac{e^{-(t-\tau)\beta}}{k}\sin k(t-\tau)\,d\tau$$

$$= \int_{0}^{t} f(\tau)\frac{e^{-\beta(t-\tau)}}{k}\sin k(t-\tau)\,d\tau, \qquad (11.11a)$$

because $f(\tau) = 0$ for $\tau < 0$. This expression can be written alternatively by shifting the origin of τ as

$$\varphi(t) = \int_{-\tau}^{t-\tau} f(\tau+\tau')\frac{e^{-(t-\tau-\tau')\beta}}{k}\sin k(t-\tau-\tau')\,d\tau'. \qquad (11.11b)$$

The mean value of φ^2 established in many experiments at a time t from the start, $\overline{\varphi(t)\varphi(t)}$ can consequently be expressed using the product of equations (11.11a) and (11.11b); it is not independent of t for some significant time.

$$\overline{\varphi(t)\varphi(t)}$$

$$= \int_{0}^{t}\int_{-\tau}^{t-\tau} \frac{\overline{f(\tau)f(\tau+\tau')}}{k^2} e^{-2\beta(t-\tau)} e^{\beta\tau'} \sin k(t-\tau)\sin k(t-\tau-\tau')\,d\tau'\,d\tau.$$

$F(\tau') = \overline{f(\tau)f(\tau+\tau')}$ is the autocorrelation of the applied force. This is a function only of τ' because it is statistically stationary and vanishes for large time differences $\tau' > T$ say.

$$\overline{\varphi(t)\varphi(t)} = \int_{0}^{t}\int_{-\tau}^{t-\tau} \frac{e^{\beta\tau'}F(\tau')}{k^2}$$

$$\times \{\sin^2 k(t-\tau)\cos k\tau' - \tfrac{1}{2}\sin 2k(t-\tau)\sin k\tau'\}\,d\tau'\,e^{-2\beta(t-\tau)}\,d\tau. \qquad (11.12)$$

The integration range for this equation is illustrated in Figure 11.3 for the case where the forcing has been active for much longer than the coherence time T. From the diagram it is clear that the effective integration range is confined to a column of approximate height t on the τ axis and of width $2T$. Now we consider the case where the contribution from the two fixed regions at the foot and at the head of the column is negligible in comparison with that from the main bulk of the column which alone increases in proportion to time t. In the limit where t/T is very large

$$\overline{\varphi(t)\varphi(t)} \underset{t/T\to\infty}{\sim} \int_{0}^{t}\int_{-T}^{T} e^{\beta\tau'}F(\tau')$$

$$\times \{\sin^2 k(t-\tau)\cos k\tau' - \tfrac{1}{2}\sin 2k(t-\tau)\sin k\tau'\}\,d\tau'$$

$$\times \frac{e^{-2\beta(t-\tau)}}{k^2}\,d\tau. \qquad (11.13)$$

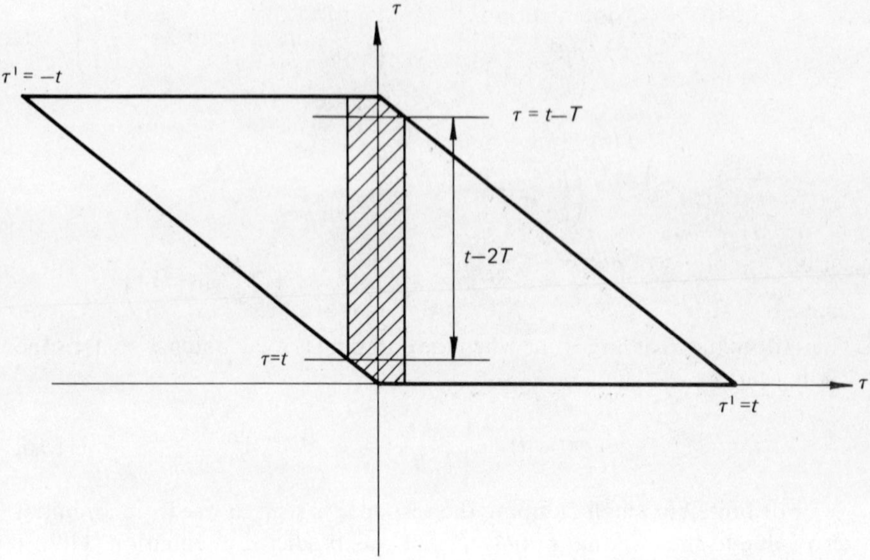

Fig. 11.3 — A diagram illustrating that the integration range in equation (11.12) is confined to a parallelogram in which only the narrow shaded pillar has non-zero integrand.

The τ' integral is a measure of the forcing elements that are coupled to the oscillator response. It is proportional to the power spectral density of the force at the resonance frequency k in the limit of zero damping.

$$\hat{F}(k) = \int_{-\infty}^{\infty} F(\tau')\, \mathrm{e}^{-ik\tau'}\, \mathrm{d}\tau' = \int_{-T}^{T} F(\tau') \cos k\tau'\, \mathrm{d}\tau', \tag{11.14}$$

Since the oscillation is only recognizable if it persists for a long time after excitation, the damping is very small in the cases of most interest in the sense that $\beta T \ll 1$. The exponential $\mathrm{e}^{\beta\tau'}$ can consequently be approximated by unity over the entire τ' range for which the correlation is non-zero and the error of this approximation is smaller than the retained term by a factor of order βT. Equation (11.13) consequently gives the eventual long time behaviour of the weakly damped excited oscillator in the form

$$\overline{\varphi(t)\varphi(t)} \underset{\substack{t/T\to\infty \\ \beta T \ll 1}}{\sim} \frac{\hat{F}(k)}{k^2} \int_0^t \sin^2 k(t-\tau)\, \mathrm{e}^{-2\beta(t-\tau)}\, \mathrm{d}\tau \tag{11.15}$$

$$= \frac{\hat{F}(k)}{k^3} \int_0^{kt} \sin^2 \alpha\, \mathrm{e}^{-2\beta\alpha/k}\, \mathrm{d}\alpha$$

$$= \frac{1}{4}\frac{\hat{F}(k)}{k^3}\frac{1}{\left(\frac{\beta^2}{k^2}+1\right)}\left[\mathrm{e}^{-2\beta\alpha/k}\left\{-\frac{2\beta}{k}\sin^2\alpha - \sin 2\alpha - \frac{k}{\beta}\right\}\right]_0^{kt}$$

$$= \frac{1}{4}\frac{\hat{F}(k)}{k^3}\frac{1}{\left(\frac{\beta^2}{k^2}+1\right)}$$

$$\times\left(\frac{k}{\beta}(1-\mathrm{e}^{-2\beta t}) - \mathrm{e}^{-2\beta t}\left(\sin 2kt + 2\frac{\beta}{k}\sin^2 kt\right)\right).$$

This is dominated at large time when the damping is weak and $\beta/k \ll 1$, by the leading term

$$\overline{\varphi(t)\varphi(t)} \sim \frac{1}{4}\frac{\hat{F}(k)}{k^2\beta}(1-\mathrm{e}^{-2\beta t}). \tag{11.16}$$

For finite but small damping the response is seen to rise from an initial zero value to the eventual $\hat{F}(k)/4k^2\beta$ that was predicted in equation (11.9), a value which sets in at a time of about β^{-1} after the forcings is initiated. If the damping is actually zero then the steady state is never reached, the limiting form of equation (11.16) for $\beta t \to 0$ being

$$\overline{\varphi(t)\varphi(t)} \sim \frac{1}{2}\frac{\hat{F}(k)}{k^2}t; \tag{11.17}$$

the mean square response of an undamped oscillator increases linearly with the duration of forcing. It is only the damping that controls the response to a steady level in the statistically stationary cases. This behaviour and the fact that it is only those elements in the force field at the resonance frequency k that determine the level of response in a weakly damped oscillator is a general characteristic of randomly forced oscillation, a characteristic illustrated in the form of equation (11.16) that is depicted in Figure 11.4.

11.2 HOMOGENEOUS RESPONSE OF A FLEXIBLE PLANE BOUNDARY

We now consider the situation illustrated in Figure 11.5 and suppose that the boundary surface is one that responds linearly to the surface pressure p, i.e. to the force applied to unit surface area.

If the response is homogeneous and linear, there is a differential equation with constant coefficients relating the response, η, to p. When Fourier transforms are taken in both space and time, that linear equation becomes an algebraic equation in which $\hat{p}(\mathbf{k}, \omega)$ is proportional to $\hat{\eta}(\mathbf{k}, \omega)$. We can call the constant of proportionality $Z(\mathbf{k}, \omega)$ and define it as

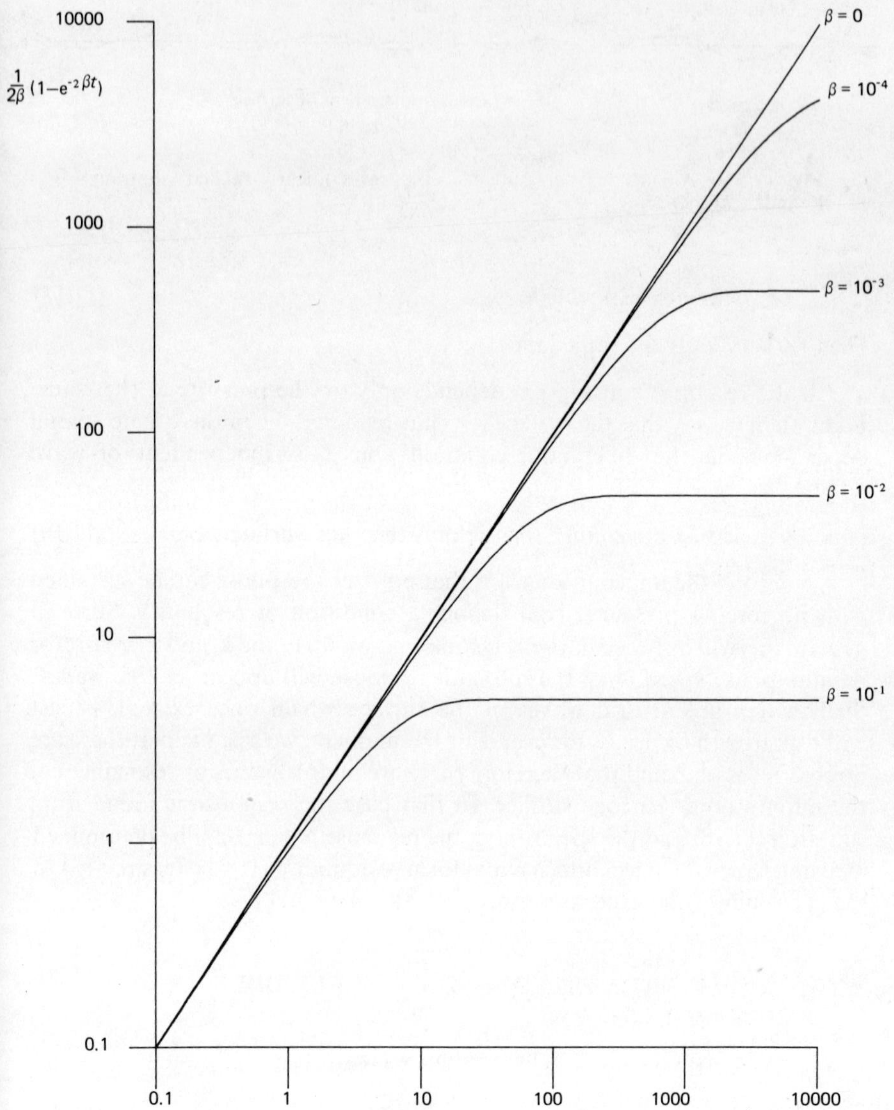

Fig. 11.4 — An illustration of the behaviour described by equation (11.16) where the response of an oscillator grows with time to a maximum level determined by the damping.

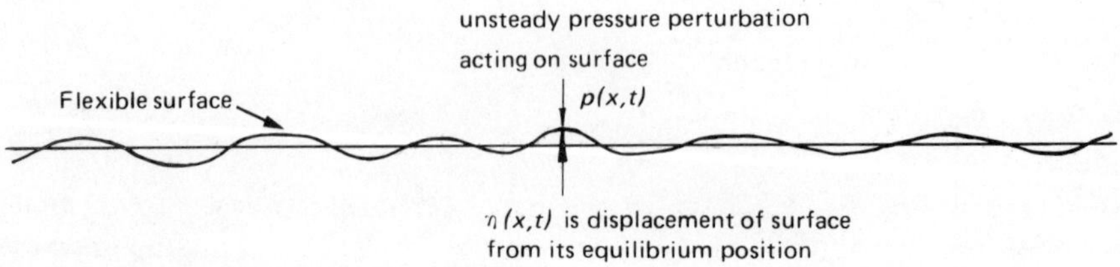

Fig. 11.5 — An unsteady pressure field induces a linearly related vibrational boundary response.

$$\hat{p}(\mathbf{k}, \omega) = Z(\mathbf{k}, \omega)\hat{\eta}(\mathbf{k}, \omega). \tag{11.18}$$

Then $iZ(\mathbf{k}, \omega)/\omega$ is the impedance.

If the response η at a point depends only on the pressure at that same point, then we say that the surface is 'point-reacting'. Then there is no spatial dependence in the differential equation and Z is independent of wave number $\mathbf{k}$.

$$Z(\mathbf{k}, \omega) = Z_p(\omega) \quad \text{for a 'point-reacting' surface.} \tag{11.19}$$

A zero in the impedance means that non-zero response can be sustained with no forcing pressure. That defines a condition of resonance. Since in general this will only occur for fixed combinations of ω and $\mathbf{k}$, and therefore for definite phase speed ω/k_i, the resonant response will appear as 'free waves'. Such waves are neutral waves of the surface which once excited, persist without growth or decay, for ever. For the moment, we assume that there are no real zeros of Z and that therefore there are no free waves or resonances of the infinite homogeneous surface. In that case the response is found from equation (11.18) and the spectrum of the response $\hat{N}(\mathbf{k}, \omega)$ can be determined by using the two (space) dimensional form of equation (10.31). It is related to the spectrum of the excitation pressure $\hat{P}(\mathbf{k}, \omega)$ as follows

$$\overline{\hat{\eta}(\mathbf{k}, \omega)\hat{\eta}(\mathbf{k}', \omega')} = \frac{\overline{\hat{p}(\mathbf{k}, \omega)\hat{p}(\mathbf{k}', \omega')}}{Z(\mathbf{k}, \omega)Z(\mathbf{k}', \omega')} \tag{11.20}$$

$$= (2\pi)^3 \hat{N}(\mathbf{k}, \omega)\, \delta(\mathbf{k} + \mathbf{k}')\, \delta(\omega + \omega') = (2\pi)^3 \frac{\hat{P}(\mathbf{k}, \omega)\, \delta(\mathbf{k} + \mathbf{k}')\, \delta(\omega + \omega')}{Z(\mathbf{k}, \omega)Z(-\mathbf{k}, -\omega)}. \tag{11.21}$$

Now $Z(-\mathbf{k}, -\omega)$ is the complex conjugate of $Z(\mathbf{k}, \omega)$. This can be seen from the fact that $\mathbf{k}$ enters the problem only through Fourier transformation where it is always associated with i. The same is true of ω. Changing the sign of both $\mathbf{k}$ and ω is fully equivalent therefore to changing the sign of i. Equation (11.21) can then be integrated over all $\mathbf{k}'$ and ω' to give

$$\hat{N}(\mathbf{k}, \omega) = \frac{\hat{P}(\mathbf{k}, \omega)}{|Z(\mathbf{k}, \omega)|^2}. \tag{11.22}$$

The power spectral density of the surface pressure, $\hat{P}(\mathbf{k}, \omega)$, is real and positive; so is $\hat{N}(\mathbf{k}, \omega)$.

A zero in Z for real $\mathbf{k}$ and real ω would mean that the surface supports an undamped free mode, or wave and, just as we saw in the case of an undamped oscillator (section 11.1), such a wave would grow algebraically in time if driven by a statistically stationary pressure field; the analysis of such as response can be approached in a similar initial value problem.

11.3 RANDOM VIBRATION OF A HOMOGENEOUS BEAM

The equations governing the motion of a beam that deflects an amount η in response to a load p/unit length is a particular case of the above in which the impedance $Z(k, \omega) = m\omega^2 - i\omega\beta - Bk^4$. This corresponds to the differential equation,

$$B\frac{\partial^4\eta}{\partial x^4} + \beta\frac{\partial\eta}{\partial t} + m\frac{\partial^2\eta}{\partial t^2} = -p, \tag{11.23}$$

where B is the bending stiffness, m is the mass of the beam/unit length and β is a mechanical damping coefficient. In terms of the Fourier transforms of η and p

$$(Bk^4 + i\omega\beta - m\omega^2)\hat{\eta}(k, \omega) = -\hat{p}(k, \omega) = -Z\hat{\eta}(k, \omega). \tag{11.24}$$

The response at wave number k and frequency ω is determined only by that component of the driving pressure at the same wave number and frequency. The spectrum can therefore be written down immediately

$$\begin{aligned}\overline{\hat{\eta}(k, \omega)\hat{\eta}(k', \omega')} &= (2\pi)^2\hat{N}(k, \omega)\,\delta(k + k')\,\delta(\omega + \omega')\\ &= \frac{\overline{\hat{p}(k, \omega)\hat{p}(k', \omega')}}{(Bk^4 + i\omega\beta - m\omega^2)(Bk'^4 + i\omega'\beta - m\omega'^2)}\\ &= \frac{(2\pi)^2\hat{P}(k, \omega)\,\delta(k + k')\,\delta(\omega + \omega')}{(Bk^4 + i\omega\beta - m\omega^2)(Bk^4 - i\omega\beta - m\omega^2)}\\ &= \frac{(2\pi)^2\hat{P}(k, \omega)\,\delta(k + k')\,\delta(\omega + \omega')}{(Bk^4 - m\omega^2)^2 + \omega^2\beta^2}.\end{aligned} \tag{11.25}$$

When this equation is integrated over all k' and ω' we obtain the response spectrum of a beam in terms of the load spectrum $\hat{P}(k, \omega)$.

$$\hat{N}(k, \omega) = \frac{\hat{P}(k, \omega)}{(Bk^4 - m\omega^2)^2 + \omega^2\beta^2}. \tag{11.26}$$

Damping ensures that the response is finite. If there were no damping, β would be zero and the impedance Z would have zeros whenever $\boldsymbol{B}k^4 = m\omega^2$. There would be a singular (actually growing) response to a statistically homogeneous driving pressure field. The unbounded response would be composed of free waves, those waves that need no pressure to sustain them and whose characteristics are defined through the dispersion equation

$$Bk^4 - m\omega^2 = 0. \tag{11.27}$$

Free flexural waves on a beam travel with a phase speed ω/k,

$$\omega/k = \sqrt{\frac{B}{m}k^2} \tag{11.28}$$

and carry energy which travels at the group velocity $\partial\omega/\partial k = 2\omega/k$, which is twice the phase velocity of free waves.

11.4 FLUID LOADING

Up to now we have assumed that the driving pressure field, or load, is given. This, together with the response equation that describes how the boundary deforms in response to its load, yields the interrelation between the response and force spectral characteristics. But the load might change as the surface responds. In fact it will inevitably change if the loads are induced by unsteady fluid pressure. Consider, for example, a point force driving a surface from below. That force will induce a general state of vibration in the surface and therefore inevitably in the nearby fluid which must move with the surface. The acceleration of the fluid can only be accomplished by unsteady pressure gradients so the surface pressure must be unsteady. This unsteady pressure represents an additional load on the surface. We call this effect 'fluid loading'.

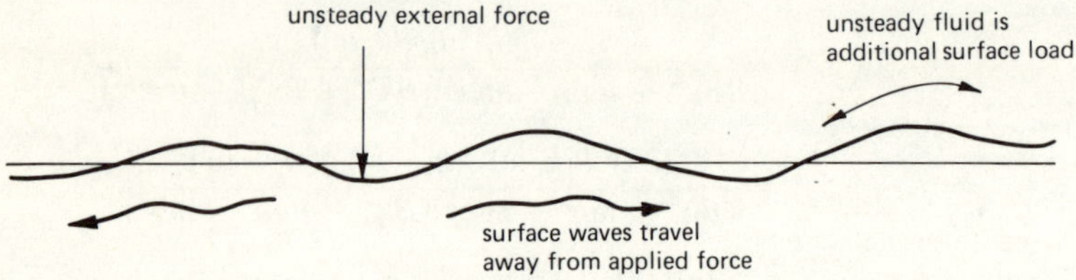

Fig. 11.6 — A force on the surface generates both a surface response and an associated unsteady flow field.

For definiteness we will consider a plane interface driven by an external force/unit area $f(x, t)$ that is independent of the third co-ordinate. The surface will deflect in response to this force, an amount $\eta(x, t)$ and this response will induce a fluid loading $p(x, t)$. We consider only weak motion in incompressible fluid, so that the equations of fluid motion can be simplified by linearisation. The motions are perturbations about a mean state of rest.

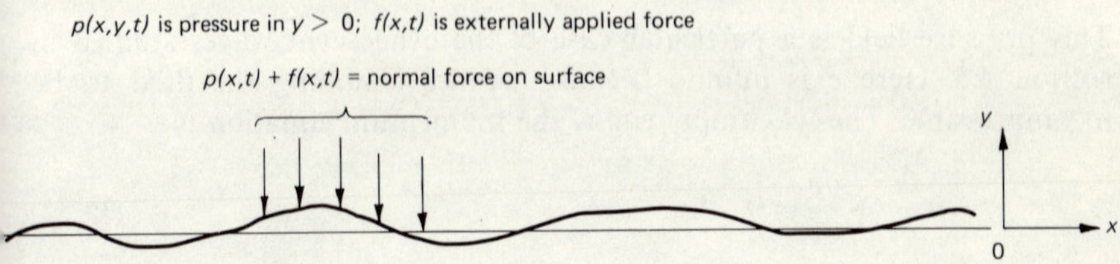

Fig. 11.7 — External force must drive both the surface response and the fluid loading.

The continuity of incompressible flow requires that the fluid moves with a solenoidal velocity

$$\frac{\partial u_i}{\partial x_i} = 0. \tag{11.29}$$

Momentum changes only because of pressure gradients in this inviscid linear motion:

$$\rho \frac{\partial u_i}{\partial t} + \frac{\partial p}{\partial x_i} = 0. \tag{11.30}$$

By taking the divergence of this equation it is evident that in incompressible flow the pressure p satisfies Laplace's equation,

$$\frac{\partial^2 p}{\partial x_i^2} = \nabla^2 p = 0$$

i.e.

$$\frac{\partial^2 p}{\partial x^2} + \frac{\partial^2 p}{\partial y^2} = 0. \tag{11.31}$$

Now we take Fourier transforms in (x, t) so that the equation governing the fluid motion is

$$\frac{\partial^2}{\partial y^2} \hat{p}(k, y, \omega) - k^2 \hat{p}(k, y, \omega) = 0 \tag{11.32}$$

and

$$\hat{p}(k, y, \omega) = A\,\mathrm{e}^{-ky} + B\,\mathrm{e}^{ky}. \tag{11.33}$$

We require that the vibration induced disturbance be confined to the vicinity of the surface at $y = 0$ and should not grow without bound away from it. This means that $\boldsymbol{B} = 0$ for $k > 0$, $A = 0$ for $k < 0$. On the surface the pressure is $p(x, t)$, which has Fourier transform $\hat{p}(k, \omega)$, so that this must be the value for $\boldsymbol{A}$ for $k > 0$ and $\boldsymbol{B}$ for $k < 0$.

$$\hat{p}(k, y, \omega) = \hat{p}(k, \omega)\,\mathrm{e}^{-|k|y}. \tag{11.34}$$

This pressure field is a particular case of the evanescent waves studied in section 4.5. Here c is infinite because we are assuming the fluid to be incompressible. The 'y' component of the momentum equation is

$$\rho\frac{\partial v}{\partial t} + \frac{\partial p}{\partial y} = 0 \tag{11.35}$$

where v is the vertical component of velocity. Fourier transformed this becomes,

$$-i\omega\rho\hat{v}(k, y, \omega) = \frac{\partial \hat{p}}{\partial y}(k, y, \omega) = -|k|\hat{p}(k, \omega)\,\mathrm{e}^{-|k|y}. \tag{11.36}$$

In particular at the surface $y = 0$ where $v = \partial\eta/\partial t$

$$-\rho(i\omega)^2\hat{\eta}(k, \omega) = -i\omega\rho\hat{v}(k, 0, \omega) = -|k|\hat{p}(k, \omega). \tag{11.37}$$

The 'surface loading' evidently has Fourier elements

$$\hat{p}(k, \omega) = -\frac{\rho\omega^2}{|k|}\hat{\eta}(k, \omega). \tag{11.38}$$

The vibration induced surface pressure is equal to $\rho/|k|$ times the surface acceleration.

It is as if a layer of fluid $1/|k|$ thick were attached to and moving with the vibrating surface, i.e. a layer $1/2\pi$ times the wavelength of the surface vibration. The fluid loading is the force required to accelerate this layer. Fluid on both sides of the surface induces twice this load.

Equation (11.18) relates the excitation force to the surface response. The excitation force is, once fluid loading is admitted,

$$\hat{p}(k, \omega) + \hat{f}(k, \omega) = Z(k, \omega)\hat{\eta}(k, \omega) \tag{11.39}$$

or, from (11.38)

$$\hat{f}(k, \omega) = \left\{Z(k, \omega) + \frac{\rho\omega^2}{|k|}\right\}\hat{\eta}(k, \omega). \tag{11.40}$$

$$\hat{\eta}(k, \omega) = \frac{\hat{f}(k, \omega)}{Z(k, w) + \rho\omega^2/|k|}. \tag{11.41}$$

In the particular example of the plate, or beam, where according to (11.24)

$$Z = m\omega^2 - i\omega\beta - Bk^4$$

$$\hat{\eta}(k, \omega) = \frac{\hat{f}(k, \omega)}{(m + \rho/|k|)\omega^2 - Bk^4 - i\omega\beta} \tag{11.42}$$

and the response spectrum follows immediately as

$$\hat{N}(k, \omega) = \frac{\hat{F}(k, \omega)}{\{(m + \rho/|k|)^2\omega^2 - Bk^4\}^2 + \omega^2\beta^2}. \tag{11.43}$$

In general the response spectrum is related to the force spectrum by the equation

$$\hat{N}(k, \omega) = \frac{\hat{F}(k, \omega)}{|Z(k, \omega) + \rho\omega^2/|k||^2}. \tag{11.44}$$

The loading induced by the small amplitude motion of incompressible fluid is an 'added mass' and is very easily accounted for by increasing the mass of the plate by the mass contained in a layer of fluid $|k|^{-1}$ thick, or twice that mass of there is fluid on both sides.

11.5 ENERGY FLOW INTO A MECHANICAL STRUCTURE

The force applied to unit area of the boundary depicted in Figure 11.7 is $(p + f)(x, t)$ and this induces the boundary displacement (in the direction of the force) $-\eta(x, t)$. The point of application of the force moves at a rate $-(\partial\eta/\partial t)(x, t)$ so that the force does work at the rate $-(p + f)(\partial\eta/\partial t)$ per unit area per unit time; i.e. there is a 'power flow' into the plate, the mean value of which P, say,

$$P = -\overline{(p + f)\frac{\partial\eta}{\partial t}} \tag{11.45}$$

$$= \frac{-1}{(2\pi)^4} \iiiint_{-\infty}^{\infty} \overline{(\hat{p} + \hat{f})(k, \omega) i\omega' \hat{\eta}(k', \omega')}$$
$$\times\, e^{i(k + k')x}\, e^{i(\omega + \omega')t}\, dk\, dk'\, d\omega\, d\omega'. \tag{11.46}$$

which, from equation (11.39), is

$$= \frac{-1}{(2\pi)^4} \iiint_{-\infty}^{\infty} \overline{(\hat{p} + \hat{f})(k, \omega) \frac{i\omega'(\hat{p} + \hat{f})(k', \omega')}{Z(k', \omega')}}$$
$$\times\, e^{i(k + k')x}\, e^{i(\omega + \omega')t}\, dk\, dk'\, d\omega\, d\omega'. \tag{11.47}$$

Now $\overline{(\hat{p} + \hat{f})(k, \omega)(\hat{p} + \hat{f})(k', \omega')}$ is real and positive definite, it being related to the spectrum of the surface force field $P\hat{F}$ say, by equation (10.31)

$$\overline{(\hat{p} + \hat{f})(k, \omega)(\hat{p} + \hat{f})(k', \omega')} = (2\pi)^2 P\hat{F}(k, \omega)\, \delta(k + k')\, \delta(\omega + \omega'). \tag{11.48}$$

Therefore,

$$P = +\int_{-\infty}^{\infty}\int_{-\infty}^{\infty} \frac{i\omega P\hat{F}(k, \omega)}{(2\pi)^2 Z(-k, -\omega)}\, \mathrm{d}k\, \mathrm{d}\omega \tag{11.49}$$

$$= \frac{1}{(2\pi)^2}\int_{-\infty}^{\infty}\int_{-\infty}^{\infty} \frac{i\omega Z(k, \omega)}{|Z(k, \omega)|^2} P\hat{F}(k, \omega)\, \mathrm{d}k\, \mathrm{d}\omega. \tag{11.50}$$

$Z(-k, -\omega)$ is the complex conjugate of $Z(k, \omega)$. This follows directly from equation (11.18) when written for the complex conjugate of $\hat{p}(k, \omega)$.

Both $P\hat{F}(k, \omega)$ and $|Z|^2$ are real even functions with diagonal symmetry in the (k, ω) plane. Similarly the *real* part of Z has the same diagonal symmetry, and the skewing influence of ω in the integral of (11.50) makes the contribution associated with the real part of Z vanish.

We write

$$Z = Z_R + iZ_I \tag{11.51}$$

$$\left.\begin{aligned} Z_R(k, \omega) &= Z_R(-k, -\omega) \\ Z_I(k, \omega) &= -Z_I(-k, -\omega) \end{aligned}\right\} \tag{11.52}$$

$$P = -\int_{-\infty}^{\infty}\int_{-\infty}^{\infty} \frac{\omega Z_I(k, \omega)}{(2\pi)^2 |Z|^2} P\hat{F}(k, \omega)\, \mathrm{d}k\, \mathrm{d}\omega. \tag{11.53}$$

The power flow is now seen to be purely real and to depend entirely on the *imaginary part of* Z. The integral possesses the diagonal symmetry that allows the integration range to be halved on each co-ordinate

$$P = \frac{-1}{\pi^2}\int_0^{\infty}\int_0^{\infty} \frac{\omega Z_I(k, \omega)}{|Z(k, \omega)|^2} P\hat{F}(k, \omega)\, \mathrm{d}k\, \mathrm{d}\omega. \tag{11.54}$$

This quantity P represents the power flow per unit area into the mechanical structure and, if the surface is dissipative, P must not be negative. $Z_I(k, \omega)$ is negative for energy absorbing surfaces.

For example, if the surface were a purely viscous resistance, then

$$p = -\alpha \frac{\partial \eta}{\partial t} \tag{11.55}$$

α being the positive resistance coefficient.

In terms of Fourier transforms,

$$\hat{p}(k, \omega) = -\alpha i\omega \hat{\eta}(k, \omega) \tag{11.56}$$

$$= Z\hat{\eta} \tag{11.57}$$

so that the Z is in this case purely imaginary

$$Z(k, \omega) = -\omega\alpha i \tag{11.58}$$

and

$$Z_I = -\omega\alpha \quad \text{is negative.}$$

A stable surface must not have a restoring force that acts in the same direction as the driving force! The steady case is included in the general formulae and corresponds to the condition $\omega = 0$. Therefore, to ensure static stability the surface response equation takes the form,

$$\hat{p}(k, \omega) = \text{negative number} \times \hat{\eta}(k, \omega) \quad \text{at } \omega = 0, \tag{11.59}$$

so that the real part of $Z(k, \omega)$ must tend to a negative number as ω tends to zero.

11.6 VIBRATION OF A SURFACE BOUNDING A STEADY INCOMPRESSIBLE FLOW

We now extend the scope of the problem to include the possibility of a uniform flow parallel to the surface and perpendicular to any wave crests. The problem is illustrated in Figure 11.8.

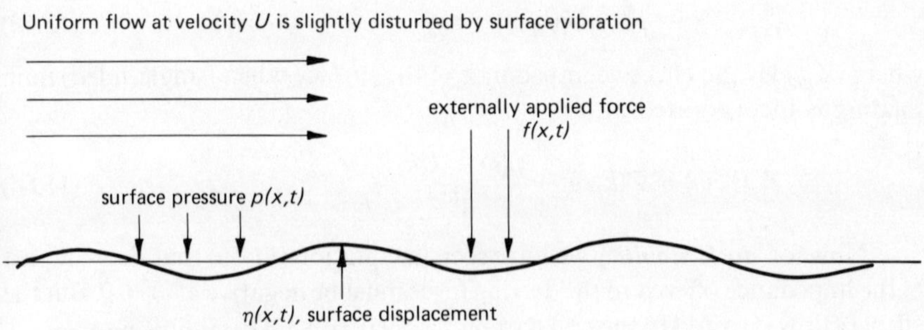

Fig. 11.8 — Uniform flow past a flat surface disturbed from rest by an unsteady force field $f(x, t)$.

The surface response calculation is unaltered by this step but the fluid problem is different. The fluid loading has changed. Equation (11.38) is no longer valid. But it would be valid in a co-ordinate system moving with the flow. We again denote by the suffic m quantities that are observed in that moving co-ordinate system, and then we can re-write equation (11.38) as

$$\hat{p}_m(k, \omega) = -\frac{\rho\omega^2}{|k|}\hat{\eta}_m(k, \omega). \tag{11.60}$$

But, as we can see from equation (10.36), convection at speed U is easily accounted for.

$$\begin{aligned}\hat{p}(k,\omega) &= \hat{p}_m(k,\omega+Uk)\\ \hat{\eta}(k,\omega) &= \hat{\eta}_m(k,\omega+Uk).\end{aligned} \tag{11.61}$$

The fluid loading equation (11.60) is therefore

$$\hat{p}(k,\omega-Uk) = -\frac{\rho\omega^2}{|k|}\hat{\eta}(k,\omega-Uk)$$

or equivalently,

$$\hat{p}(k,\omega) = -\frac{\rho(\omega+Uk)^2}{|k|}\hat{\eta}(k,\omega). \tag{11.62}$$

By incorporating this 'moving fluid loading' expression into the basic response equation,

$$\hat{p}(k,\omega) + \hat{f}(k,\omega) = Z(k,\omega)\hat{\eta}(k,\omega),$$

equation (11.40) is modified in the moving fluid case to:

$$\begin{aligned}\hat{f}(k,\omega) &= \left\{Z(k,\omega) + \frac{\rho(\omega+Uk)^2}{|k|}\right\}\hat{\eta}(k,\omega)\\ &= Z_f(k,\omega)\hat{\eta}(k,\omega), \quad \text{say}\end{aligned} \tag{11.63}$$

where iZ_f/ω is the effective impedance of the surface when (single sided) fluid loading is incorporated in it.

$$Z_f(k,\omega) = Z(k,\omega) + \frac{\rho(\omega+Uk)^2}{|k|}. \tag{11.64}$$

Now for *static stability* we have seen in equation (11.59) that the real part of the impedance offered to the driving force must be negative at $\omega = 0$. But the effect of flow is to add to the real part of Z a positive, i.e. a destabilising element. Flow is equivalent to a negative spring stiffness.

$$Z_f(k,0) = Z(k,0) + \frac{\rho U^2 k^2}{|k|}.$$

i.e.

$$Z_f(k,0) = Z(k,0) + \rho U^2|k|. \tag{11.65}$$

The surface will be statically unstable when

$$\rho U^2|k| > |Z_R(k,0)|, \tag{11.66}$$

at which condition the 'moving fluid loading' has reduced the effective surface stiffness to below zero and into the negative range. A slightly deformed surface would then experience a force that deformed it further; the deformation would then grow exponentially. The deformation would diverge and this kind of static instability is often referred to as *divergence*. Of course $Z_f(k,\omega)$ might well vanish for non-zero frequencies in which case there would be a free wave condition in which resonant modes could exist without any excitation. The zeros of Z_f which define the free wave condition may occur for complex values of the frequency and/or wave number. In these cases the waves will either grow or decay depending on the sign of the imaginary part of ω (or k). If the waves grow then the free waves are a manifestation of *flutter* instability which is commonly found when fluid flows past flexible surfaces. The most common example is the fluttering of a flag, the instability there being an example of a flow first defined by Kelvin and Helmholtz.

11.7 KELVIN–HELMHOLTZ INSTABILITY

Mechanical systems are unstable if they can sustain motions of growing amplitude when the driving force is zero. This is the topic we expand now. Because the amplitude grows, the technique of analysis constructed for statistically stationary processes is inadequate. But the basic Fourier analysis remains valid—provided we are dealing with disturbances started from rest and confine our attention to the region in time, or space, where the amplitudes remain bounded.

Accordingly we examine the possibility that the fluid loading term is of a type that can sustain an otherwise unforced vibration of the bounding surface.

The fluid loading pressure on the surface is,

$$\hat{p}(k, \omega) = \frac{-\rho(\omega + Uk)^2}{|k|} \hat{\eta}(k, w). \tag{11.67}$$

From equation (11.63) this must also be equal to $Z(k, \omega)\hat{\eta}$ when there is no other driving force. Motion is possible provided

$$|k|Z(k, \omega) = -\rho(\omega + Uk)^2. \tag{11.68}$$

This condition relating wave number k to frequency ω is a *dispersion* equation.

Consider as an example the surface consisting of a limp passive mass (such as a curtain, or flag) of mass $\boldsymbol{m}$/unit area. Then the equation of a surface motion is

$$p = -m \frac{\partial^2 \eta}{\partial t^2}$$

$$\hat{p}(k, \omega) = m\omega^2 \hat{\eta}(k, \omega); \quad Z(k, \omega) = Z(\omega) = m\omega^2. \tag{11.69}$$

Equation (11.68) is, in that case,

$$m\omega^2|k| = -\rho(\omega + Uk)^2, \tag{11.70}$$

so that ω, or k, or both must be complex for motion to exist.

We solve equation (11.70) for $\omega/k = -C$, the phase speed.

$$C^2 + \frac{\rho}{m|k|}(C - U)^2 = 0 \tag{11.71}$$

or, equivalently,

$$C = \frac{\rho U}{\rho + m|k|} \pm i\frac{\rho U\sqrt{(m|k|/\rho)}}{\rho + m|k|} \tag{11.72}$$

$$C = C_R + iC_I, \quad \text{say.}$$

The phase speed is complex. A disturbance with real wave number k must then take the form

$$\eta \sim e^{ikx}\, e^{i\omega t} = e^{ik(x - Ct)} \tag{11.73}$$

$$= e^{ik(x - C_R t)}\, e^{kC_I t}. \tag{11.74}$$

The wave of positive wave number k will decay exponentially with time if C_I is negative. But if C_I is positive, and we see from (11.72) that in this case it can be, the wave will grow exponentially in time; it is then said to be *temporally* unstable. If ω, the frequency is fixed and real then it will be seen that the wave can grow exponentially with downstream distance x; the wave is then said to be a *spatial* instability. The wave moves downstream with phase speed C_R as its amplitude grows to produce a degeneration of the steady flow into waves that grow and break and may eventually become turbulence.

The Kelvin–Helmholtz problem concerns the stability of a vortex sheet that separates fluids of density ρ^+ moving at speed U^+ from fluid of density ρ^- moving in the same direction at speed U^-. The two fluids are separated at an initially plane surface that is disturbed from rest with a displacement proportional to $e^{i\omega t}\, e^{ikx}$. The pressure and displacement of the interface is continuous, and since the pressure is related to the displacement by the fluid loading equation (11.62),

$$\hat{p}^+ = -\frac{\rho^+(\omega + U^+k)^2}{|k|}\hat{\eta}$$

$$\hat{p}^- = +\frac{\rho^-(\omega + U^-k)^2}{|k|}\hat{\eta}. \tag{11.75}$$

The signs of the two fluid loading terms are different because positive displacements are measures in the direction leading *into* the fluids concerned, which are of course on different sides of the interface.

This equation can be solved for $\omega/k = -C$ to give

$$C = \frac{\rho^+U^+ + \rho^-U^-}{\rho^+ + \rho^-} \pm i\frac{\sqrt{(\rho^+\rho^-(U^+ - U^-)^2)}}{\rho^+ + \rho^-}, \tag{11.76}$$

i.e., again we find that $C = C_R + iC_I$.

Waves at the interface between two fluids travel with a mass averaged speed $(\rho^+U^+ + \rho^-U^-)/(\rho^+ + \rho^-)$ and, because C is complex, have an exponentially varying amplitude. Vortex sheets are unstable and this 'Kelvin–Helmholtz' instability is the main reason why jet and wake flows often break down into the chaotic motions we call turbulence.

It is a relatively simple matter to investigate the growth of other wave forms in this case because because the complex phase speed is independent of the wave scale. This is bound to be so because the vortex sheet has itself no scale on which to measure those effects.

Consider for example a vortex sheet disturbed from its plane position by an initial disturbance proportional to,

$$\eta_0(x) = \frac{1}{(d^2 + x^2)}. \tag{11.77}$$

This is a disturbance comprising a distribution of wavelets in a 'wave packet', the strength of each element being the Fourier transform of equation (11.77)

$$\hat{\eta}_0(k) = 2\int_0^\infty \frac{\cos kx}{(d^2 + x^2)}\,\mathrm{d}x$$

which may be found from tables or worked out by contour integration to be

$$\eta_0(k) = \frac{\pi}{d}\,\mathrm{e}^{-kd}. \tag{11.78}$$

One possible disturbance that can exist on the vortex sheet is the particular collection of wavelets admissible by equation (11.76),

$$\eta(x, t) = \frac{1}{\pi}\int_0^\infty \hat{\eta}_0(k)\cos k(x - C_R t)\,\mathrm{e}^{kC_I t}\,\mathrm{d}k$$

$$\eta(x, t) = \frac{1}{d}\int_0^\infty \cos k(x - C_R t)\,\mathrm{e}^{k(C_I t - d)}\,\mathrm{d}k. \tag{11.79}$$

This integral exists only as long as $C_I t < d$ and shows the interesting effect that wave packets can actually disintegrate at some particular time when they propagate on infinitely thin vortex sheets. But before that disintegration time the integral in (11.79) has a perfectly definite value which may again be calculated by contour integration or taken from tables of Fourier transforms,

$$\int_0^\infty \cos ka\,\mathrm{e}^{-kb}\,\mathrm{d}k = \frac{b}{a^2 + b^2},$$

so that according to equation (11.79) one (only one because C_I can be either positive or negative) admissable vortex sheet disturbance with the initial value given in equation (11.77) is proportional to

$$\eta(x, t) = \frac{(1 - C_I t/d)}{(1 - C_I t/d)^2 + (x/d - C_R t/d)^2}. \tag{11.80}$$

This wave packet whose basic scale is set by the parameter d, and its lifetime by the ratio d/C_I is illustrated in Figure 11.9.

EXERCISES FOR CHAPTER 11

1. A circular walled open ended tube of length L is placed in an open space where the background noise spectrum if $P_i(\omega)$. The background consists of sound waves travelling perpendicular to the axis of the tube.

(a) Show that the maximum spectral sound level inside the tube is $2P_i(\omega)/(1 + \cos \omega L/c)$, and
(b) What is the sound level at the ends of the tube?

2. A sphere of mass M and radius a is tethered far below the free surface of water by a spring of stiffness K. Long waves on the water surface cause the sphere to move, and would exert on a rigidly tethered sphere a force $F \cos \Omega t$. Viscous and damping effects are negligible and the water motion around the sphere is potential and incompressible.

Determine the frequency at which the unsteady force exerted on the sphere by the water is (a) a maximum and (b) a minimum.

Determine the displacement of the sphere from its equilibrium position given that it is at rest at times $t = 0$ and $2\pi/\Omega$.

3. Determine the effective normal impedance of the free surface of deep water disturbed from rest. Surface tension may be neglected. A homogeneous turbulent boundary layer is generated on the water surface by a wind which starts at time $t = 0$ and thereafter exerts on the surface an unsteady pressure field. The Fourier transform of the surface pressure at wave number $\mathbf{k}$ and frequency ω is $\hat{p}(\mathbf{k}, \omega)$. Determine the Fourier transform of the surface response and describe the development of the water wave system as it absorbs energy from the wind according to this model.

4. Determine the compressible fluid version of equation (11.44) relating the response spectrum to that of the applied force. In particular show that

$$\hat{N}(k, \omega) = \frac{\hat{F}(k, \omega)}{|Z(k, \omega) + L|^2}$$

where the fluid loading term L for fluid of density ρ_1 and sound speed c_1 is given by:

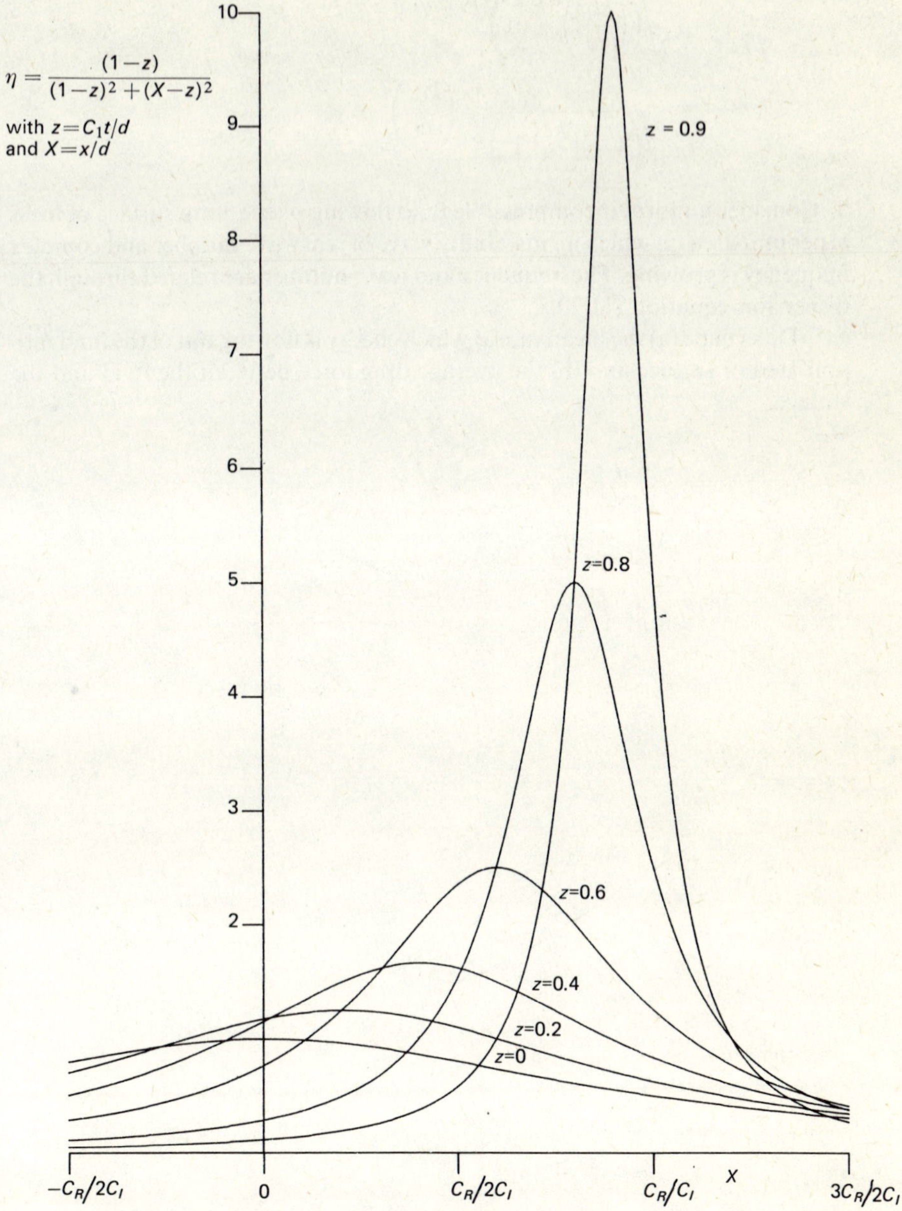

Fig. 11.9 — A wave packet growing on a vortex sheet and disintegrating at time d/C_1.

$$\boldsymbol{L} = \frac{\omega\rho_1 c_1}{\sqrt{\left(\left(\frac{kc_1}{\omega}\right)^2 - 1\right)}}; \quad \left(\frac{kc_1}{\omega}\right)^2 > 1$$

$$= \frac{-i\omega\rho_1 c_1}{\sqrt{\left(1 - \left(\frac{kc_1}{\omega}\right)^2\right)}}; \quad \left(\frac{kc_1}{\omega}\right)^2 \angle 1.$$

5. Consider uniform incompressible fluid flowing over a limp surface of mass $\boldsymbol{m}$ per unit area on which an instability wave of real wave number and complex frequency is growing. The frequency and wave number are related through the dispersion equation (11.70).

Determine (a) the mean rate at which energy is flowing out of the fluid into unit area of surface and (b) the average drag force between the fluid and the surface.

Solutions to Exercises

SOLUTIONS TO EXERCISES FOR CHAPTER 1

$\neq$Qu1. If the pressure perturbation produced by one tone is $p_1 \cos \omega t$, then 80 tones of different frequencies produce a rms pressure perturbation $\sqrt{40}\, p_1$:

$$\text{SPL} = 20 \log_{10}\left(\frac{p_1}{2\sqrt{2}\,10^{-5}}\right) + 10 \log_{10}(80) = \underline{79.03 \text{ dB}}.$$

$\neq$Qu2. The SPL due to printing press alone = 90 dB, which is a rms pressure perturbation $p_1' = 0.632 \text{ N/m}^2$. The ambient SPL = 87 dB, which corresponds to a rms pressure $p_2' = 0.45 \text{ N/m}^2$. Since the two sound sources are not coherent the resultant rms pressure $= (p_1'^2 + p_2'^2)^{\frac{1}{2}} = 0.77 \text{ N/m}^2 = \underline{92 \text{ dB}}$.

$\neq$Qu3. Let p_B' be the rms background pressure, p_M' be the rms measured pressure with the source switched on, p_S' be the rms pressure caused by the source alone, which we wish to find.

If the difference between the SPL with and without source is X then $20 \log_{10}(p_M'/p_B') = X$, i.e. $p_M'/p_B' = 10^{X/20}$. Since the sound fields are not coherent $p_M'^2 = p_B'^2 + p_S'^2$. so that

$$p_S'^2 = p_M'^2 - p_B'^2 = p_M'^2\left(1 - \frac{p_B'^2}{p_M'^2}\right) = p_M'^2(1 - 10^{-X/10})$$

$$20 \log_{10}\left(\frac{p_S'}{p_{\text{ref}}}\right) = 20 \log_{10}\left(\frac{p_M'}{p_{\text{ref}}}\right) + 10 \log_{10}(1 - 10^{-X/10})$$

SPL of source alone = measured SPL + correction

where the correction is $10 \log_{10}(1 - 10^{-X/10})$.
When $X = 10$, the correction is $\underline{-0.46, \text{ negligible}}$.
$X = 3$, the correction is $\underline{-3 \text{ dB}}$.

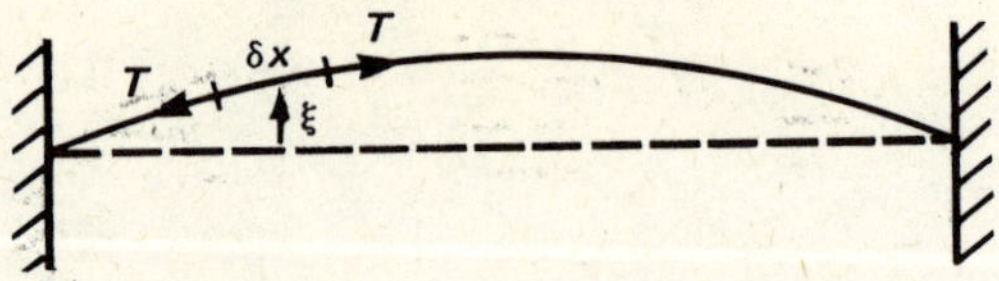

Fig. A.1.1 — A tensioned string.

≠Qu4. Consider a length δx of the string. Then

$$m\,\delta x\,\frac{\partial^2\xi}{\partial t^2} = \delta\left(T\,\frac{\partial\xi}{\partial x}\right)$$

$$m\,\frac{\partial^2\xi}{\partial t^2} = T\,\frac{\partial^2\xi}{\partial x^2}.$$

By comparison with (1.8) and (1.9) the general solution is

$$\xi(x, t) = f(x - ct) + g(x + ct),$$

where the wave speed $c = \sqrt{T/m}$.

≠Qu5. For SPL of 100 dB, $p'_{\text{rms}} = 2\ \text{N/m}^2$, and $p'_{\text{peak}} = 2\sqrt{2}\ \text{N/m}^2$. Then

$$u_{\text{peak}} = \frac{p'_{\text{peak}}}{\rho_0 c} = \begin{cases} 7 \times 10^{-3}\ \text{m/s in air} \\ 2 \times 10^{-6}\ \text{m/s in water} \end{cases}$$

$$\eta_{\text{peak}} = \frac{u_{\text{peak}}}{|\omega|} = \begin{cases} 1.1\ 10^{-6}\ \text{m in air} \\ 3.1\ 10^{-10}\ \text{m in water.} \end{cases}$$

≠Qu6. A wave is generated ahead of the train in which $u = 60$ mph,

$$p' = \rho_0 c u = 1.2 \times 340\,\frac{60 \times 1.61 \times 10^3}{60 \times 60} = \underline{11\ \text{kN/m}^2}.$$

≠Qu7. (a) $\lambda = \dfrac{2\pi c}{\omega} = \dfrac{1450}{20.10^3} = \underline{0.0725\ \text{m}}$

(b) $\text{Intensity} = \dfrac{p'^2_{\text{rms}}}{\rho_0 c} = \dfrac{100}{\pi(0.25)^2}$

$p'_{\text{rms}} = 2.7\ 10^4\ \text{N/m}^2 = \underline{183\ \text{dB}}$

(c) $u = \sqrt{2}\,\dfrac{p'_{\text{rms}}}{\rho_0 c} = \underline{0.027\ \text{m/s}}$

(d) $\eta = \dfrac{|u|}{\omega} = \underline{2.1\ 10^{-7}\ \text{m}}.$

≠Qu8. (a) Intensity $= \dfrac{\overline{p'^2}}{\rho_0 c}$

$$\frac{I_{\text{air}}}{I_{\text{water}}} = \frac{(\rho_0 c)_{\text{water}}}{(\rho_0 c)_{\text{air}}} = \underline{3554}$$

(b) Intensity $= \overline{u^2}\,\rho_0 c$

$$\frac{I_{\text{air}}}{I_{\text{water}}} = \frac{(\rho_0 c)_{\text{air}}}{(\rho_0 c)_{\text{water}}} = \frac{1}{3554}.$$

SOLUTIONS TO EXERCISES FOR CHAPTER 2

≠Qu1. The radiated power, P, can be calculated by integrating $p'u$ over a large sphere. In the far-field $u = p'/\rho_0 c$ and

$$\overline{p'u}(\mathbf{x}) = \frac{p_0^2 l^{2n+2}}{2\rho_0 c r^2}\left(\frac{\omega}{c}\right)^{2n} \cos^{2n}\theta,$$

where θ is the angle between $\mathbf{x}$ and the 1-axis. Hence

$$P = 2\pi \frac{p_0^2 l^{2n+2}}{2\rho_0 c}\left(\frac{\omega}{c}\right)^{2n} \int_0^\pi \cos^{2n}\theta \sin\theta \, \mathrm{d}\theta = \frac{2\pi p_0^2 l^2}{\rho_0 c(2n+1)}\left(\frac{\omega l}{c}\right)^{2n}.$$

≠Qu2. The pressure perturbation radiated by the pulsating sphere is

$$p'(r, t) = \frac{a}{r}\frac{i\omega a/c}{1 + i\omega a/c}\rho_0 c \hat{u}_a \,\mathrm{e}^{i\omega(t-(r-a)/c)},$$

and hence the radiated power

$$\begin{aligned} P = \frac{2\pi\rho_0 \varepsilon^2 a^4 \omega^4}{c(1 + (\omega a/c)^2)} &= \frac{1.79625 \times 10^{-10}\omega^4}{1 + \omega^2 \times 7.785 \times 10^{-7}} \\ &= 2.8\ \text{mW} \quad \text{at} \quad 10\ \text{Hz} \\ &= 21.4\ \text{W} \quad \text{at} \quad 100\ \text{Hz} \\ &= 8.8\ \text{kW} \quad \text{at} \quad 1\ \text{kHz}. \end{aligned}$$

≠Qu3. (a) *For the plane boundary* $p' = \rho_0 c u$. Therefore $|p'| = 1.2 \times 340 \times \omega \times 10^{-3}\ \text{N/m}^2$ and the

$$\begin{aligned} \text{SPL} &= 139\ \text{dB} \quad \text{at } 100\ \text{Hz} \\ &= 179\ \text{dB} \quad \text{at } 10\ \text{kHz}. \end{aligned}$$

(b) *For the cylinder* $p'(\mathbf{x}, t) = \dfrac{\rho_0 \varepsilon \omega c \cos\theta}{H_0^{(2)\prime\prime}(\omega a/c)} H_0^{(2)\prime}\left(\dfrac{\omega r}{c}\right) \mathrm{e}^{i\omega t}.$

For $0 < X \ll 1$ $H_0^{(2)}(X) \sim \dfrac{-2i}{\pi}\log_e X$ and hence $H_0^{(2)\prime\prime}(X) \sim \dfrac{2i}{\pi X^2}$.

For $X \gg 1$ a comparison of equations (2.62) and (2.64) shows that

$$H_0^{(2)}(X) \sim \sqrt{\frac{2}{\pi X}}\,e^{-iX + i\pi/4}.$$

At a frequency of 100 Hz with $a = 5$ cm, $r = 10$ m $\omega a/c$ is much less than unity and $\omega r/c$ is much greater than unity. Hence $|p'(\mathbf{x}, t)|$ simplifies to

$$|p'(\mathbf{x}, t)| = \rho_0 \varepsilon \omega c|\cos\theta|\left(\frac{\omega a}{c}\right)^2\sqrt{\frac{\pi c}{2\omega r}}.$$

$$\text{SPL} = 87 + 20\log_{10}|\cos\theta|\ \text{dB}.$$

At a frequency of 10 kHz both $\omega a/c$ and $\omega r/c$ are large in comparison with unity. Then

$$|p'(\mathbf{x}, t)| \sim \rho_0 \varepsilon \omega c\sqrt{\frac{a}{r}}\,|\cos\theta|$$

and the

$$\text{SPL} = 156 + 20\log_{10}|\cos\theta|\ \text{dB}.$$

(c) *For the sphere*

$$p'(\mathbf{x}, t) = \frac{a}{r}\,\frac{i\omega a/c}{1 + i\omega a/c}\,\rho_0 c\hat{u}_a\, e^{i\omega(t-(r-a)/c)}.$$

At a frequency of 100 Hz SPL = 72.4 dB.
At a frequency of 10 kHz SPL = 133 dB.

≠Qu4. For this impulsive sound, $p' = \rho_0 cu$ and hence the maximum pressure in the sound wave $= \rho_0 c \times 10 = 4080\ \text{N/m}^2$.

≠Qu5. Let the velocity potential have the form

$$\varphi(r, t) = \frac{f(ct - r)}{r} + \frac{g(ct + r)}{r}.$$

Finiteness conditions at $r = 0$ give $f(ct) = -g(ct)$ and hence

$$\varphi(r, t) = \frac{f(ct - r)}{r} - \frac{f(ct + r)}{r}.$$

At $t = 0$ there is no velocity and therefore $f(-r) = f(r)$. Also

$$p'(r, 0) = \frac{\rho_0 c}{r}\{-f'(-r) + f'(r)\} = \frac{2\rho_0 c f'(r)}{r} = \begin{cases} \dfrac{2T}{a} & \text{for } 0 < r < a \\ 0 & \text{for } r > a \end{cases}$$

Hence

$$f'(r) = \begin{cases} \dfrac{Tr}{\rho_0 ca} & \text{for } 0 < r < a \\ 0 & \text{for } r > a, \end{cases}$$

and for $r > a$,

$$p'(r, t) = \frac{\rho_0 c}{r}\{f'(ct + r) - f'(ct - r)\}$$

$$= \begin{cases} 0 & \text{for } r > ct + a \text{ or } r < ct - a \\ \dfrac{T}{ra}(r - ct) & \text{for } ct + a > r > ct - a. \end{cases}$$

The total radiated energy is equal to the initial energy which is all potential and equal to

$$\tfrac{2}{3}\pi a^3\left(\frac{2T}{a}\right)^2 \frac{1}{\rho_0 c^2} = \frac{8\pi a T^2}{3\rho_0 c^2}.$$

≠Qu6. $\varphi(r, t) = \dfrac{f(t - r/c)}{r}$

$$u(r, t) = -\frac{f(t - r/c)}{r^2} - \frac{f'(t - r/c)}{rc}$$

$u = 0$ for $t < 0$, hence $f(t) = 0$ for $t < 0$

$u = 1$ on $r = t$ for $0 < t < 10^{-3}$, and so

$$-\frac{f(t - t/c)}{t^2} - \frac{f'(t - t/c)}{tc} = 1$$

$c \gg 1, t$ and so this leads to $f(t) = -t^2$ for $0 < t < 10^{-3}$.

$$p'(1, t) = -\rho_0 f'(t - 1/c) = \begin{cases} 0 & \text{for } t < c^{-1} \\ 2\rho_0(t - 1/c) & \text{for } c^{-1} < t < c^{-1} + 10^{-3}. \end{cases}$$

SOLUTIONS TO EXERCISES FOR CHAPTER 3

≠Qu1. Frequency of fundamental $= \dfrac{2\pi\, 4000}{60} \times 2$ rad/sec.

Hence the minimum effective length $= \dfrac{\pi c}{2\omega} = 0.64$ m.

≠Qu2. On the end $x = l + d$, $u = 0$ and so

$$B = C\, e^{2i\omega(l+d)/c}. \tag{A.3.1}$$

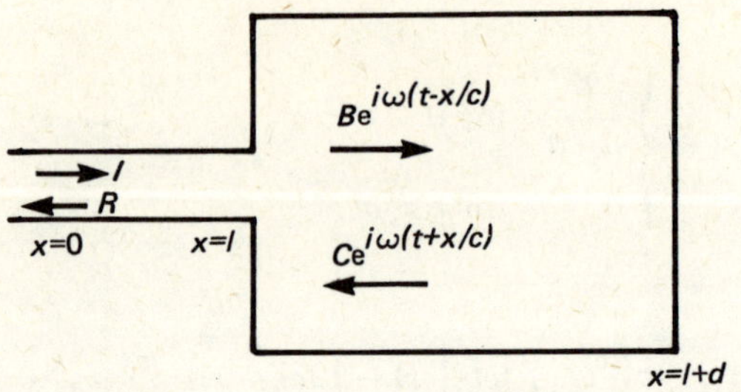

Fig. A.3.1 — The waves in the vessel.

Continuity at $x = l$ gives

$$I\,e^{-i\omega l/c} + R\,e^{i\omega l/c} = B\,e^{-i\omega l/c} + C\,e^{i\omega l/c} \tag{A.3.2}$$

$$A_1(I\,e^{-i\omega l/c} - R\,e^{i\omega l/c}) = A_2(B\,e^{-i\omega l/c} - C\,e^{i\omega l/c}). \tag{A.3.3}$$

After solving equations (A.3.1)–(A.3.3) to give R in terms of I, we find

$$R = I\,\frac{A_1 - iA_2\tan\left(\dfrac{\omega d}{c}\right)}{A_1 + iA_2\tan\left(\dfrac{\omega d}{c}\right)}\,e^{-2i\omega l/c}.$$

At $x = 0$, $p' = (I + R)\,e^{i\omega t}$ and $Q = (A_1/c)(I - R)\,e^{i\omega t}$. It therefore follows that

$$\frac{p'}{Q} = \frac{c}{A_1 i}\,\frac{A_1 - A_2\tan\left(\dfrac{\omega l}{c}\right)\tan\left(\dfrac{\omega d}{c}\right)}{A_1\tan\left(\dfrac{\omega l}{c}\right) + A_2\tan\left(\dfrac{\omega d}{c}\right)}.$$

For $\omega d/c \ll 1$, $\omega l/c \ll 1$, and $A_1 \ll A_2$

$$\frac{p'}{Q} \sim \frac{c^2}{i\omega V}\left(1 - \frac{V\omega^2 l}{c^2 A_1}\right).$$

≠Qu3. When the valve W closes abruptly at $t = 0$ pressure waves f, g propagate along the main and supply pipes. After a time $t = 2h/c$ the wave r (reflected from the end D) also influences the conditions at B and the strengths of the waves f and g are changed. For $t < 2h/c$, continuity of pressure at B gives $f(t) = g(t)$, and continuity of mass flux at B shows

$$A_1\left(U - \frac{f(t)}{\rho_0 c}\right) = \frac{g(t)}{\rho_0 c}A_2$$

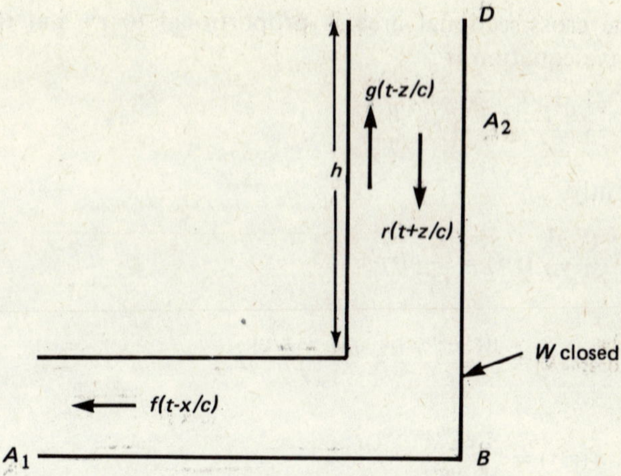

Fig. A.3.2 — Pressure waves in the pipe.

when the valve W is closed. Hence

$$g(t) = 0 \qquad \text{for } t < 0$$

$$= \frac{A_1}{A_1 + A_2}\rho_0 c U \qquad \text{for } 0 < t < 2h/c.$$

For $t > 2h/c$, $g(t)$ is changed from $(A_1/[A_1 + A_2])\rho_0 c U$ due to the influence of the reflected wave r. Now $p' = 0$ at $z = h$ for all time, hence $g(t - h/c) + r(t + h/c) = 0$, and therefore $r(t) = -g(t - 2h/c)$. u, the velocity of the fluid delivered to the dwelling,

$$= \frac{1}{\rho_0 c}(g(t - h/c) - r(t + h/c)) = \frac{2}{\rho_0 c} g(t - h/c)$$

$$= 0 \qquad \text{for } t < h/c$$

$$= \frac{2UA_1}{A_1 + A_2} \qquad \text{for } \frac{h}{c} < t < \frac{3h}{c}.$$

After $t = 3h/c$ the flow velocity at $z = h$ changes. Therefore the steady velocity $2UA_1/(A_1 + A_2)$ persists for a time $2h/c$, and the volume flow through the end D during this time

$$= 2\frac{10^{-2}}{10^{-2} + 10^{-3}}\frac{2 \times 15}{1450} 10^{-3} = \underline{3.8\ 10^{-5}\ \text{m}^3}.$$

≠Qu4. The cross-sectional area is proportional to r^2 and therefore the relevant wave equation is

$$\frac{r^2}{c^2}\frac{\partial^2 p'}{\partial t^2} = \frac{\partial}{\partial r}\left(r^2 \frac{\partial p'}{\partial r}\right),$$

or equivalently

$$\frac{1}{c^2}\frac{\partial^2}{\partial t^2}(rp') = \frac{\partial^2}{\partial r^2}(rp').$$

Hence

$$p'(r, t) = \frac{f(t - r/c)}{r} + \frac{g(t + r/c)}{r}.$$

Here

$$p'(r, t) = \frac{A\,e^{i\omega(t - r/c)}}{r}$$

$$|u| = \frac{|p|}{\rho_0 c}\left\{1 + \left(\frac{c}{\omega r}\right)^2\right\}^{\frac{1}{2}} = \frac{2\sqrt{2}}{1.2 \times 340}\left\{1 + \left(\frac{340}{2\pi \times 10^3 \times 5 \times 10^{-2}}\right)^2\right\}^{\frac{1}{2}}$$

$$= \underline{10^{-2}\ \text{m/s}}.$$

SOLUTIONS TO EXERCISES FOR CHAPTER 4

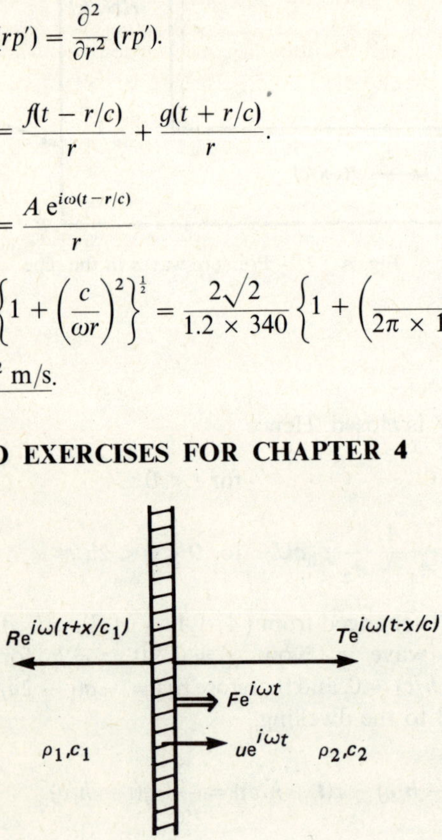

Fig. A.4.1 — The vibration of the plate generates waves.

≠Qu1. Continuity of velocity at the plate $u = -(R/\rho_1 c_1) = (T/\rho_2 c_2)$. Equation of motion of the plate $i\omega m u = F + R - T$. Substituting for R and T

$$u = \frac{F}{i\omega m + \rho_1 c_1 + \rho_2 c_2},$$

$$|u| = \frac{F}{\{(\omega m)^2 + (\rho_1 c_1 + \rho_2 c_2)^2\}^{\frac{1}{2}}} = 6.8\ 10^{-4}\ \text{m/s}.$$

$$\text{Amplitude of vibration} = \frac{|u|}{\omega} = \underline{1.1\ 10^{-6}\ \text{m}}.$$

$$\text{The sound intensity, } I = \overline{p'u} = \frac{|T|^2}{2\rho_2 c_2} = \tfrac{1}{2}|u|^2\rho_2 c_2.$$

$$\text{Intensity level} = 10\log_{10}\left(\frac{I}{10^{-12}}\right) = \underline{115\ \text{dB}}.$$

[This question illustrates a difficulty encountered with the design of ships and submarines. If the machinery in the vessel produces a fluctuating force on parts of the hull, the hull will vibrate. Although the amplitude of vibration of the hull may be very small, it will still radiate a strong sound field into the water, advertising the vessel's position!]

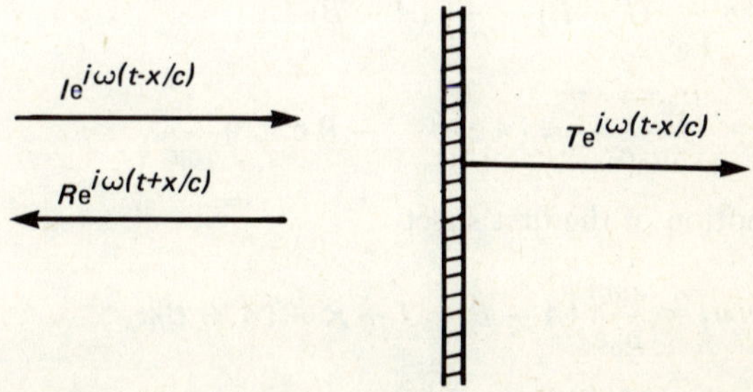

Fig. A.4.2 — The transmission of sound across a single pane.

≠Qu2. Let $u\,e^{i\omega t}$ be the velocity of the glass sheet. Continuity of velocity $u = (I - R)(1/\rho_0 c) = (T/\rho_0 c)$. Equation of motion of the glass $i\omega m u = I + R - T$. Eliminating R

$$T = \frac{2I}{2 + i\beta} \quad \text{where } \beta = \frac{\omega m}{\rho_0 c}$$

$$|T|^2 = \frac{|I|^2}{1 + \beta^2/4}.$$

$$\text{Transmission loss} = 10\log_{10}\left\{\frac{|I|^2}{|T|^2}\right\} = 10\log_{10}\{1 + \beta^2/4\}$$

$$= \underline{17.8\ \text{dB}}.$$

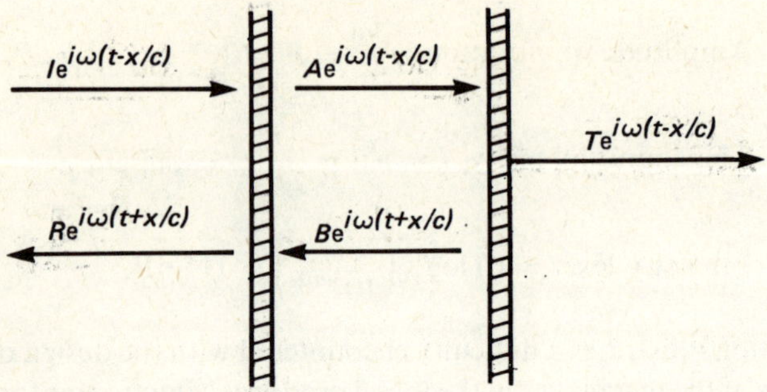

Fig. A.4.3 — The transmission of sound through double glazing.

Let $u_1\ \mathrm{e}^{i\omega t}$ be the velocity of the first sheet of glass. Let $u_2\ \mathrm{e}^{i\omega t}$ be the velocity of the second sheet of glass. Then continuity of velocity:

$$u_1 = \frac{1}{\rho_0 c}(I - R) = \frac{1}{\rho_0 c}(A - B)$$

$$u_2 = \frac{T\,\mathrm{e}^{-i\omega d/c}}{\rho_0 c} = (A\,\mathrm{e}^{-i\omega d/c} - B\,\mathrm{e}^{i\omega d/c})\frac{1}{\rho_0 c}.$$

Equation of motion of the first sheet

$$i\omega m u_1 = \frac{i\omega m}{\rho_0 c}(A - B) = I + R - (A + B).$$

Equation of motion of the second sheet

$$i\omega m u_2 = \frac{i\omega m}{\rho_0 c} T\,\mathrm{e}^{-i\omega d/c} = A\,\mathrm{e}^{-i\omega d/c} + B\,\mathrm{e}^{i\omega d/c} - T\,\mathrm{e}^{-i\omega d/c}.$$

These equations give

$$T = \frac{I}{1 - \frac{\beta^2}{4}\left\{1 - \cos\left(\frac{2\omega d}{c}\right)\right\} + i\beta\left\{1 - \frac{\beta}{4}\sin\left(\frac{2\omega d}{c}\right)\right\}}$$

$$\frac{|I|^2}{|T|^2} = \left\{1 - \frac{\beta^2}{4} + \frac{\beta^2}{4}\cos\left(\frac{2\omega d}{c}\right)\right\}^2 + \beta^2\left\{1 - \frac{\beta}{4}\sin\left(\frac{2\omega d}{c}\right)^2\right\}^2$$

$$= \left(1 + \frac{\beta^2}{4}\right)^2 + \frac{\beta^4}{16} - \frac{\beta^2}{2}\left(\frac{\beta^2}{4} + 1\right)\cos\left(\frac{2\omega d}{c} - \alpha\right)$$

where $\cos\alpha = \dfrac{\beta^2 - 4}{\beta^2 + 4}$, $\sin\alpha = \dfrac{4\beta}{\beta^2 + 4}$.

The transmission loss is maximum when $\cos[(2\omega d/c) - \alpha] = -1$, i.e. when

$$\frac{2\omega d}{c} = \alpha + \pi.$$

For 500 Hz, $\beta = 15.4$, $\alpha = 0.26$ and the transmission loss is maximum when $d \approx 0.18$ m, then

$$\frac{|I|^2}{|T|^2} = 1 + \beta^2 + \frac{\beta^4}{4},$$

and transmission loss $= 10 \log_{10}\left(\frac{|I|^2}{|T|^2}\right) = \underline{42 \text{ dB}}$.

If $d \approx 0.014$ m then

$$\frac{2\omega d}{c} \approx \alpha, \quad \text{and} \quad \frac{|I|^2}{|T|^2} \approx 1$$

and transmission loss $= 10 \log_{10}\left(\frac{|I|^2}{|T|^2}\right) = \underline{0 \text{ dB}}$.

$\neq$Qu3. $R = \dfrac{Z - \rho_0 c/\cos\theta}{Z + \rho_0 c/\cos\theta} I$

$$= \frac{b - \dfrac{\rho_0 c}{\cos\theta} + i\left(\omega m - \dfrac{k}{\omega}\right)}{b + \dfrac{\rho_0 c}{\cos\theta} + i\left(\omega m - \dfrac{k}{\omega}\right)} I.$$

Fraction of incident energy absorbed

$$= 1 - \frac{|R|^2}{|I|^2}$$

$$= \frac{\left(b + \dfrac{\rho_0 c}{\cos\theta}\right)^2 - \left(b - \dfrac{\rho_0 c}{\cos\theta}\right)^2}{\left(b + \dfrac{\rho_0 c}{\cos\theta}\right)^2 + \left(\omega m - \dfrac{k}{\omega}\right)^2}$$

$$= \frac{4b\dfrac{\rho_0 c}{\cos\theta}}{\left(b + \dfrac{\rho_0 c}{\cos\theta}\right)^2 + \left(\omega m - \dfrac{k}{\omega}\right)^2}.$$

This is maximised for any θ if $\omega^2 = k/m$, $\underline{\omega = \pm\sqrt{k/m}}$.

Then the fraction of energy absorbed $= E_{ab} = \dfrac{4b\cos\theta\rho_0 c}{(b\cos\theta + \rho_0 c)^2}$.

The maximum absorption occurs at $b \cos\theta = \rho_0 c$ i.e.

$$\theta = \cos^{-1}(\rho_0 c/b) \quad \text{if } \rho_0 c < b.$$

If $\rho_0 c > b$ then the maximum absorption occurs when $b\cos\theta$ is as large as possible i.e. at $\theta = 0°$.

≠Qu4. $$R = \frac{\dfrac{\rho_1 c_1}{\cos\varphi} - \dfrac{\rho_0 c_0}{\cos\theta}}{\dfrac{\rho_1 c_1}{\cos\varphi} + \dfrac{\rho_0 c_0}{\cos\theta}} I$$

where $\dfrac{\sin\varphi}{c_1} = \dfrac{\sin\theta}{c_0}$.

For 100% transmission $R = 0$.
Hence

$$\frac{\rho_1 c_1}{\cos\varphi} = \frac{\rho_0 c_0}{\cos\theta}$$

i.e.

$$\cos\theta\rho_1 c_1 = \rho_0 c_0 \cos\varphi = \rho_0 c_0\left\{1 - \sin^2\theta\frac{c_1^2}{c_0^2}\right\}^{\frac{1}{2}}.$$

Squaring

$$\cos^2\theta(\rho_1 c_1)^2 = (\rho_0 c_0)^2 - \sin^2\theta(c_1^2\rho_0^2)$$

$$(\rho_1 c_1)^2 = (\rho_0 c_0)^2\sec^2\theta - \tan^2\theta(c_1^2\rho_0^2)$$

$$(\rho_1 c_1)^2 = [(\rho_0 c_0)^2 - (c_1^2\rho_0^2)]\tan^2\theta + (\rho_0 c_0)^2$$

$$\tan\theta = \frac{1}{\rho_0}\left[\frac{\rho_1^2 c_1^2 - \rho_0^2 c_0^2}{c_0^2 - c_1^2}\right]^{\frac{1}{2}}.$$

≠Qu5. $$\text{Transmission loss} = 10\log_{10}\left\{\left|\frac{I}{T}\right|^2\right\}$$

$$= 10\log_{10}\left[1 + \frac{1}{4}\left(\frac{\cos\theta}{\rho_0 c_0}\frac{\rho_1 c_1}{\alpha} + \frac{\rho_0 c_0}{\cos\theta}\frac{\alpha}{\rho_1 c_1}\right)^2 \sinh^2\left(\frac{h\omega}{c_1}\alpha\right)\right]$$

where $\alpha = \sqrt{\dfrac{c_1^2}{c_0^2}\sin^2\theta - 1}$

$$= 10\log_{10}[1 + 71.2\sinh^2(17.4\,\mathrm{h}/\lambda)].$$

When $h = \lambda$, Transmission loss ~ 164 dB, transmitted wave negligible.

When $h = 0.1 \times \lambda$, Transmission loss ~ 27 dB.

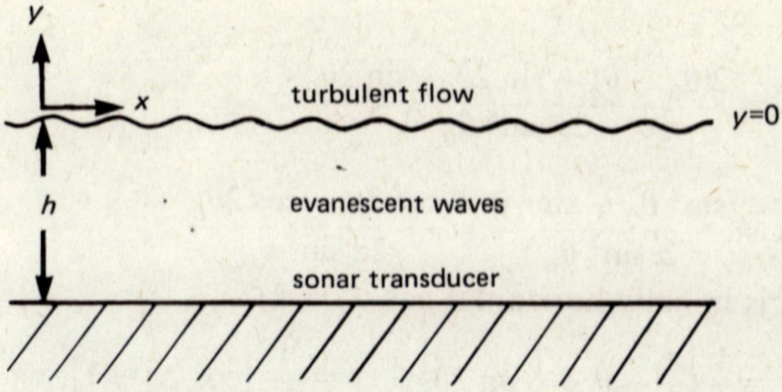

Fig. A.4.4 — The geometry of the cover over the sonar.

≠Qu6. On $y = 0$ $p'_{\text{noise}} = p_0\, e^{i\omega(t - x/V)}$ where $V = 0.8 \times 15$ m/s.
In $-h < y < 0$ $p'_{\text{noise}} = A\, e^{i\omega(t - x/V) + \gamma y} + B\, e^{i\omega(t - x/V) - \gamma y}$

where $\gamma = \dfrac{\omega}{c}\sqrt{\dfrac{c^2}{V^2} - 1}$.

On $y = -h$, $\dfrac{\partial p'_{\text{noise}}}{\partial y} = 0$.

Hence $p'_{\text{noise}} = p_0 \dfrac{\cosh \gamma(y + h)}{\cosh \gamma h} e^{i\omega(t - x/V)}$ in $-h < y < 0$.

p'_{noise} on $y = -h$ is 60 dB less than on $y = 0$ if $\cosh \gamma h = 10^3$ i.e. $\gamma h = 7.6$ and $h = \underline{1.5 \text{ cm}}$.

SOLUTIONS TO EXERCISES FOR CHAPTER 5

≠Qu1. $c = c_0(1 - \alpha x)^{\frac{1}{2}}$ where c_0 is the sound speed at ground level and so Snell's law leads to

$$\sin \theta = \frac{\mathrm{d}y/\mathrm{d}x}{\{(\mathrm{d}y/\mathrm{d}x)^2 + 1\}^{\frac{1}{2}}} = \sin \theta_0 (1 - \alpha x)^{\frac{1}{2}}$$

i.e.

$$\frac{\mathrm{d}y}{\mathrm{d}x} = \pm \frac{\sin \theta_0 (1 - \alpha x)^{\frac{1}{2}}}{\{1 - \sin^2 \theta_0 (1 - \alpha x)\}^{\frac{1}{2}}}$$

$$y = \int_0^x \frac{\sin \theta_0 (1 - \alpha x')^{\frac{1}{2}}\, \mathrm{d}x'}{\{1 - \sin^2 \theta_0 (1 - \alpha x')\}^{\frac{1}{2}}}.$$

Rewriting the integral in terms of θ where $\sin \theta = \sin \theta_0 (1 - \alpha x)^{\frac{1}{2}}$ we find

$$y = -\frac{1}{\alpha \sin^2 \theta_0} \int_{\theta_0}^{\theta} (1 - \cos 2\theta)\, \mathrm{d}\theta = \frac{-1}{\alpha \sin^2 \theta_0} \{\theta - \theta_0 - \tfrac{1}{2}(\sin 2\theta - \sin 2\theta_0)\}$$

i.e.

$$y = \frac{2(\theta_0 - \theta) + \sin 2\theta - \sin 2\theta_0}{2\alpha \sin^2 \theta_0}$$

and

$$x = \frac{\sin^2 \theta_0 - \sin^2 \theta}{\alpha \sin^2 \theta_0} = \frac{\cos 2\theta - \cos 2\theta_0}{2\alpha \sin^2 \theta_0}.$$

The ray that is initially horizontal has $\theta_0 = \pi/2$.

$$y = \frac{1}{\alpha}\left\{\frac{\pi}{2} - \theta + \tfrac{1}{2}\sin 2\theta\right\}, \quad x = \frac{1}{2\alpha}\{\cos 2\theta + 1\}.$$

At the origin $\theta = \pi/2$ and it subsequently decreases to 0. The maximum value of y is $\pi/2\alpha$.

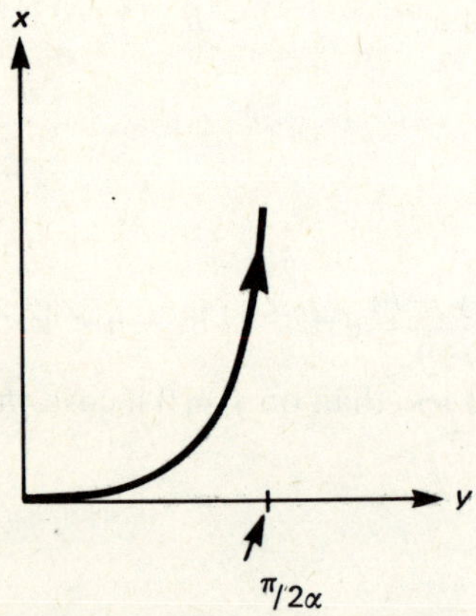

Fig. A.5.1 — The sound rays are refracted upwards.

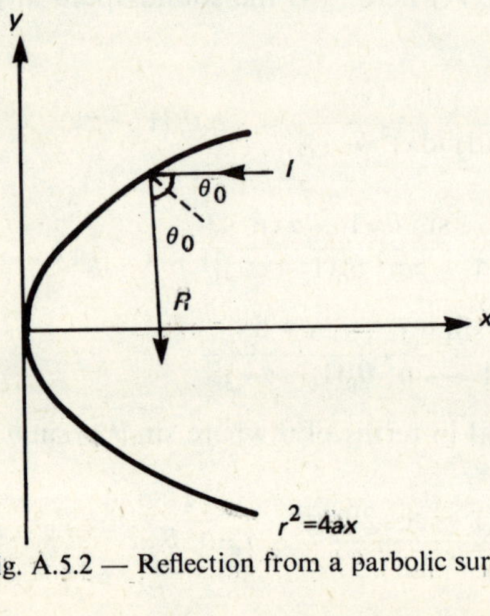

Fig. A.5.2 — Reflection from a parbolic surface.

≠Qu2. Consider a ray parallel to the x-axis which meets the parabolic reflector and choose the (x, y) plane to contain this ray and the focus of the paraboloid. Suppose this ray meets the paraboloid at the point $(y_0^2/4a, y_0, 0)$. Then $\tan\theta_0 = y_0/2a$, and the reflected wave has slope $-\tan 2\theta_0$ i.e. the equation of the reflected ray is $z = 0$,

$$y - y_0 = -\frac{2\tan\theta_0}{1-(\tan\theta_0)^2}\left(x - \frac{y_0^2}{4a}\right),$$

which simplifies to

$$y\left(1 - \frac{y_0^2}{4a^2}\right) = y_0\left(1 - \frac{x}{a}\right).$$

We see that for any value of y_0 the ray passes through $(a, 0, 0)$, and hence all reflected rays pass through the focus.

≠Qu3. Along a ray $\dfrac{\sin\theta}{c} = \dfrac{\sin\theta_0}{c_0} = \frac{1}{1500}$ for a ray which is horizontal at the surface. At its maximum depth the ray is again horizontal. Then $\theta = 90°$, $c = 1500$ and so the maximum depth is 3 km. Take the x-axis to be vertically

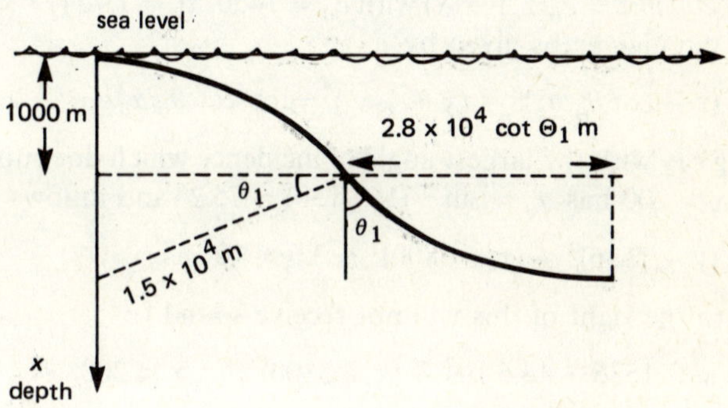

Fig. A.5.3 — The path of the ray that is horizontal at the surface.

downwards. In $0 < x < 1$ km, $c = c_0(1 - \alpha_0 x)$ where $c_0 = 1500$, $\alpha_0 = (1.5 \times 10^4)^{-1}$. The ray path in $0 < x < 1$ km is therefore part of a circle with radius 1.5×10^4 m. When the ray reaches a depth of 1 km the angle $\theta_1 = \sin^{-1}(14/15) = 69°$ and the horizontal distance $= \cos\theta_1\, 1.5 \times 10^4$ m $=$ 5385 m.

In the region $x > 1$ km the sound speed can be written as $c = c_1(1 + \alpha_1(x - x_1))$ where $(x_1, y_1) = (1000, 5385)$, $c_1 = 1400$ and $\alpha_1 = (2.8 \times 10^4)^{-1}$. In this region the ray follows a different circular path; one with a centre at $(x_1 - 2.8 \times 10^4$ m, $y_1 + \cot\theta_1\, 2.8 \times 10^4$ m). The maximum depth is attained at $y = y_1 + 2.8 \times 10^4 \cot \theta_1$ m = 16155 m. The ray then follows a similar path back to the surface, and from symmetry it meets the surface of the sea a distance 2 × 16155 = 32310 m from its starting point.

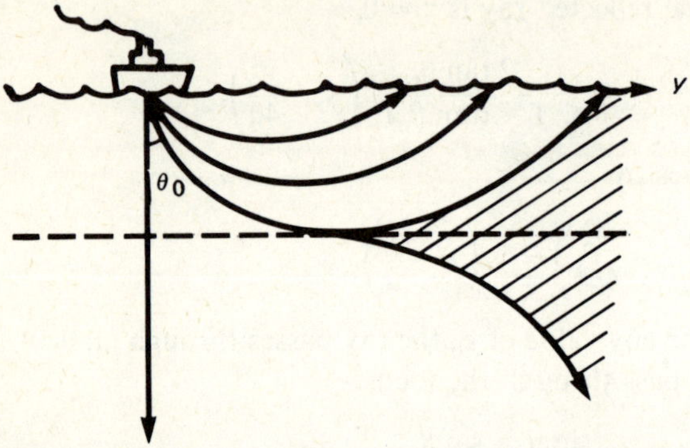

Fig. A.5.4 — The shaded region is a 'zone of silence' into which the ship's sonar cannot penetrate.

≠Qu4. If we trace the sonar rays from the ship and none of them meet the submarine, the submarine will remain undetected by the ship's sonar.

In $x < 200$ *m* $c = c_0(1 + \alpha_0 x)$ with $c_0 = 1450$, $\alpha_0 = 1.72413 \times 10^{-4}$. The rays follow circular paths given by

$$(y - \cot \theta_0/\alpha_0)^2 + (x + 1/\alpha_0)^2 = \operatorname{cosec}^2 \theta_0/\alpha_0^2.$$

The limiting ray with the largest angle of incidence which does not penetrate the region $x > 200$ has $\theta_0 = \sin^{-1}(145/150) = 75.2°$ and follows the curve

$$(y - 1536)^2 + (x + 5800)^2 = 3.6 \times 10^7,$$

the region to the right of this will not receive sound i.e.

$$y > 1536 + \{3.6\ 10^7 - (x + 5800)^2\}^{\frac{1}{2}}, \quad x < 200$$

is not illuminated by sonar rays.

In $x > 200$ *m* $c = c_1(1 - \alpha_1 x)$, $c_1 = 1500$, $\alpha_1 = 8.333 \times 10^{-5}$. These rays now follow circular paths

$$(Y + \cot \theta_1/\alpha_1)^2 + (X - 1/\alpha_1)^2 = \operatorname{cosec}^2 \theta_1/\alpha_1^2,$$

where X, Y are co-ordinates measured from the position of the ray as it enters the region $x > 200$ m. The most shallow ray to enter the region $x > 200$ does so at (200, 1536) at an angle of incidence $\pi/2$, and subsequently follows the path

$$(y - 1536)^2 + (x - 200 - 1.2\ 10^4)^2 = (1.2\ 10^4)^2$$

$$y = 1536 + \{1.44\ 10^8 - (x - 1.22\ 10^4)^2\}^{\frac{1}{2}}.$$

The region to the right of this will not receive sound i.e.

$$y > 1536 + \{1.44\ 10^8 - (x - 1.22\ 10^4)^2\}^{\frac{1}{2}}, \quad x > 200$$

is not illuminated by sonar rays.

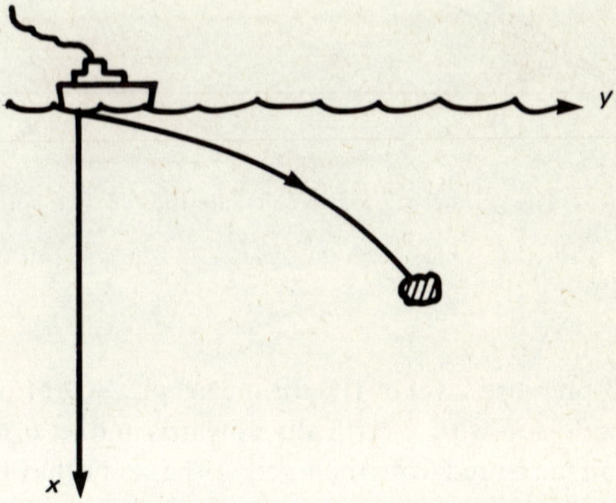

Fig. A.5.5 — The reflection of the ray by the shoal of fish.

≠Qu5. The sound speed $c = c_0(1 - \alpha x)$ with $c_0 = 1500$, $\alpha = 6.67 \times 10^{-5}$. The ray therefore follows the circular path

$$y^2 + (x - 1.5 \times 10^4\ \mathrm{m})^2 = (1.5 \times 10^4\ \mathrm{m})^2. \qquad \text{(A.5.1)}$$

The ray passes from ship to shoal in 1 s. Hence

$$1 = \int_{x=0}^{x} \frac{\mathrm{d}s}{c} = \int_{x=0}^{x} \left\{\left(\frac{\mathrm{d}y}{\mathrm{d}x}\right)^2 + 1\right\}^{\frac{1}{2}} \frac{\mathrm{d}x}{c}$$

$$= \frac{1}{c_0} \int_{x=0}^{x} \frac{\mathrm{d}x}{\{1 - (1 - \alpha x)^2\}^{\frac{1}{2}}(1 - \alpha x)}$$

since a rearrangement of Snell's law gives

$$\frac{\mathrm{d}y}{\mathrm{d}x} = \frac{1 - \alpha x}{\{1 - (1 - \alpha x)^2\}^{\frac{1}{2}}}.$$

Making a substitution $\sin\theta = 1 - \alpha x$ leads to

$$1 = -\frac{1}{\alpha c_0} \int_{\pi/2}^{\theta} \operatorname{cosec}\theta\, \mathrm{d}\theta = -\frac{1}{\alpha c_0} \ln(\tan\theta/2).$$

Hence $\theta = 84°$.
$x = (1 - \sin\theta)/\alpha = 75$ m and from equation (A.5.1) $y = 1495$ m.

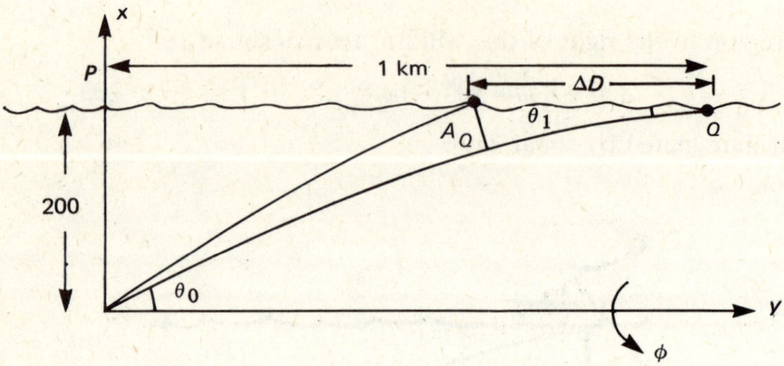

Fig. A.5.6 — The geometry used to calculate the radiated sound power.

≠Qu6. A sound pressure level of 100 dB means $p'_{\text{rms}} = 2\ \text{N/m}^2$. Choose axes centred on the explosion with x vertically upwards and let θ_0 be the direction of ray propagation measured from the y axis, φ the azimuthal angle around the y axis. If W = sound power output from explosion

$$\frac{W}{4\pi} \sin\theta_0 \, d\theta \, d\varphi = \frac{\overline{p'^2} A}{\rho_0 c}$$

where $\boldsymbol{A}$ is the area of the ray tube consisting of rays with initial directions between (θ_0, φ) and $(\theta_0 + \delta\theta, \varphi + \delta\varphi)$. The sound speed $c = c_0(1 + \alpha x)$ with $c_0 = 1500$ m/s, $\alpha = 6.67 \times 10^{-5}$, and so the ray paths are circles:

$$(y - \tan\theta_0/\alpha)^2 + (x + 1/\alpha)^2 = \sec^2\theta_0/\alpha^2$$

i.e.

$$\frac{2y}{\alpha}\tan\theta_0 = y^2 + x^2 + \frac{2x}{\alpha}.$$

The ray that goes through (200 m, 1 km) has $\theta_0 = 13.2°$, and at $x = 200$ m $\theta = \theta_1 = \theta\cos^{-1}(\cos\theta_0\ 152/150) = 9.4°$. $A_Q = 200\ \delta\varphi\ \Delta D \sin\theta_1$, where D is the value of y at $x = 200$, and

$$\Delta D = D(\theta_0 + \delta\theta) - D(\theta_0) = \frac{\mathrm{d}D}{\mathrm{d}\theta_0}\delta\theta.$$

Now

$$D^2 - \frac{2D}{\alpha}\tan\theta_0 + 200^2 + \frac{2 \times 200}{\alpha} = 0. \tag{A.5.2}$$

Hence

$$\frac{\mathrm{d}D}{\mathrm{d}\theta_0} = \frac{D\sec^2\theta_0}{D\alpha - \tan\theta_0} \tag{A.5.3}$$

and

$$A_Q = \frac{200.\delta\varphi\, \delta\theta D \sec^2\theta_0 \sin\theta_1}{|D\alpha - \tan\theta_0|}.$$

Since the pressure at $Q = 2\ \mathrm{N/m^2}$ we find

$$W = \frac{4\pi\, 2^2\, 200\, D \sec^2\theta_0 \sin\theta_1}{\rho_1 c_1 \sin\theta_0 |D\alpha - \tan\theta_0|}.$$

The vertical ray has $\theta_0 = \pi/2$ and $(\mathrm{d}D/\mathrm{d}\theta_0)$ is indeterminate from (A.5.3). But $D(\pi/2) = 0$ and $D(\pi/2 - \delta)$ can be easily evaluated from (A.5.2). It gives

$$D\left(\frac{\pi}{2} - \delta\right) = \frac{\alpha\delta}{2}\left(x^2 + \frac{2x}{\alpha}\right) + 0(\delta^2)$$

and hence

$$\frac{\mathrm{d}D}{\mathrm{d}\theta} = -\frac{\alpha}{2}\{(200)^2 + 400 \times 15000\} \quad \text{at } P,$$

$$A_P = 200\, \delta\varphi\, \delta\theta \frac{\alpha}{2}\{(200)^2 + 400 \times 15000\}.$$

The pressure at P can now be calculated from

$$\overline{p'^2} = \frac{W\rho_1 c_1\, \delta\theta\, \delta\varphi}{4\pi A_P}$$

$$= \underline{113\ \mathrm{dB}}.$$

SOLUTIONS TO EXERCISES FOR CHAPTER 6

$\neq$Qu1. The normal modes are

$$p'(\mathbf{x}, t) = A_{mnq} \cos\left(\frac{m\pi x_1}{l_1}\right) \cos\left(\frac{n\pi x_2}{l_2}\right) \cos\left(\frac{q\pi x_3}{l_3}\right)$$

where $\dfrac{\omega}{c} = \pi\left\{\dfrac{m^2}{l_1^2} + \dfrac{n^2}{l_2^2} + \dfrac{q^2}{l_3^2}\right\}^{\frac{1}{2}}$.

Hence for the (1, 0, 0) mode the resonance frequency = 28 Hz, for the (0, 1, 0) mode 34 Hz, and for the (1, 1, 0) mode 44 Hz.

$\neq$Qu2. $p'(\mathbf{x}, t) = \dfrac{f(t - r/c)}{r} + \dfrac{g(t + r/x)}{r}$

satisfies both the wave equation and the condition of no normal velocity into the walls of the tube. The condition that p' be finite when $r = 0$ implies $f(t) = -g(t)$. Hence for disturbances of frequency ω

$$p'(\mathbf{x}, t) = \frac{A}{r}\{e^{i\omega(t-r/c)} - e^{i\omega(t+r/c)}\}$$

$$= -\frac{2Ai}{r}e^{i\omega t}\sin\left(\frac{\omega r}{c}\right).$$

$p' = 0$ when $\boldsymbol{r} = l$ leads to $(\omega l/c) = n\pi$, i.e. the resonance frequencies are equal to $n\pi c/l$, where n is an integer.

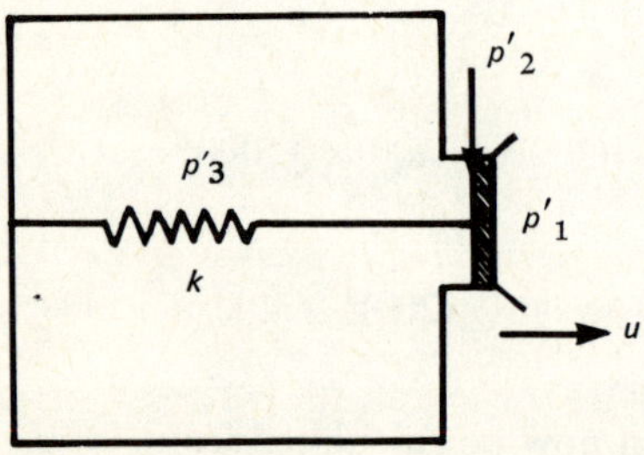

Fig. A.6.1 — The loudspeaker cabinet and cone.

≠Qu3. (a) The Helmholtz resonator frequency $= \sqrt{\dfrac{c^2S}{Vl}} = 124$ Hz.

(b) Let u be the velocity of the loudspeaker cone. Then the equation of motion of the cone vibrating with frequency ω is

$$i\omega mu + \frac{ku}{i\omega} = S(p_2' - p_1').$$

But from equation (6.8)

$$p_2' = -\left(\frac{c^2}{Vi\omega} + \frac{i\omega l}{S}\right)\rho_0 Su,$$

and therefore

$$-Sp_1' = \left\{i\omega(m + \rho_0 lS) + \frac{1}{i\omega}\left(k + \frac{\rho_0 c^2 S^2}{V}\right)\right\}u.$$

The resonant frequency is now

$$\sqrt{\frac{k + \rho_0 c^2 S^2/V}{m + \rho_0 lS}} = 78 \text{ Hz}.$$

(c) The radiated power $= S\overline{p'_1 u} = S\rho_0 c\overline{u^2}$ for a plane wave. The radiated power is therefore maximised when u is maximised. We can calculate u from the equation of motion of the cone. If the applied force is $\boldsymbol{F}\,e^{i\omega t}$ then

$$F\,e^{i\omega t} = (\rho_0 cS + iX)u \quad \text{where } X = \omega\left\{m + \rho_0 lS - \frac{1}{\omega^2}\left(k + \frac{\rho_0 c^2 S^2}{V}\right)\right\},$$

and hence

$$|u| = \frac{|F|}{((\rho_0 cS)^2 + X^2)^{\frac{1}{2}}}.$$

$|u|$ is therefore maximised when $X = 0$ i.e. at the resonance frequency found in (b). At this frequency $|u| = |F|/\rho_0 cS$, and the radiated acoustic power $= \overline{F^2}/\rho_0 cS = 15.6$ watts.

$\neq$Qu4. For plane waves of frequency ω

$$p'(x, t) = A\,e^{i\omega(t - x/c)} + B\,e^{i\omega(t + x/c)}.$$

The condition of no velocity at $x = 0$ leads to $\boldsymbol{A} = B$, and then the open end condition implies $\cos(\omega L/c) = 0$ i.e. the resonant frequencies are $\omega = \pi c/2L$, $3\pi c/2L$, $5\pi c/2L$, ... etc.

At these resonant frequencies

$$p' = 2A\,e^{i\omega t}\cos(\Omega x/c), \quad u = -(2Ai/\rho_0 c)\,e^{i\omega t}\sin(\Omega x/c)$$

and so $\hat{u} = -(i/\rho_0 c)\tan(\Omega x/c)\hat{p}$ for the resonant frequency Ω. Therefore, at the fundamental frequency, u lags p' by $\pi/2$, and if the heat release lags the particle velocity by α, it must lag the pressure perturbation by $\alpha + \pi/2$. Now if $\overline{p'q}$ is positive, energy is fed into the acoustic waves in each cycle and the system is unstable. We see then that if $-\pi < \alpha < 0$ the system is unstable.

In practice the pressure waves radiate energy from the open end. The waves therefore only grow in magnitude if the heat release rate is sufficiently large that the waves gain more energy from the unsteady heating that they lose by radiation.

$\neq$Qu5. Let p' be the pressure perturbation due to the bubble pulsation. Within the bubble $p' = p'|_{r=a_0} = p_a$ say.

For adiabatic disturbances $(p_0 + p_a)a^\gamma$ is constant, which leads to $p_a = -(3\gamma p_0/a_0)(a - a_0)(t)$.

In $\boldsymbol{r} > a$ $p'(r, t)$ satisfies the wave equation and so

$$p'(r, t) = -\frac{3\gamma p_0}{r}(a - a_0)\left(t - \frac{r - a_0}{c}\right) \quad \text{in } r > a.$$

The momentum equation states that

$$\rho_0 \frac{\mathrm{d}^2 a}{\mathrm{d}t^2} = -\left.\frac{\partial p}{\partial r}\right|_{r=a_0}$$

i.e.

$$\rho_0 \frac{\mathrm{d}^2 a}{\mathrm{d}t^2} + \frac{3\gamma p_0}{a_0 c}\frac{\mathrm{d}a}{\mathrm{d}t} + \frac{3\gamma p_0}{a_0^2}(a - a_0) = 0.$$

To find the resonance frequency we look for a solution of this equation with $a - a_0$ proportional to e^{mt}. Then

$$m = -\frac{3\gamma p_0}{2\rho_0 a_0 c} \pm \frac{1}{2\rho_0}\sqrt{\left(\frac{3\gamma p_0}{a_0 c}\right)^2 - \frac{12\gamma p_0 \rho_0}{a_0^2}},$$

and writing $m = -\alpha + i\omega$ leads to a resonance frequency of 3.3 kHz.

$\neq$Qu6. A reverberation time of 1.5 s means that the total sound absorption in the empty workshop is 108 metric sabins. Hence when the machines are installed the sound absorption will be 128 metric sabins.

If W is the sound power output of each machine the balance of energy in the diffuse sound field gives

$$20W = Ia = \frac{ec}{4}a = \frac{\overline{p'^2}\, a}{4\rho_0 c}.$$

When the sound pressure level is 90 dB, $W = 1.6$ m$W = 92$ dB. The maximum sound power output of each machine is therefore 92 dB.

SOLUTIONS TO EXERCISES FOR CHAPTER 7

$\neq$Qu1. $p'(\mathbf{x}, t) = \dfrac{p_0}{r}\sin\omega(t - r/c) - \dfrac{p_0}{r'}\sin\omega(t - r'/c)$,

where $r = (x_1^2 + x_2^2 + x_3^2)^{\frac{1}{2}}$ and $r' = ((x_1 - l)^2 + x_2^2 + x_3^2)^{\frac{1}{2}} \approx r - l\cos\theta$ for large r. Therefore for large r,

$$p'(\mathbf{x}, t) = \frac{p_0}{r}\left\{\sin\omega\left(t - \frac{r}{c}\right) - \sin\omega\left(t - \frac{r}{c} + \frac{l\cos\theta}{c}\right)\right\}$$

$$= \frac{2p_0}{r}\cos\omega\left(t - \frac{r}{c} + \frac{\omega l\cos\theta}{2c}\right)\sin\left(\frac{\omega l\cos\theta}{2c}\right)$$

$$\overline{p'^2} = \frac{2p_0^2}{r^2}\sin^2\left(\frac{\omega l\cos\theta}{2c}\right)$$

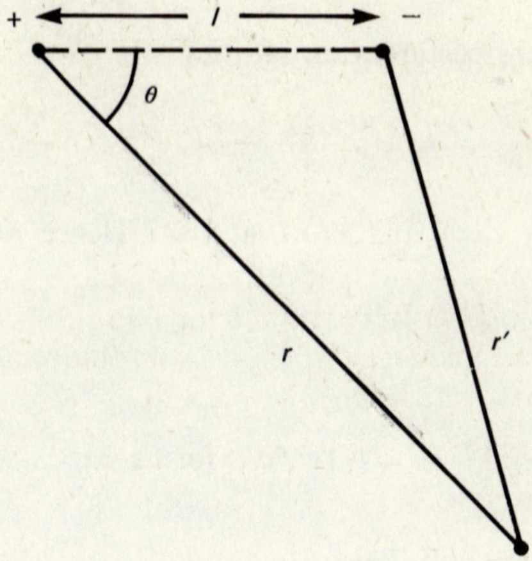

Fig. A.7.1 — Equal and opposite sources separated by a distance l produce a pressure field
$p'(x, t) = \frac{p_0}{r} \sin \omega(t - r/c) - \frac{p_0}{r} \sin \omega(t - r'/c)$.

and the radiated power $= (1/\rho_0 c) \int \overline{p'^2}\, \mathrm{d}S$ where the integral is evaluated over the surface of a sphere of radius r.

$$\text{Radiated power} = 4\pi \frac{p_0^2}{\rho_0 c} \int_0^\pi \sin^2\left(\frac{\omega l \cos\theta}{2c}\right) \sin\theta \, \mathrm{d}\theta$$

$$= \frac{4\pi p_0^2}{\rho_0 c}\left[1 - \frac{c}{\omega l} \sin\left(\frac{\omega l}{c}\right)\right].$$

In isolation the power radiated by each source is $2\pi p_0^2/\rho_0 c$. As l tends to infinity the power radiated by the two sources is just the sum of the power they would radiate in isolation. As l tends to zero the power radiated by the sources also tends to zero. The anti-source creates an environment around the first source which prevents it doing work on the surrounding fluid. Energy is never generated when $\omega l \ll c$, and therefore the question is not sensible.

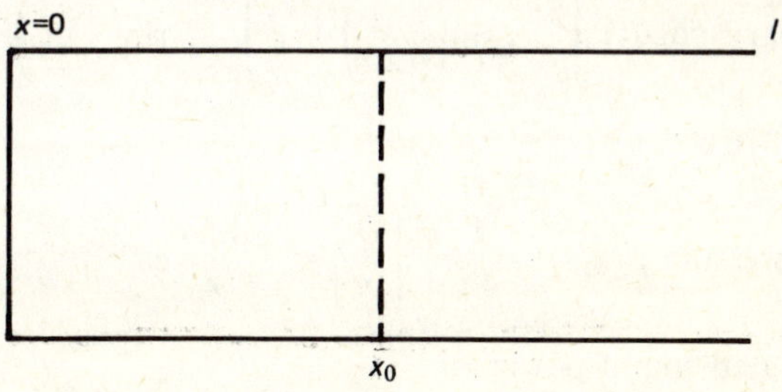

Fig. A.7.2 — Heat is added at x_0 to fluid in a pipe with one open and one closed end.

≠Qu2. The pressure perturbation satisfies

$$\frac{\partial^2 p'}{\partial x^2} + \frac{\omega^2}{c^2} p' = \frac{-i\omega\rho_0(\gamma - 1)}{c^2} h_0 \, \mathrm{e}^{i\omega t} \, \delta(x - x_0) \tag{A.7.1}$$

with $(\partial p'/\partial x) = 0$ at $x = 0$ and $p' = 0$ at $x = l$. Hence

$$p' = A \cos\left(\frac{\omega x}{c}\right) \mathrm{e}^{i\omega t} \qquad \text{for } x < x_0$$

and

$$p' = B \sin\left(\frac{\omega}{c}(x - l)\right) \mathrm{e}^{i\omega t} \qquad \text{for } l > x > x_0.$$

Integrating (A.7.1) we find that

$$\left[\frac{\partial p'}{\partial x}\right]_{x = x_0^-}^{x_0^+} = \frac{-i\omega\rho_0(\gamma - 1)h_0}{c^2} \mathrm{e}^{i\omega t} \quad \text{and} \quad [p']_{x = x_0^-}^{x_0^+} = 0,$$

which gives

$$A = B \frac{\sin \omega(x_0 - l)/c}{\cos \omega x_0/c} = \frac{-i\rho_0(\gamma - 1)h_0}{c} \frac{\sin \omega(x_0 - l)/c}{\cos \omega l/c}$$

$$p' = \frac{i\rho_0(\gamma - 1)h_0}{c} \frac{\sin \omega(l - x_0)/c}{\cos \omega l/c} \cos\left(\frac{\omega x}{c}\right) \mathrm{e}^{i\omega t} \quad \text{for } x < x_0$$

$$= \frac{i\rho_0(\gamma - 1)h_0}{c} \frac{\cos \omega x_0/c}{\cos \omega l/c} \sin\left(\frac{\omega(l - x)}{c}\right) \mathrm{e}^{i\omega t} \quad \text{for } x_0 < x < l.$$

≠Qu3. The compactness ratio $\omega D/c = 0.18$, and so the drum is compact.

$$Q = \rho_0 \int U_n \, \mathrm{d}S$$

$$= \rho_0 i\omega \, \mathrm{e}^{i\omega t}(0.01) \int_0^{0.25} \sin\left(\frac{\pi x_1}{0.25}\right) \mathrm{d}x_1 \int_0^{0.25} \sin\left(\frac{\pi x_2}{0.25}\right) \mathrm{d}x_2$$

$$= i \, 0.076 \, \mathrm{e}^{i\omega t}.$$

The sound pressure $p'(\mathbf{x}, t) = \dfrac{1}{4\pi} \dfrac{\partial}{\partial t}\left[\dfrac{Q}{r}\right] = -\dfrac{1.53}{r} \mathrm{e}^{i\omega(t - r/c)}$

and the radiated sound power $= \dfrac{4\pi r^2 \overline{p^2}}{\rho_0 c}$

$= \underline{0.036 \text{ watts}}.$

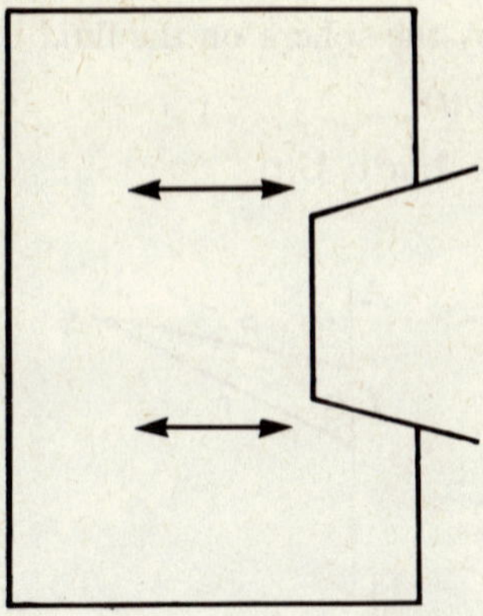

Fig. A.7.3 — The sound generated by a compact loudspeaker is given by $p'(r, t) = \dfrac{1}{4\pi r}\dfrac{\partial Q}{\partial t}$.

≠Qu4. The compactness ratio $= \omega a/c = 0.09$, and so the loudspeaker is compact. Hence

$$p' = \frac{1}{4\pi r}\frac{\partial Q}{\partial t}$$

and

$$p'_{\mathrm{rms}} = \frac{0.21}{r}\,\mathrm{N\,m^{-2}}.$$

The radiated power is therefore 1.35 milliwatts.

≠Qu5. (i) For a compact pulsating sphere, from (2.34),

$$p'(r, t) = \frac{\rho_0 a^2}{r}\frac{\partial u_a}{\partial t}(t - r/c)$$

$$= \frac{\rho_0}{4\pi r}\frac{\partial^2 V}{\partial t^2}(t - r/c).$$

i.e. The sound field is equivalent to that generated by a monopole of strength $\rho_0 \ddot{V}(t)$.

(ii) For a compact oscillating sphere from (2.40)

$$p'(\mathbf{x}, t) = -\tfrac{1}{2} i\omega\rho_0 U a^3 \frac{\partial}{\partial x_1}\left\{\frac{e^{i\omega(t - r/c)}}{r}\right\}$$

i.e. the sound field is equivalent to that generated by a dipole of strength $2\pi a^3 \rho_0 \dot{\mathbf{U}}(t)$.

The force exerted by the sphere on the fluid is from (2.43),

$$\mathbf{F} = \tfrac{2}{3}\pi a^3 \rho_0 \dot{\mathbf{U}}(t)$$

Dipole strength $= \mathbf{F}(t) + \tfrac{4}{3}\pi a^3 \rho_0 \dot{\mathbf{U}}(t)$.

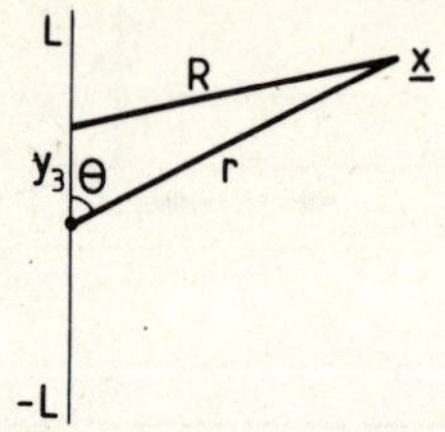

Fig. A.7.4 — The sound generated by the vibrating wire is heard at x.

≠Qu6. The source distribution is a collection of dipoles of strength $-L\,\mathrm{e}^{i\omega t}\,\mathrm{d}y_3$ pointing in the y_1-direction.

$$p'(\mathbf{x}, t) = \frac{\partial}{\partial x_1} \int_{-l}^{l} \frac{L\,\mathrm{e}^{i\omega(t-R/c)}}{4\pi R}\,\mathrm{d}y_3.$$

In the far-field $R = (r^2 - 2ry_3\cos\theta + y_3^2)^{\frac{1}{2}} \sim r - y_3\cos\theta$, and

$$p'(\mathbf{x}, t) \sim \frac{\partial}{\partial x_1}\left\{\frac{L\,\mathrm{e}^{i\omega(t-r/c)}}{4\pi r}\int_{-l}^{l} \mathrm{e}^{i\omega l\cos\theta y_3/c}\,\mathrm{d}y_3\right\}$$

$$= \frac{-x_1 i\omega L}{4\pi r^2 c}\,\mathrm{e}^{i\omega(t-r/c)}\,\frac{2\sin(\omega l x_3/rc)}{\omega x_3/rc}.$$

If $(\omega l/c) \ll 1$,

$$p'(\mathbf{x}, t) \sim \frac{-x_1 i\omega 2Ll}{4\pi r^2 c}\,\mathrm{e}^{i\omega(t-r/c)},$$

the field due to a dipole of strength $-2Ll\,\mathrm{e}^{i\omega t}$ which is equal to the total force on the fluid.

≠Qu7. $D_1 = 2D_2$ say.

For equal thrust, $\rho_e U_1^2 D_1^2 = \rho_e U_2^2 D_2^2$

$$U_2 = 2U_1.$$

Sound power for engine 1 $= kU_1^8 D_1^2$,

and sound power for engine 2 $= kU_2^8 D_2^2 = kU_1^8 D_1^2 \times 2^6$.

The smaller diameter engine is $10\log_{10}(2^6)$ dB louder, i.e. 18 dB.

SOLUTIONS TO EXERCISES FOR CHAPTER 8

≠Qu1. $$p'(\mathbf{x}, t) = \frac{i\omega\rho_0\sqrt{2m}}{4\pi r}\,\mathrm{e}^{i\omega(t-r/c)} - \frac{i\omega\rho_0\sqrt{2m}}{4\pi r'}\,\mathrm{e}^{i\omega(t-r'/c)},$$

where $r = \{x_1^2 + x_2^2 + (x_3 - h)^2\}^{\frac{1}{2}}$, $r' = \{x_1^2 + x_2^2 + (x_3 + h)^2\}^{\frac{1}{2}}$.

We can determine the radiated acoustic power by calculating the energy flux through a large hemisphere. For large $|\mathbf{x}|$

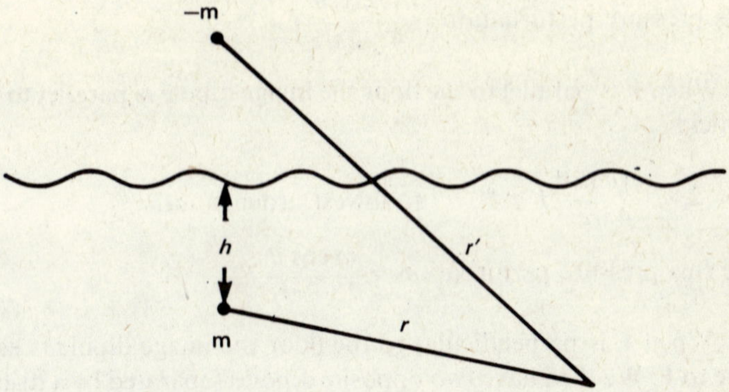

Fig. A.8.1 — The reflection of sound in a free surface is equivalent to an image source of opposite sign.

$$p'(\mathbf{x}, t) = \frac{i\omega\rho_0\sqrt{2m}}{4\pi|\mathbf{x}|} e^{i\omega(t-|\mathbf{x}|/c)}\{e^{i\omega h \sin\theta/c} - e^{-i\omega h \sin\theta/c}\}$$

$$= -\frac{\omega\rho_0\sqrt{2m}}{2\pi|\mathbf{x}|} e^{i\omega(t-|\mathbf{x}|/c)} \sin\left(\frac{\omega h \sin\theta}{c}\right).$$

Integration of $p'u$ around a hemisphere centred on the origin gives the radiated power, P,

$$= \left(\frac{\omega\rho_0 m}{2\pi}\right)^2 \frac{1}{\rho_0 c}\int_{\phi=0}^{2\pi}\int_{\theta=0}^{\pi/2} \sin^2\left(\frac{\omega h\cos\theta}{c}\right)\sin\theta\, d\theta\, d\varphi$$

$$= \frac{\rho_0\omega^2 m^2}{4\pi c}\left[1 - \frac{c}{2\omega h}\sin\left(\frac{2\omega h}{c}\right)\right].$$

In deep water $P = \rho_0\omega^2 m^2/4\pi c$, P has half this value when

$$1 - \frac{c}{2\omega h}\sin\left(\frac{2\omega h}{c}\right) = \tfrac{1}{2} \quad \text{i.e. when } \sin\left(\frac{2\omega h}{c}\right) = \frac{\omega h}{c}.$$

Numerical solution of $\sin x = 0.5 \times x$ gives $x \sim 1.9$ and hence $\mathbf{h} = 2.2$ m.

$\neq$Qu2. Suppose **F** acts in the 1-direction then in the absence of surfaces

$$p'(\mathbf{x}, t) = \frac{-\partial}{\partial x_1}\left\{\frac{F\, e^{i\omega(t-r/c)}}{4\pi r}\right\},$$

which in the far-field simplifies to

$$p'(\mathbf{x}, t) = \frac{i\omega\cos\theta}{4\pi r c} F\, e^{i\omega(t-r/c)},$$

where θ is the angle between the observer and the x_1-axis.

The rms pressure perturbation $= \dfrac{\omega \cos\theta}{4\sqrt{2}\pi rc} F.$

(i) When **F** is parallel to the floor the image dipole is parallel to **F** and in the far-field

$$p'(\mathbf{x}, t) = 2 \times \frac{i\omega \cos\theta}{4\pi rc} F\, e^{i\omega(t - r/c)} \text{ to lowest order in } \omega L/c,$$

and the rms pressure perturbation $= \dfrac{\omega \cos\theta}{2\sqrt{2}\pi rc} F.$

(ii) When **F** is perpendicular to the floor the image dipole is equal and opposite to **F**. We then have two opposite dipoles separated by a distance $2L$. The resultant pressure perturbation for small L is therefore

$$p'(\mathbf{x}, t) = 2L \frac{\partial^2}{\partial x_1^2} \left\{ \frac{F\, e^{i\omega(t - r/c)}}{4\pi r} \right\},$$

which in the far-field simplifies to

$$p'(\mathbf{x}, t) = \frac{-2L\omega^2 \cos^2\theta}{4\pi rc^2} F\, e^{i\omega(t - r/c)},$$

and the rms pressure perturbation $= \dfrac{\omega^2 L \cos^2\theta F}{2\sqrt{2}\pi rc^2}.$

Qu3. On the axis $p'(\mathbf{x}, t) = \frac{1}{2}(\rho_0 i\omega a^2/R) U_0\, e^{i\omega(t - R/c)}$ and a SPL of 100 dB when $R = 10$ m implies $U_0 = 0.17$ m/s.

The major lobe is confined within $\theta < \sin^{-1}(3.83c/\omega a) = 43.7°$. On the wall where $R = 10$ m

$$p'(\mathbf{x}, t) = 2\sqrt{2}\, \frac{2J_1(\omega a/c)}{\omega a/c}\, e^{i\omega t} \text{ N/m}^2.$$

which leads to a sound pressure level of 82 dB.

≠Qu4. $p'(\mathbf{x}, t) = \dfrac{\rho_0 i\omega U_0}{2\pi} \displaystyle\int_0^a \int_0^a \frac{e^{i\omega(t - r/c)}}{r}\, dy_2\, dy_3.$

For large $|\mathbf{x}|$ this becomes (cf. equation 8.21)

$$p'(\mathbf{x}, t) = \frac{\rho_0 i\omega U_0}{2\pi R} e^{i\omega(t - R/c)} \int_0^a e^{i\omega x_2 y_2/Rc}\, dy_2 \int_0^a e^{i\omega x_3 y_3/Rc}\, dy_3$$

where $R = |\mathbf{x}|$ i.e.

$$p'(\mathbf{x}, t) = \frac{\rho_0 i\omega U_0 a^2}{2\pi R} e^{i\omega(t - R/c)}\, e^{i\omega a(x_2 + x_3)/2Rc} \frac{\sin\left(\dfrac{\omega a x_2}{2Rc}\right)}{\dfrac{\omega a x_2}{2Rc}} \frac{\sin\left(\dfrac{\omega a x_3}{2Rc}\right)}{\dfrac{\omega a x_3}{2Rc}}.$$

For $\omega a/c \ll 1$, $p'(\mathbf{x}, t) \sim \dfrac{\rho_0 i\omega U_0 a^2}{2\pi R} e^{i\omega(t-R/c)}$

an omnidirectional radiation pattern.

For $\omega a/c \gg 1$, $p'(\mathbf{x}, t) = 0$ when $\sin(\omega a x_2/2Rc) = 0$, $x_2 \neq 0$, or $\sin(\omega a x_3/2Rc) = 0$, $x_3 \neq 0$, i.e. the pressure perturbation vanishes whenever $(x_2/R) = (2n\pi c/\omega a)$ or $(x_3/R) = (2m\pi c/\omega a)$, with m and n non-zero integers and there is a pattern of lobes.

≠Qu5. The incident pressure perturbation at the fish,

$$p_i(t) = \frac{-\rho_0 \omega^2 Q}{4\pi s} e^{i\omega(t-s/c)}.$$

The incident pressure perturbation causes the bladder to contract, thus producing a scattered field $p'(r, t) = (b/r)p_b(t - r/c)$ where $\boldsymbol{r}$ is measured from the centre of the fish. The linearised momentum equation implies $\rho_0(\partial u/\partial t) = -(\partial p'/\partial r)$ i.e.

$$\rho_0 \frac{\partial^2 b}{\partial t^2} = \frac{p_b(t)}{b}$$

for a compact fish. Hence

$$b - b_0 = -\frac{1}{\rho_0 \omega^2 b} p_b(t).$$

Now since the fluid in the bladder is isothermal

$$(p_0 + p_i + p_b)b^3 = \text{constant}$$

and so

$$(p_i + p_b)b_0 + 3p_0(b - b_0) = 0$$

i.e.

$$p_i = -p_b - \frac{3p_0}{b}(b - b_0) = p_b\left\{-1 + \frac{3p_0}{\rho_0 \omega^2 b^2}\right\}$$

and

$$p_b = -\frac{p_i}{1 - 3p_0/\rho_0 \omega^2 b^2}.$$

The scattered sound field $p'(r, t)$ is therefore given by

$$p'(r, t) = \frac{-b}{r} \frac{p_i(t - r/c)}{1 - 3p_0/\rho_0 \omega^2 b^2} = \frac{b\rho_0 \omega^2 Q\, e^{i\omega(t-(s+r)/c)}}{4\pi rs\{1 - 3p_0/\rho_0 \omega^2 b^2\}}.$$

The magnitude of the field scattered to the sonar $= \dfrac{b\rho_0 \omega^2 |Q|}{4\pi s^2 |1 - 3p_0/\rho_0 \omega^2 b^2|}.$

≠Qu6. We wish to find the distant sound field radiated by a quadrupole near a surface. The reciprocal problem involves evaluating the field, φ_2, near the surface due to a distance point source at $\mathbf{x}_2$. The waves incident on the surface due to this distance source are nearly plane and are reflected with the plane wave reflection coefficient given in equation (4.12). Hence

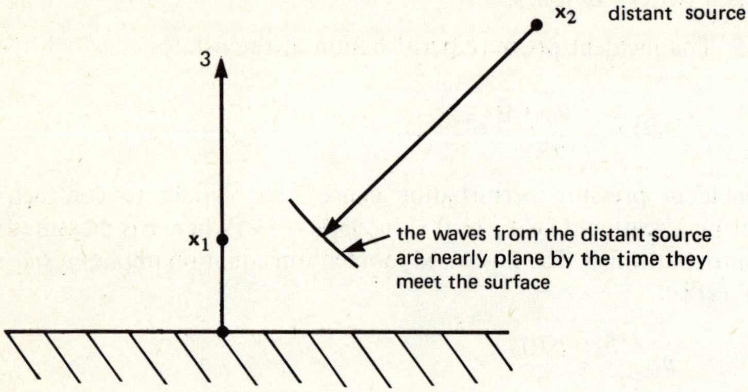

Fig. A.8.2 — The reciprocal problem.

$$\varphi_2(\mathbf{x}) = -\frac{e^{i\omega|\mathbf{x}-\mathbf{x}_2|/c}}{4\pi|\mathbf{x}-\mathbf{x}_2|} - R\frac{e^{-i\omega|\mathbf{x}-\mathbf{x}_2^+|/c}}{4\pi|\mathbf{x}-\mathbf{x}_2^+|}$$

where $\mathbf{x}_2 = (x_{21}, x_{22}, x_{23})$ and $\mathbf{x}_2^+$ is its image point $(x_{21}, x_{22}, -x_{23})$.

$$R = \frac{Z - \rho_0 c/\cos\theta}{Z + \rho_0 c/\cos\theta} \quad \text{with } \cos\theta = x_{23}/|\mathbf{x}_2|.$$

At the quadrupole $\mathbf{x} = \mathbf{x}_1$ and

$$\frac{\partial^2\varphi_2}{\partial x_i\,\partial x_j} = \frac{\omega^2}{4\pi|\mathbf{x}_2|^3c^2}\,e^{-i\omega|\mathbf{x}_2|/c}\{x_{2i}x_{2j} + Rx_{2i}^+x_{2j}^+\}.$$

(We have assumed that $|\mathbf{x}_2| \gg |\mathbf{x}_1|$.) The sound field radiated by the quadrupole near the surface is

$$-T_{ij}\frac{\partial^2\varphi_2}{\partial x_i\,\partial x_j}(\mathbf{x}_1) = \frac{-T_{ij}\omega^2}{4\pi|\mathbf{x}_2|^3c^2}\,e^{-i\omega|\mathbf{x}_2|/c}\{x_{2i}x_{2j} + Rx_{2i}^+x_{2j}^+\}$$

i.e. in the real problem

$$p'(\mathbf{x}, t) = \frac{-T_{ij}\omega^2}{4\pi|\mathbf{x}|^3c^2}\,e^{i\omega(t-|\mathbf{x}|/c)}\{x_ix_j + Rx_i^+x_j^+\}.$$

We see that the sound field = $(1 \pm R) \times$ the field that would be radiated by the quadrupole in free space. For a sound absorbing surface $ReZ > 0$ and so $|R| \leqslant 1$. It therefore follows that the sound field radiated by turbulence near a plane sound absorbing surface is at most twice the field that would be radiated by the turbulence in free space. The adjacent surface does not therefore substantially increase the sound generated by turbulence.

SOLUTIONS TO EXERCISES FOR CHAPTER 9

$\neq$Qu1. From equation (9.7)

$$p'(\mathbf{x}, t) = \frac{1}{4\pi} \int_{-l}^{l} Q \frac{e^{i\omega\tau^*}}{r|1 - M_r|} d\eta_2$$

where

$$\tau^* = \left(t - \frac{|\mathbf{x}|}{c} + \frac{x_i \eta_i}{|\mathbf{x}|c}\right) \frac{1}{1 - M \cos\theta} \quad \text{for } |\mathbf{x}| \gg l, |\mathbf{U}\tau^*|$$

i.e.

$$p'(\mathbf{x}, t) = \frac{Q}{4\pi|\mathbf{x}|(1 - M\cos\theta)}$$

$$\times \int_{-l}^{l} \exp i\omega\left[\left(t - \frac{|\mathbf{x}|}{c} + \frac{x_2\eta_2}{|\mathbf{x}|c}\right)\frac{1}{1 - M\cos\theta}\right] d\eta_2$$

$$= \frac{c}{2\pi\omega x_2} \exp\left[\frac{i\omega(t - |\mathbf{x}|/c)}{1 - M\cos\theta}\right] \sin\left(\frac{\omega x_2 l}{|\mathbf{x}|c(1 - M\cos\theta)}\right).$$

For $\omega l \ll c(1 - M\cos\theta)$

$$p'(\mathbf{x}, t) \sim \frac{2lQ}{4\pi|\mathbf{x}|(1 - M\cos\theta)} \exp\left[\frac{i\omega(t - |\mathbf{x}|/c)}{1 - M\cos\theta}\right],$$

and the sound field is equivalent to that generated by a moving point source of strength $2lQ\, e^{i\omega t}$.

$\neq$Qu2. Rewriting (9.2) in terms of moving co-ordinates

$$p'(\mathbf{x}, t) = Q \int \frac{e^{i\omega\tau}\, \delta(t - \tau - |\mathbf{x} - \boldsymbol{\eta} - \mathbf{U}\tau|/c)\, \delta(\eta_1)\, \delta(\eta_3)}{4\pi|\mathbf{x} - \boldsymbol{\eta} - \mathbf{U}\tau|} d^3\boldsymbol{\eta}\, d\tau.$$

The argument of the δ-function is

$$g = t - \tau - \frac{|\mathbf{x} - \boldsymbol{\eta} - \mathbf{U}\tau|}{c}, \quad \frac{\partial g}{\partial \eta_2} = \frac{x_2 - \eta_2}{c|\mathbf{x} - \boldsymbol{\eta} - \mathbf{U}\tau|}$$

and g vanishes at $x_2 - \eta_2 = \pm\{c^2(t-\tau)^2 - (x_1 - U\tau)^2 - x_3^2\}^{\frac{1}{2}}$. Hence evaluation of the $\boldsymbol{\eta}$-integrals gives

$$p'(\mathbf{x}, t) = \frac{Qc}{2\pi}\int \frac{e^{i\omega\tau} H[c(t-\tau) - \{(x_1 - U\tau)^2 + x_3^2\}^{\frac{1}{2}}]}{[c^2(t-\tau)^2 - \{(x_1 - U\tau)^2 + x_3^2\}]^{\frac{1}{2}}}\, d\tau$$

$$c^2(t-\tau)^2 - (x_1 - U\tau)^2 - x_3^2 = (c^2 - U^2)\left(\tau - \frac{c^2 t - x_1 U}{c^2 - U^2}\right)^2 - \frac{(Ut - x_1)^2 + (1 - M^2)x_3^2}{1 - M^2}$$

$$= (\xi^2 - 1)\frac{(Ut - x_1)^2 + (1 - M^2)x_3^2}{1 - M^2}$$

where $\xi = \dfrac{c(1 - M^2)}{\{(Ut - x_1)^2 + (1 - M^2)x_3^2\}^{\frac{1}{2}}}\left(\dfrac{t - x_1 U/c^2}{1 - M^2} - \tau\right)$.

Expressing p' in terms of an integral over ξ gives

$$p'(\mathbf{x}, t) = \frac{Q}{2\pi(1 - M^2)^{\frac{1}{2}}}\exp\left[i\omega\left(\frac{t - x_1 U/c^2}{1 - M^2}\right)\right] \times \int_1^\infty \frac{1}{(\xi^2 - 1)^{\frac{1}{2}}}\exp\left[-i\xi\omega\frac{\{(Ut - x_1)^2 + (1 - M^2)x_3^2\}^{\frac{1}{2}}}{c(1 - M^2)}\right] d\xi$$

$$= \frac{-iQ}{4(1 - M^2)^{\frac{1}{2}}}\exp\left[i\omega\left(\frac{t - x_1 U/c^2}{1 - M^2}\right)\right] \times H_0^{(2)}\left(\omega\frac{\{(Ut - x_1)^2 + (1 - M^2)x_3^2\}^{\frac{1}{2}}}{c(1 - M^2)}\right).$$

$\neq$Qu3. From equation (9.7)

$$p'(\mathbf{x}, t) = \frac{m}{4\pi r|1 - M_r|},$$

where for large $|\mathbf{x}|$ and $x_2 = 0$, $c(t - \tau^*) \approx |\mathbf{x}| - (ax_3/|\mathbf{x}|)\sin\omega\tau^*$, $(1/r) \approx |\mathbf{x}|^{-1}$ and $1 - M_r \approx 1 - \omega a\cos\omega\tau^*(x_3/c|\mathbf{x}|)$. Hence

$$p'(\mathbf{x}, t) = \frac{m}{4\pi|\mathbf{x}|[1 - \omega a\cos\omega\tau^* x_3/c|\mathbf{x}|]}.$$

Writing $\theta = \omega\tau^*$, we have

$$\theta = \omega\left(t - \frac{|\mathbf{x}|}{c}\right) + \omega\frac{ax_3}{c|\mathbf{x}|}\sin\theta$$

with

$$\frac{\mathrm{d}\theta}{\mathrm{d}t} = \frac{\omega}{1 - \omega a \cos \theta x_3/c|\mathbf{x}|}.$$

Hence

$$p'(\mathbf{x}, t) = \frac{m}{4\pi\omega|\mathbf{x}|} \frac{\mathrm{d}\theta}{\mathrm{d}t}.$$

≠Qu4. From equation (8.19)

$$\begin{aligned}
p'(\mathbf{x}, t) &= \frac{\partial}{\partial t} \int_S \left[\frac{\rho_0}{2\pi r} \frac{\partial \xi}{\partial \tau}\bigg|_{\mathbf{y}} \right] \mathrm{d}y_1\, \mathrm{d}y_2 \\
&= \frac{\partial}{\partial t} \int \frac{\rho_0}{2\pi r} \frac{\partial \xi}{\partial \tau}\bigg|_{\mathbf{y}} \delta(t - \tau - |\mathbf{x} - \mathbf{y}|/c)\, \mathrm{d}y_1\, \mathrm{d}y_2\, \mathrm{d}\tau \\
&= -\frac{\partial}{\partial t} \int \frac{\rho_0}{2\pi r} \xi \frac{\partial}{\partial \tau} \delta(t - \tau - |\mathbf{x} - \mathbf{y}|/c)\, \mathrm{d}y_1\, \mathrm{d}y_2\, \mathrm{d}\tau \\
&= \frac{\partial^2}{\partial t^2} \int \frac{\rho_0}{2\pi r} \xi\, \delta(t - \tau - |\mathbf{x} - \mathbf{y}|/c)\, \mathrm{d}y_1\, \mathrm{d}y_2\, \mathrm{d}\tau.
\end{aligned}$$

Rewriting this integral in terms of $\boldsymbol{\eta} = \mathbf{y} + \mathbf{U}\tau$ gives

$$p'(\mathbf{x}, t) = \frac{\partial^2}{\partial t^2} \int \frac{\rho_0 \xi_0\, \mathrm{e}^{i\omega\tau^*}}{2\pi r|1 + M_r|} \cos\left(\frac{\pi\sigma}{2a}\right) \mathrm{d}\eta_1\, \mathrm{d}\eta_2,$$

with $\sigma = (\eta_1^2 + \eta_2^2)^{\frac{1}{2}}$. For a compact source and an observer in the far-field this becomes

$$p'(\mathbf{x}, t) = \frac{\partial^2}{\partial t^2} \left[\frac{\rho_0 \xi_0\, \mathrm{e}^{i\omega\tau_0^*}}{2\pi|\mathbf{x}|(1 + M \cos \theta)}\, 2\pi \int_0^a \cos\left(\frac{\pi\sigma}{2a}\right) \sigma\, \mathrm{d}\sigma \right]$$

where $\cos \theta = x_1/|\mathbf{x}|$, and $\tau_0 = \dfrac{t - |\mathbf{x}|/c}{1 + M \cos \theta}$

$$\begin{aligned}
&= \frac{\partial^2}{\partial t^2} \left[\frac{(2\pi - 4)}{\pi^2} \frac{\rho_0 \xi_0 a^2\, \mathrm{e}^{i\omega\tau_0^*}}{|\mathbf{x}|(1 + M \cos \theta)} \right] \\
&= -\left(\frac{2\pi - 4}{\pi^2}\right) \frac{\rho_0 \xi_0 a^2 \omega^2}{|\mathbf{x}|(1 + M \cos \theta)^3} \exp i\omega\left[\frac{t - |\mathbf{x}|/c}{1 + M \cos \theta}\right].
\end{aligned}$$

SOLUTIONS TO EXERCISES FOR CHAPTER 10

≠Qu1. If the pressure perturbation produced by one speaker is $[p_s(t - r/c)/r]$, then the observer hears

$$p(t) = \frac{p_s(t - d/c)}{d} + \frac{p_s(t - \sqrt{2}d/c)}{\sqrt{2}d},$$

and

$$\hat{p}_{ob}(\omega) = \hat{p}_s(\omega)\left(\frac{e^{-i\omega d/c}}{d} + \frac{e^{-i\omega\sqrt{2}d/c}}{\sqrt{2}d}\right).$$

The spectrum of the sound heard by the observer,

$$\hat{P}_{ob}(\omega) = \left|\frac{e^{-i\omega d/c}}{d} + \frac{e^{-i\omega\sqrt{2}d/c}}{\sqrt{2}d}\right|^2 \hat{P}_s(\omega).$$

White noise has a uniform power spectrum and so

$$\hat{P}_{ob}(\omega) = C\left|\tfrac{3}{2} + \sqrt{2}\cos\left(\frac{\omega d}{c}(\sqrt{2}-1)\right)\right| \quad \text{where } C \text{ is a constant.}$$

≠Qu2. From equation (3.6)

$$\hat{p}_T(\omega) = \hat{p}_i(\omega)\frac{e^{i\omega l/c}}{\cos\left(\frac{\omega l}{c}\right) + \frac{i}{2}\left(\frac{A_2}{A_1} + \frac{A_1}{A_2}\right)\sin\left(\frac{\omega l}{c}\right)}$$

$$\hat{P}_T(\omega) = \hat{P}_I(\omega)\left|\frac{e^{i\omega l/c}}{\cos\left(\frac{\omega l}{c}\right) + \frac{i}{2}\left(\frac{A_2}{A_1} + \frac{A_1}{A_2}\right)\sin\left(\frac{\omega l}{c}\right)}\right|^2$$

$$\hat{P}_T(\omega) = \hat{P}_I(\omega)\frac{1}{1 + \frac{1}{4}\left(\frac{A_2}{A_1} - \frac{A_1}{A_2}\right)^2 \sin^2\left(\frac{\omega l}{c}\right)}.$$

≠Qu3. Near the bubble the pressure is

$$p_0 + p_i(t) + \frac{a}{r}p_s\left(t - \frac{r-a}{c}\right).$$

Since the bubble is isothermal,

$$p_i(t) + p_s(t) = -3p_0\frac{\delta a}{a}.$$

For disturbances of frequency ω, the linear momentum equation gives

$$-\rho_0\omega^2\,\delta a(\omega) = \left(\frac{1}{a} + \frac{i\omega}{c}\right)\hat{p}_s(\omega).$$

It follows from these three equations that

$$\hat{\delta a}(\omega) = \frac{\hat{p}_i(\omega)\left(1 + \frac{i\omega a}{c_0}\right)}{\left(\rho_0\omega^2 a - \frac{3p_0}{a}\right) - \frac{3i\omega p_0}{c}},$$

and hence

$$A(\omega) = \frac{P(\omega)}{\left(\dfrac{3p_0}{a} - \rho_0 a\omega^2\right)^2 + \dfrac{9\omega^2 p_0^2}{c^2}}, \quad \text{for } \left|\frac{\omega a}{c}\right| \ll 1.$$

$\neq$Qu4. Denote the sound pressure at a distance of 100 m away from the source in free space by $p(t)$. Then the pressure at the observer,

$$p_{ob} = \frac{100}{r} p(t - (r - 100)/c) + \frac{100}{r'} p(t - (r' - 100)/c)$$

$$= p(T) + p(T - \tau_0) \quad \text{to within } 1\%.$$

Where $T = t - 0.398/c$

$$\tau_0 = \frac{0.398}{340} = 1.17\ 10^{-3}.$$

The mean square pressure $= \overline{\{(p(T) + p(T - \tau_0)\}\{p(T) + p(T - \tau_0)\}}$

$$= 2\overline{p^2} + 2\overline{p(T)p(T + \tau_0)}.$$

By definition (i.e. equation 10.14)

$$\overline{p(T)p(T + \tau_0)} = \frac{1}{2\pi}\int_{-\infty}^{\infty} W(\omega)\, e^{i\omega\tau_0}\, d\omega$$

$$= \frac{\overline{p^2}}{10^3\sqrt{\pi}}\int_{-\infty}^{\infty} e^{-\omega^2 10^{-6} - i\omega\tau_0}\, d\omega$$

$$= \overline{p^2} \exp\left\{\frac{-\tau_0^2}{4 \times 10^{-6}}\right\} = \overline{p^2} \times 1.41$$

mean square pressure $= \overline{p^2}\{2 + 2 \times 1.41\} = 4.82\overline{p^2}$.

Autocorrelation at observer

$$P_0(\tau) = \overline{\{p(T) + p(T - \tau_0)\}\{p(T + \tau) + p(T - \tau_0 + \tau)\}}$$

$$= 2P(\tau) + P(\tau - \tau_0) + P(\tau + \tau_0)$$

where P is the autocorrelation at 100 m in free space.

$$\text{Spectrum} = \int_{-\infty}^{\infty} \{2P(\tau) + P(\tau - \tau_0) + P(\tau + \tau_0)\}\, e^{-i\omega\tau}\, d\tau$$

$$= 2W(\omega) + W(\omega)\, e^{-i\omega\tau_0} + W(\omega)^{i\omega\tau_0}$$

$$= W(\omega) 2[1 + \cos\omega\tau_0]$$

$$= \frac{4\pi\overline{p^2}}{10^3\sqrt{\pi}} \exp\left\{\frac{-\omega^2}{10^6}\right\}[1 + \cos(\omega \times 1.17\ 10^{-3})].$$

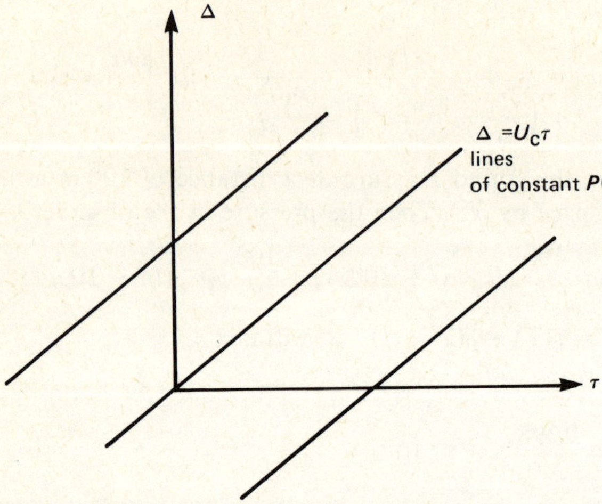

Fig. A.10.1 — The contours of the cross-correlation function for eddies conforming with Taylor's hypothesis.

≠Qu5. Taylor's hypothesis is that turbulent eddies just convect with speed U. Therefore the pressure (or any other flow variable has the form $p(x, t) = f(x - Ut)$.

$$P(\Delta, \tau) = \overline{f(x - Ut + \Delta - U\tau)f(x - Ut)} = F(\Delta - U\tau)$$

i.e. $P(\Delta, \tau)$ is const. along 'lines' $\Delta - U\tau = \text{const}$, and is maximal when $\Delta = U\tau$

$$\begin{aligned}\hat{P}(k,\omega) &= \iint P(\Delta, \tau)\, e^{-ik\Delta - i\omega\tau}\, d\Delta\, d\tau \\ &= \iint F(\Delta - U\tau)\, e^{-ik\Delta - i\omega\tau}\, d\Delta\, d\tau \\ &= \hat{F}(k) 2\pi\, \delta(\omega + Uk)\end{aligned}$$

where $\hat{F}(k) = \displaystyle\int F(\Delta - U\tau)\, e^{-ik(\Delta - U\tau)}\, d(\Delta - U\tau)$.

≠Qu6. $$\begin{aligned}V(\Delta, \tau) &= \overline{v(x, t)v(x + \Delta, t + \tau)} \\ &= \overline{\frac{\partial^2}{\partial x\, \partial t} p(x, t) \frac{\partial^2}{\partial \Delta\, \partial \tau} p(x + \Delta, t + \tau)}.\end{aligned}$$

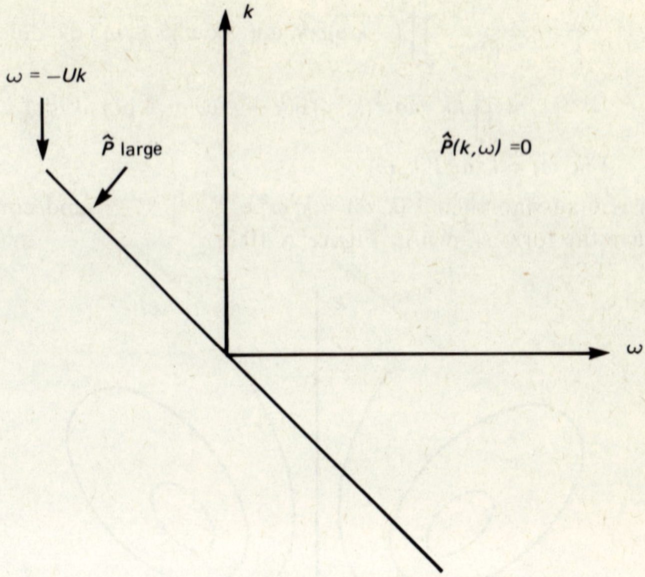

Fig. A.10.2 — The cross-spectral density for eddies conforming with Taylor's hypothesis.

Now

$$\overline{\frac{\partial}{\partial t}p(x,t)p(x+\Delta,t+\tau)} = \frac{1}{2T}\int_{-T}^{T}\frac{\partial p}{\partial t}(x,t)p(x+\Delta,t+\tau)\,\mathrm{d}t$$

$$= \frac{1}{2T}\left[p(x,t)p(x+\Delta,t+\tau)\right]$$

$$-\frac{1}{2T}\int_{-T}^{T}p(x,t)\frac{\partial}{\partial t}p(x+\Delta,t+\tau)\,\mathrm{d}\tau$$

$$= -\overline{p(x,t)\frac{\partial p}{\partial \tau}(x+\Delta,t+\tau)}.$$

The x-derivative can be treated similarly

$$V(\Delta,\tau) = \overline{p(x,t)\frac{\partial^4}{\partial\Delta^2\,\partial\tau^2}p(x+\Delta,t+\tau)}$$

$$V(\Delta,\tau) = \frac{\partial^4}{\partial\Delta^2\,\partial\tau^2}P(\Delta,\tau)$$

$$\hat{V}(k,\omega) = \frac{1}{(2\pi)^2}\iint\overline{\hat{v}(k,\omega)\hat{v}(k',\omega')}\,\mathrm{d}k'\,\mathrm{d}\omega'$$

$$= \frac{1}{(2\pi)^2} \iint \overline{(-k\omega)\hat{p}(k,\omega)(-k'\omega')\hat{p}(k',\omega')}\, \mathrm{d}k'\, \mathrm{d}\omega'$$

$$= \iint kk'\omega\omega' \hat{P}(k,\omega)\, \delta(k+k')\, \delta(\omega+\omega')\, \mathrm{d}k'\, \mathrm{d}\omega'$$

$$\hat{V}(k,\omega) = k^2\omega^2 \hat{P}(k,\omega).$$

If $\hat{P}(k,\omega)$ is Gaussian then $\hat{V}(k,\omega) = k^2\omega^2\, \mathrm{e}^{-(k^2/k_0^2)-(\omega^2/\omega_0^2)}$ and contours of $\hat{V}(k,\omega)$ have the form shown in Figure A 10.3.

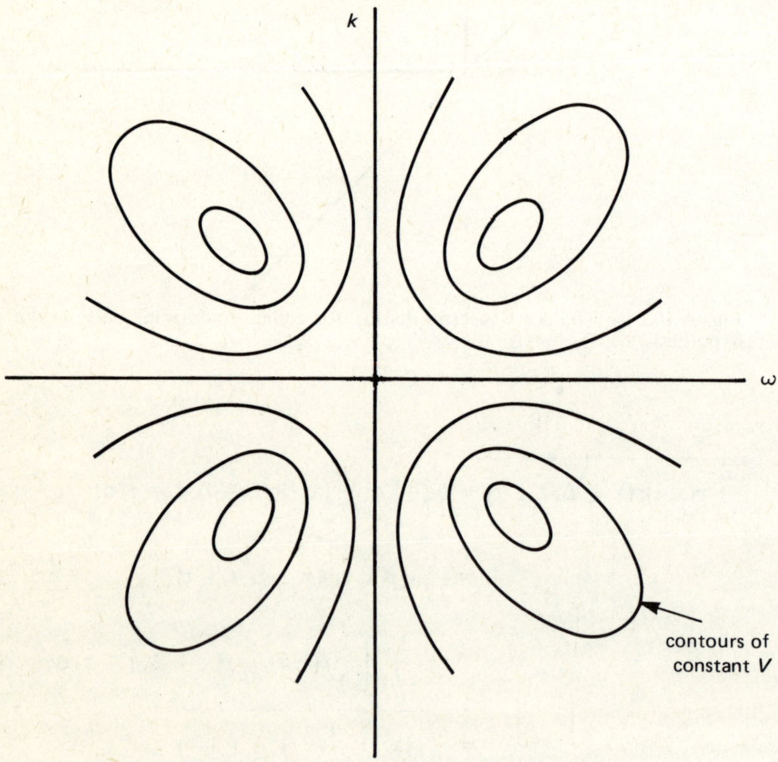

Fig. A.10.3 — Contours of the cross-power spectral density of v.

≠Qu7. $$\overline{f\frac{\partial f}{\partial t}} = \lim_{T\to\infty} \frac{1}{2T}\int_{-T}^{T} f\frac{\partial f}{\partial t}\,\mathrm{d}t = \lim_{T\to\infty}\frac{1}{4T}\,[f^2]_{-T}^{T} = 0$$

$$\overline{f\frac{\partial^2 f}{\partial t^2}} = \lim_{T\to\infty}\frac{1}{2T}\int_{-T}^{T} f\frac{\partial^2 f}{\partial t^2}\,\mathrm{d}t$$

$$= \lim_{T\to\infty}\left\{\frac{1}{2T}\left[f\frac{\partial f}{\partial t}\right]_{-T}^{T} - \frac{1}{2T}\int_{-\infty}^{\infty}\left(\frac{\partial f}{\partial t}\right)^2 \mathrm{d}t\right\} = -\overline{\left(\frac{\partial f}{\partial t}\right)^2}.$$

Now

$$p'(\mathbf{x}, t) = \frac{x_1}{r}\left\{\frac{1}{rc}\frac{\partial^2 f}{\partial t^2} + \frac{1}{r^2}\frac{\partial f}{\partial t}\right\}$$

$$u_i = -\frac{1}{\rho_0}\frac{\partial}{\partial x_i}\left\{\frac{x_1}{r}\left(\frac{1}{rc}\frac{\partial f}{\partial t} + \frac{1}{r^2}f\right)\right\}$$

$$= \frac{1}{\rho_0}\left[-\frac{\partial}{\partial x_i}\left(\frac{x_1}{r}\right)\left(\frac{1}{rc}\frac{\partial f}{\partial t} + \frac{1}{r^2}f\right)\right.$$

$$\left. + \frac{x_1 x_i}{r^2}\left(\frac{1}{rc^2}\frac{\partial^2 f}{\partial t^2} + \frac{2}{r^2 c}\frac{\partial f}{\partial t} + \frac{2f}{r^3}\right)\right].$$

Hence

$$\overline{p'u_i} = \frac{x_1^2 x_i}{\rho_0 r^3}\left[\frac{1}{r^2c^3}\overline{\left(\frac{\partial^2 f}{\partial t^2}\right)^2} + \frac{2}{r^4 c}\overline{\left(\frac{\partial f}{\partial t}\right)^2} + \frac{2}{r^4 c}\overline{f\frac{\partial^2 f}{\partial t^2}}\right]$$

$$= \frac{x_1^2 x_i}{\rho_0 r^5 c^3}\overline{\left(\frac{\partial^2 f}{\partial t^2}\right)^2} = \bar{I}_i,$$

and

$$\bar{\mathbf{I}} = \overline{\left(\frac{\partial^2 f}{\partial t^2}\right)^2}\,\frac{x_1^2\mathbf{x}}{\rho_0 c^3 r^5}.$$

$$\frac{\partial}{\partial x_i}\bar{I}_i = \overline{\left(\frac{\partial^2 f}{\partial t^2}\right)^2}\,\frac{1}{\rho_0 c^3}\frac{\partial}{\partial x_i}\left(\frac{x_1^2 x_i}{r^5}\right)$$

and

$$\frac{\partial}{\partial x_i}\left(\frac{x_1^2 x_i}{r^5}\right) = \frac{2x_1\,\delta_{i1}x_i + 3x_1^2}{r^5} - \frac{x_1^2 x_i 5x_i}{r^7}$$

$$= \frac{5x_1^2}{r^5} - \frac{5x_1^2}{r^5} = 0.$$

≠Qu8. $$\overline{\hat{p}(\omega)\hat{q}(\omega')} = \int_{-\infty}^{\infty}\int_{-\infty}^{\infty}\overline{p(t)q(t+\tau)}\,\mathrm{e}^{-i\omega t - i\omega'(t+\tau)}\,\mathrm{d}t\,\mathrm{d}\tau$$

$$= \int_{-\infty}^{\infty} C_{pq}(\tau)\,\mathrm{e}^{-i\omega'\tau}\,\mathrm{d}\tau\int_{-\infty}^{\infty}\mathrm{e}^{-i(\omega+\omega')t}\,\mathrm{d}t$$

$$= 2\pi\hat{C}_{pq}(\omega')\,\delta(\omega+\omega').$$

For

$$\hat{q}(\omega) = T(\omega)\hat{p}(\omega) + \hat{n}(\omega),$$

$$\hat{C}_{pq}(\omega') = \frac{1}{2\pi}\int_{-\infty}^{\infty}\left(T(\omega')\overline{\hat{p}(\omega)\hat{p}(\omega')} + \overline{\hat{p}(\omega)\hat{n}(\omega')}\right)\mathrm{d}\omega$$

$$= T(\omega')\hat{P}(\omega') \quad \text{since } \overline{\hat{p}(\omega)\hat{n}(\omega')} = 0,$$

and similarly

$$\hat{Q}(\omega') = \frac{1}{2\pi}\int_{-\infty}^{\infty} (\overline{T(\omega)T(\omega')\overline{\hat{p}(\omega)\hat{p}(\omega')} + \overline{\hat{n}(\omega)\hat{n}(\omega')}})\,\mathrm{d}\omega$$

$$= |T(\omega')|^2 \hat{P}(\omega') + \hat{N}(\omega').$$

Hence

$$\gamma^2 = \frac{|T(\omega)|^2 \hat{P}(\omega)}{|T(\omega)|^2 \hat{P}(\omega) + \hat{N}(\omega)} = \frac{\hat{S}(\omega)}{\hat{S}(\omega) + \hat{N}(\omega)}$$

and

$$\frac{\hat{S}(\omega)}{\hat{N}(\omega)} = \frac{\gamma^2}{1-\gamma^2}.$$

SOLUTIONS TO EXERCISES FOR CHAPTER 11

≠Qu1. The one-dimensional waves inside an organ pipe will be

$$p' = A\,\mathrm{e}^{i\omega(t - x/c)} + B\,\mathrm{e}^{-i\omega(t + x/c)}.$$

The background incident noise will be $p_i(t)$ at both ends of the pipe because the pipe lies parallel to the wave fronts.

As a first approximation we neglect the scattered field, as we did in the organ pipe calculation and set the interior pressure equal to the incident pressure at the open ends.

$$p_i(\omega)\,\mathrm{e}^{i\omega t} = (A + B)\,\mathrm{e}^{i\omega t} = (A\,\mathrm{e}^{-i\omega L/c} + B\,\mathrm{e}^{i\omega L/c})\,\mathrm{e}^{i\omega t}$$

i.e.,

$$B = \mathrm{e}^{-i\omega L/c} A$$

so that

$$p_i(\omega) = (1 + \mathrm{e}^{-i\omega L/c})A$$

giving the pressure inside the pipe as

$$p(\omega) = \frac{p_i(\omega)}{(1 + \mathrm{e}^{-i\omega L/c})}\{\mathrm{e}^{i\omega(t - x/c)} + \mathrm{e}^{-i\omega L/c}\,\mathrm{e}^{i\omega(t + x/c)}\}.$$

The spectrum is formed by multiplying this by its complex conjugate and averaging.

$$P(\omega) = \frac{P_i(\omega)}{\left(1 + \cos\frac{\omega L}{c}\right)^2 + \sin^2\frac{\omega L}{c}}$$

$$\times \{\mathrm{e}^{-i\omega x/c} + \mathrm{e}^{-i\omega L/c}\,\mathrm{e}^{i\omega x/c}\}\{\mathrm{e}^{i\omega x/c} + \mathrm{e}^{i\omega L/c}\,\mathrm{e}^{-i\omega x/c}\}$$

$$= \frac{P_i(\omega)}{2\left(1 + \cos\frac{\omega L}{c}\right)}\{1 + \mathrm{e}^{-i\omega L/c}\,\mathrm{e}^{i\omega 2x/c} + \mathrm{e}^{i\omega L/c}\,\mathrm{e}^{-i\omega 2x/c} + 1\}$$

$$= \frac{P_i(\omega)}{2\left(1 + \cos\dfrac{\omega L}{c}\right)}\left\{2 + 2\cos\frac{\omega L}{c}\cos 2\frac{\omega x}{c} + 2\sin\frac{\omega L}{c}\sin 2\frac{\omega x}{c}\right\}$$

$$= \frac{P_i(\omega)}{\left(1 + \cos\dfrac{\omega L}{c}\right)}\left(1 + \cos\frac{\omega}{c}(2x - L)\right).$$

The maximum pressure is at the centre of the tube where $x = L/2$.

$$P(\omega) = \frac{2P_i(\omega)}{\left(1 + \cos\dfrac{\omega L}{c}\right)}.$$

The pressure at the tube ends is to this approximation equal to the incident field, but we saw while discussing end corrections in Section 6.4 that there is actually a small difference between the pressure at the open end and the incident field. An organ pipe is effectively just a little bit longer than its physical length. The pressure at the end therefore has a little of the 'internal' pressure supplementing the incident field, and since the internal field can be very loud indeed near the even resonances where $[1 + \cos(\omega L/c)]$ vanishes, a loud harmonic note can be heard at the open end. This is the explanation of the commonly observed fact that resonators placed close to the ear seem to contain a musical sound; sea shells evoke the sound of the sea.

≠Qu2. The force on a sphere of volume V due to the disturbance it causes by moving relative to its ($c = \infty$) environment at speed $U(t)$ was determined in equation (2.43) to be a drag of magnitude $\frac{1}{2}\rho_0 V\dot{U}$. If the environment itself is accelerating there must be additionally a force equal to $\rho_0 V$ times that acceleration acting on every volume; this is the case in the wave field where particles are displaced a distance η from their equilibrium position. If the displacement of the sphere from its equilibrium position is ξ then the force F exerted on the sphere by the fluid is in total

$$F(t) = -\tfrac{1}{2}\rho_0 V(\ddot{\xi} - \ddot{\eta}) + \rho_0 V\ddot{\eta};$$

thus force must equal $M\ddot{\xi} + K\xi$.

The force on a tethered (i.e. $\xi = 0$) sphere is given as $F\cos\Omega t = \frac{3}{2}\rho_0 V\ddot{\eta}$ so that the force on the displaced sphere is

$$F(t) = F\cos\Omega t - \tfrac{1}{2}\rho_0 V\ddot{\xi} \tag{1}$$

$$= M\ddot{\xi} + K\xi$$

i.e.

$$(M + \tfrac{1}{2}\rho_0 V)\ddot{\xi} + K\xi = F\cos\Omega t. \tag{2}$$

The mass term is increased by the 'fluid loading' to $M + \frac{1}{2}\rho_0 V$. The solution to equation (2) such that $\dot{\xi} = 0$ at $t = 0$ and $t = 2\pi/\Omega$ is

$$\xi(t) = \frac{F \cos \Omega t}{K - \Omega^2(M + \frac{1}{2}\rho_0 V)}.$$

Equation (1) gives the force on the sphere as

$$F(t) = F \cos \Omega t \left\{ 1 + \frac{\frac{1}{2}\rho_0 V \Omega^2}{K - \Omega^2(M + \frac{1}{2}\rho_0 V)} \right\}$$

$$= \frac{K - \Omega^2 M}{K - \Omega^2(M + \frac{1}{2}\rho_0 V)} F \cos \Omega t.$$

This is a maximum (infinite) at the fluid loaded resonance frequency

$$\{K/(M + \tfrac{1}{2}\rho_0 V)\}^{\frac{1}{2}}$$

and is a minimum (zero) at the 'in vacuo' resonance frequency $\sqrt{(K/M)}$.

$\neq$Qu3. Surface disturbances on deep water disturbed from rest conform with a potential field $\nabla^2 \varphi = 0$. Therefore at wave number $\mathbf{k}$ in the plane of the surface

$\varphi = e^{kz} e^{i\mathbf{k}\cdot\mathbf{x}}$ is the solution that vanishes at large depth; $k = |\mathbf{k}|$.

The surface displaced vertically upwards an amount η generates at the $z = 0$ position of the undisturbed surface a hydrostatic pressure increase equal to $\rho g \eta$. In unsteady flow the pressure perturbation $p_s + \rho g \eta$ must be equal to $-\rho_0(\partial\varphi/\partial t) = -\rho i\omega\varphi$ for a single Fourier component at frequency ω. Therefore

$$p_s + \rho g \eta = -i\omega\rho\varphi. \tag{1}$$

The normal velocity at the surface is

$$\frac{\partial \eta}{\partial t} = i\omega\eta = \frac{\partial \varphi}{\partial z} = k\varphi. \tag{2}$$

Equations (1) and (2) thus relate the surface pressure p' to the vertically upwards surface displacement η.

$$p_s + \rho g \eta = -i\omega\rho\left(\frac{i\omega}{k}\right)\eta; \quad p_s = \rho\left(\frac{\omega^2}{k} - g\right)\eta = \rho\left(g - \frac{\omega^2}{k}\right)(-\eta)$$

so that the effective surface impedance is $[g - \omega^2/k]\rho/(i\omega)'$.

The surface response (measured positive upward) Fourier element at wave number **k** and frequency ω is therefore

$$\hat{\eta}(\mathbf{k}, \omega) = \frac{\hat{p}(\mathbf{k}, \omega)}{\rho\left(\dfrac{\omega^2}{k} - g\right)}.$$

This response is unbounded at the free wave condition $\omega^2 = gk$. Surface waves satisfying this dispersion equation will therefore grow with time as the surface absorbs energy from the wind.

≠Qu4. In section 4.5 the ratio of surface pressure to the normal velocity at the plane boundary of homogeneous fluid was found to be such that

$$u = \frac{\gamma p}{i\omega\rho_1} \quad \text{where } \gamma = \frac{\omega}{c_1}\sqrt{\left(\frac{c_1^2}{U^2} - 1\right)} \tag{1}$$

where ρ_1, c_1 are respectively the mean density and speed of sound in the medium driven by the surface motion and U is the phase speed of waves in the plane of the boundary. For a particular harmonic component at surface wave number **k** and frequency ω.

$$U = \frac{\omega}{k}, \quad k = |\mathbf{k}|,$$

so that

$$\gamma = \sqrt{(k^2 - \omega^2/c_1^2)} \quad \text{provided that } k^2 > \omega^2/c_1^2.$$

In that case equation (1) gives

$$p = \frac{i\rho_1 c_1 u}{\sqrt{\left(\left(\dfrac{kc_1}{\omega}\right)^2 - 1\right)}}; \quad \left(\frac{kc_1}{\omega}\right)^2 > 1 \tag{2}$$

and,

$$p = \frac{\rho_1 c_1 u}{\sqrt{\left(1 - \left(\dfrac{kc_1}{\omega}\right)^2\right)}} \text{ is the appropriate form when } \left(\frac{kc_1}{\omega}\right)^2 < 1. \tag{3}$$

The normal velocity is $u = -i\omega\eta$ (when η is measured positive out of the fluid) so that the fluid loading for a compressible fluid is given by equations (2) and (3) the forms being appropriate according to whether the phase speed of the wave along the surface is either less than or greater than the speed of sound.

The fluid loading term is therefore

$$L = \frac{\omega\rho_1 c_1}{\sqrt{\left(\left(\dfrac{kc_1}{\omega}\right)^2 - 1\right)}}; \quad \left(\frac{kc_1}{\omega}\right)^2 > 1$$

$$= \frac{-i\omega\rho_1 c_1}{\sqrt{\left(1-\left(\frac{kc_1}{\omega}\right)^2\right)}}; \quad \left(\frac{kc_1}{\omega}\right)^2 < 1.$$

$\neq$Qu5. According to equation (11.72)

$$-\frac{\omega}{k} = C = \frac{\rho U}{\rho + m|k|} \pm i\frac{\rho U\sqrt{(m|k|/\rho)}}{\rho + m|k|} = C_R + iC_i \tag{1}$$

so that the instability wave is proportioned to

$$e^{ik(x - C_R t)}\, e^{kC_I t}. \tag{2}$$

According to equation (11.67) the harmonic wave fluid loading pressure on a surface displaced by η from its equilibrium position is for $k > 0$,

$$\hat{p} = \frac{-\rho(\omega + Uk)^2}{|k|}\hat{\eta} = -\rho(-iC_i - C_R + U)^2 k\hat{\eta}$$
$$= -\rho((U - C_R)^2 - C_I^2 + 2i(C_R - U)C_I)k\hat{\eta}$$

in the case of an instability wave. Written in terms of real quantities this is

$$-p(x, t) = \rho((U - C_R)^2 - C_I^2)k\eta(\mathbf{x}, t) + 2\rho(C_R - U)C_I\frac{\partial\eta}{\partial x}(x, t). \tag{3}$$

To be specific, let

$$\eta(x, t) = \cos k(x - C_R t)\, e^{kC_I t}$$

$$\frac{\partial\eta}{\partial x}(x, t) = -k\sin k(x - C_R t)\, e^{kC_I t}$$

$$\frac{\partial\eta}{\partial t}(x, t) = kC_I\cos k(x - C_R t)\, e^{kC_I t} + kC_R\sin k(x - C_R t)\, e^{kC_I t}$$

so that

$$p(x, t) = -\{\rho((U - C_R)^2 - C_I^2)k\cos k(x - C_R t)$$
$$-2k\rho C_I(C_R - U)\sin k(x - C_R t)\}\, e^{kC_I t}.$$

The power flow out of the fluid, equation (11.45), is

$$P = -\overline{p\frac{\partial\eta}{\partial t}}$$
$$= \tfrac{1}{2}\rho\, e^{2kC_I t}k^2\{C_I((U - C_R)^2 - C_I^2) - 2C_R C_I(C_R - U)\}$$
$$= \tfrac{1}{2}\rho k^2\, e^{2kC_I t}C_I\{U^2 + C_R^2 - 2C_R U - C_I^2 - 2C_R^2 + 2C_R U\}$$

i.e.

$$P = \tfrac{1}{2}\rho k^2\, e^{2kC_I t} C_I\{U^2 - C_R^2 - C_I^2\}. \tag{4}$$

$$C_R^2 + C_I^2 = \frac{\rho U^2}{\rho + mk}$$

so that

$$P = \frac{1}{2}\frac{\rho m k^3 U^2}{\rho + mk} C_I\, e^{2kC_I t}. \tag{5}$$

The power flow out of the fluid into the limp surface is positive if C_I is positive and the instability wave is growing but negative if it is decaying. The flow provides the power required to energise the limp surface into a state of increasing vibrational amplitude.

The drag force per unit surface area between the surface and the fluid is

$$D = \overline{p\frac{\partial \eta}{\partial x}} = -\rho k^2 C_I(C_R - U)\, e^{2kC_I t} = \frac{\rho m k^3 U}{\rho + mk} C_I\, e^{2kC_I t}. \tag{6}$$

The energy density in the surface $= e_s = \tfrac{1}{2}m\overline{(\partial\eta/\partial t)^2} = \tfrac{1}{4}mk^2(C_I^2 + C_R^2)\, e^{2kC_I t}$, i.e.

$$e_s = \frac{1}{4}\frac{\rho m k^2 U^2}{\rho + mk} e^{2kC_I t} \tag{7}$$

The energy density/unit surface area in the fluid adjacent to that area is

$$e_f = \int_0^\infty \tfrac{1}{2}\rho\overline{v^2}\, dz = \int_0^\infty \rho\overline{v_z^2}\, dz = \int_0^\infty \rho\overline{v_z^2}|_{z=0}\, e^{-2kz}\, dz = \left.\frac{\rho\overline{v_z^2}}{2k}\right|_{z=0}. \tag{8}$$

The motion decays away from the surface exponentially and in a wave energised in incompressible flow by a plane surface, $\overline{v_x^2} = \overline{v_z^2}$.

$$e_f = \frac{\rho}{2k}\overline{\left(\frac{\partial\eta}{\partial t} + U\frac{\partial\eta}{\partial x}\right)^2}$$

$$= \frac{\rho k^2}{2k}\overline{(C_I \cos k(x - C_R t) + (C_R - U)\sin k(x - C_R t))^2}\, e^{2kC_I t}$$

$$= \frac{\rho}{4}k(C_I^2 + C_R^2 + U^2 - 2C_R U)\, e^{2kC_I t}$$

i.e.

$$e_f = \frac{1}{4}\frac{\rho m k^2 U^2}{\rho + mk} e^{2kC_I t}. \tag{9}$$

There is evidently an equipartition of disturbance kinetic energy between the surface material and the fluid.

The rate of increase of the surface energy is

$$\frac{\partial}{\partial t} e_s = \frac{1}{2} \frac{\rho m k^3 U^2 C_I}{\rho + mk} e^{2kC_I t}$$

which we have already determined to be P from equation (5). The total energy requirement from the mean flow to energise both the surface and the fluid disturbance is twice this, a requirement that is met by the mean flow working against the surface drag as can be checked from equation (6).

$$\frac{\partial}{\partial t}(e_s + e_f) = \frac{\rho m k^3 U^2 C_I}{\rho + mk} e^{2kC_I t} = UD. \tag{10}$$

Appendix: Useful Data and Definitions

PHYSICAL PROPERTIES

Atmospheric pressure at 15°C and 1 atmosphere = 1 bar = 10^5 N/m^2.

Gravitational acceleration	9.81 m/s^2
Speed of sound in air	340 m/s
Speed of sound in water	1465 m/s
Density of air	1.2 kg/m^3
Density of water	10^3 kg/m^3
Impedance of air $\rho_0 c$	408 kg/m^2 s
Impedance of water	1.456×10^6 kg/m^2 s

UNITS OF SOUND MEASUREMENT

$$\text{SPL (sound pressure level)} = 20 \log_{10}\left(\frac{p'_{\text{rms}}}{2 \times 10^{-5}\ \text{N/m}^2}\right) \text{dB}.$$

$$\text{IL (intensity level)} = 10 \log_{10}\left(\frac{\text{intensity}}{10^{-12}\ \text{watts/m}^2}\right) \text{dB}.$$

$$\text{PWL (power level)} = 10 \log_{10}\left(\frac{\text{sound power output}}{10^{-12}\ \text{watts}}\right) \text{dB}.$$

PBL (pressure band level) = sound pressure level in dB in a specified frequency band.

PSL (pressure spectrum level) = sound pressure level in dB in a frequency band of width 1 Hz.

IBL (intensity band level) = sound intensity level in dB in a specified frequency band.

ISL (intensity spectrum level) = sound intensity level in dB in a frequency band of width 1 Hz.

L_{eq} = average ***A***-weighted sound level over a sufficiently long sample of noise.

Loudness level: a tone has a loudness level of N phons if it is judged by the ear to be as loud as a pure tone of frequency 1 kHz at an intensity level of N dB.

Loudness: a tone has a loudness of N sones if it is judged by the ear to be N times as loud as a sound whose loudness level is 40 phons.

DEFINITIONS

Surface impedance, Z_s, relates the pressure perturbation applied to a surface, p', to its normal velocity v_n; $p' = Z_s v_n$.

Normal impedance of a fluid $\rho_0 c$.

Specific impedance of a surface $Z_s/\rho_0 c$.

Transmission loss $= 10 \log_{10}\left(\dfrac{\text{incident sound power}}{\text{transmitted sound power}}\right)$.

Absorption coefficient of a sound absorber $= \dfrac{\text{sound power absorbed}}{\text{incident sound power}}$.

Sound absorption (in metric sabins) $= \sum \alpha_i S_i$ where S_i is surface area (in metres^2) with absorption coefficient α_i.

Reverberation time of a room = time taken for the sound intensity level in the room to drop from 60 dB to the threshold of hearing.

Wavelength, λ, for sound waves with angular frequency ω

$$\lambda = \frac{2\pi c}{\omega}.$$

Wavenumber $k = 2\pi/\lambda$.

Phase speed $= \omega/k$ *Group velocity* $= \dfrac{\partial \omega}{\partial k}$.

Helmholtz number (or *compactness ratio*) $= D/\lambda$ where D is a typical dimension of the source.

Strouhal number $= \omega D/2\pi U$ for sound of radian frequency ω, produced in a flow with speed U, length scale D.

An octave is a frequency interval over which the frequency doubles.

A 1/3-octave is the frequency interval between two frequencies having the ratio $2^{1/3}$:1.

Heaviside function:

$$H(t-\tau) = 1 \quad t > \tau$$
$$= 0 \quad t < \tau.$$

Sgn function

$$\operatorname{sgn}(t-\tau) = 1 \quad t > \tau$$
$$= -1 \quad t < \tau$$
$$\operatorname{sgn}(t) = 2H(t) - 1.$$

Modulus

$$|t-\tau| = (t-\tau) \qquad t > \tau$$
$$= -(t-\tau) \quad t < \tau$$
$$|t| = t\,\operatorname{sgn}(t) = 2tH(t) - t$$
$$\frac{\mathrm{d}H}{\mathrm{d}t} = \delta(t)$$
$$\frac{\mathrm{d}|t|}{\mathrm{d}t} = \operatorname{sgn}(t)$$
$$\frac{\mathrm{d}\,\operatorname{sgn}(t)}{\mathrm{d}t} = 2\,\delta(t).$$

Impulse or δ-function

In one dimension: $\delta(t) = 0$ for $t \neq 0$

$$f(t)\,\delta(t-\tau) = f(\tau)\,\delta(t-\tau)$$

and $\int_{-\infty}^{\infty} \delta(t-\tau) f(t)\,\mathrm{d}t = f(\tau)$ for any function $f(t)$.

In three-space dimensions:

$$\delta(\mathbf{x}) = \delta(x_1)\,\delta(x_2)\,\delta(x_3)$$
$$\delta(\mathbf{x}) = 0 \quad \text{for } |\mathbf{x}| \neq 0$$
$$f(\mathbf{x})\,\delta(\mathbf{x}-\mathbf{y}) = f(\mathbf{y})\,\delta(\mathbf{x}-\mathbf{y})$$

and $\int_{\infty} \delta(\mathbf{x}-\mathbf{y}) f(\mathbf{x})\,\mathrm{d}^3\mathbf{x} = f(\mathbf{y})$ for any function $f(\mathbf{x})$.

BASIC EQUATIONS OF LINEAR ACOUSTICS

Conservation of mass $\dfrac{\partial \rho}{\partial t} + \rho_0 \operatorname{div} \mathbf{v} = 0.$

Conservation of momentum $\rho_0 \dfrac{\partial \mathbf{v}}{\partial t} + \text{grad}\, p' = 0$

$$c^2 = \frac{\mathrm{d}p}{\mathrm{d}\rho}.$$

These equations combine to give the *wave equation*

$$\frac{1}{c^2}\frac{\partial^2 p'}{\partial t^2} - \nabla^2 p' = 0.$$

Energy density $e = \frac{1}{2}\rho_0 v^2 + \frac{1}{2}p'^2/\rho_0 c^2$.

Mean intensity $\mathbf{I} = \overline{p'\mathbf{v}}$

$\text{div}\,\mathbf{I} = 0$ for statistically stationary (in time) sound fields.

Waves and sources

p' is a wave field when

$$\frac{1}{c^2}\frac{\partial^2 p'}{\partial t^2} - \nabla^2 p' = 0.$$

The source field q is defined when this equation is not true, i.e.

$$\frac{1}{c^2}\frac{\partial^2 p'}{\partial t^2} - \nabla^2 p' = q.$$

For example if volume is added at the point $\mathbf{x}$ at a rate $\beta(\mathbf{x}, t)$ per unit volume and a force $f(\mathbf{x}, t)$ is applied per unit volume at $\mathbf{x}$ then

$$q = \rho_0 \frac{\partial \beta}{\partial t} - \text{div}\,\mathbf{f}.$$

SIMPLE WAVE FIELDS

One-dimensional or plane wave

The general solution of the 1-D wave equation is

$$p'(x_1, t) = f(x_1 - ct) + g(x_1 + ct)$$

where f and g are arbitrary functions. In a plane wave propagating to the right $p' = \rho_0 cu$. In a plane wave propagating to the left $p' = -\rho_0 cu$, u being the particle velocity. The one-dimensional outward propagating response to an impulse satisfies

$$\left\{\frac{1}{c^2}\frac{\partial^2}{\partial t^2} - \frac{\partial^2}{\partial x_1^2}\right\} p'(x_1, t) = \delta(x_1)\,\delta(t)$$

and is

$$p'(x_1, t) = \frac{c}{2} \quad \text{for } ct > |x_1|$$

$$= 0 \quad \text{for } ct < |x_1|.$$

Two-dimensional waves

The two-dimensional outward propagating response to an impulse satisfies

$$\left\{\frac{1}{c^2}\frac{\partial^2}{\partial t^2} - \left(\frac{\partial^2}{\partial x_1^2} + \frac{\partial^2}{\partial x_2^2}\right)\right\} p'(x_1, x_2, t) = \delta(x_1)\,\delta(x_2)\,\delta(t)$$

and is

$$p'(x_1, x_2, t) = \frac{1}{2\pi}\frac{1}{[t^2 - (x_1^2 + x_2^2)/c^2]^{\frac{1}{2}}} \quad \text{if } ct > (x_1^2 + x_2^2)^{\frac{1}{2}}$$

$$= 0 \quad \text{if } ct < (x_1^2 + x_2^2)^{\frac{1}{2}}.$$

Three-dimensional waves

The three-dimensional outward propagating response to an impulse satisfies

$$\left\{\frac{1}{c^2}\frac{\partial^2}{\partial t^2} - \nabla^2\right\} p'(\mathbf{x}, t) = \delta(\mathbf{x})\,\delta(t)$$

and is

$$p'(\mathbf{x}, t) = \frac{\delta(t - |\mathbf{x}|/c)}{4\pi|\mathbf{x}|}.$$

A point-monopole field is the sound field produced by a point source at the origin. The sound field is a function of $|\mathbf{x}|$ and time only and

$$p'(\mathbf{x}, t) = \frac{\rho_0}{4\pi r}\frac{\mathrm{d}q(t - r/c)}{\mathrm{d}t}$$

where $q(t)$ = rate of volume outflow from the source, and $r = |\mathbf{x}|$. The field at radius r is therefore simply related to the field at r_0:

$$p'(r, t) = \frac{r_0}{r}\, p'\left(r_0, t - \frac{(r - r_0)}{c}\right).$$

A point dipole field is

$$p'(\mathbf{x}, t) = -\frac{1}{4\pi}\frac{\partial}{\partial x_i}\left[\frac{F_i(t - r/c)}{r}\right]$$

where F_i is the dipole strength and $r = |\mathbf{x}|$

MORE COMPLICATED THREE-DIMENSIONAL WAVE FIELDS

The outward propagating solution of

$$\left\{\frac{1}{c^2}\frac{\partial^2}{\partial t^2} - \nabla^2\right\}p'(\mathbf{x}, t) = q(\mathbf{x}, t)$$

is

$$p'(\mathbf{x}, t) = \int \frac{q(\mathbf{y}, t - |\mathbf{x} - \mathbf{y}|/c)}{4\pi|\mathbf{x} - \mathbf{y}|}\, d^3\mathbf{y}.$$

A source distribution is said to degenerate into a *dipole* distribution if the source has the form $-\mathrm{div}\,\mathbf{f}$ for some function $\mathbf{f}$, then

$$\left\{\frac{1}{c^2}\frac{\partial^2}{\partial t^2} - \nabla^2\right\}p'(\mathbf{x}, t) = -\mathrm{div}\,\mathbf{f} = -\frac{\partial f_i}{\partial x_i}$$

and the solution is

$$p'(\mathbf{x}, t) = -\frac{\partial}{\partial x_i}\int \frac{f_i(\mathbf{y}, t - |\mathbf{x} - \mathbf{y}|/c)}{4\pi|\mathbf{x} - \mathbf{y}|}\, d^3\mathbf{y}.$$

A source distribution is said to degenerate into a *quadrupole* distribution if the source has the form $(\partial^2 T_{ij}/\partial x_i\,\partial x_j)$ for some function T_{ij}. Then

$$\left\{\frac{1}{c^2}\frac{\partial^2}{\partial t^2} - \nabla^2\right\}p'(\mathbf{x}, t) = \frac{\partial^2 T_{ij}}{\partial x_i\,\partial x_j}$$

and the solution is

$$p'(\mathbf{x}, t) = \frac{\partial^2}{\partial x_i\,\partial x_j}\int \frac{T_{ij}(\mathbf{y}, t - |\mathbf{x} - \mathbf{y}|/c)}{4\pi|\mathbf{x} - \mathbf{y}|}\, d^3\mathbf{y}.$$

Lighthill's acoustic analogy shows that jet noise is generated by quadrupoles with strength T_{ij}

$$T_{ij} = \rho v_i v_j + p_{ij} - c^2\rho'\,\delta_{ij}$$

where $p_{ij} = p'\,\delta_{ij}$ + viscous terms. Then

$$\left\{\frac{\partial^2}{\partial t^2} - c^2\nabla^2\right\}\rho'(\mathbf{x}, t) = \frac{\partial^2 T_{ij}}{\partial x_i\,\partial x_j}$$

and the sound field is given by

$$\rho'(\mathbf{x}, t) = \frac{\partial^2}{\partial x_i\,\partial x_j}\int \frac{T_{ij}(\mathbf{y}, t - |\mathbf{x} - \mathbf{y}|/c)}{4\pi c^2|\mathbf{x} - \mathbf{y}|}\, d^3\mathbf{y}.$$

Exact solution of Lighthill's equations in a region V, exterior to a stationary control surface S,

$$4\pi c^2 \rho'(\mathbf{x}, t) = \frac{\partial^2}{\partial x_i\, \partial x_j} \int_V \left[\frac{T_{ij}}{r}\right] d^3\mathbf{y} - \frac{\partial}{\partial x_j} \int_S n_i \left[\frac{\rho v_i v_j + p_{ij}}{r}\right] dS + \frac{\partial}{\partial t} \int_S \left[\frac{\rho \mathbf{v}\cdot\mathbf{n}}{r}\right] dS$$

where $r = |\mathbf{x} - \mathbf{y}|$ and the functions within the square brackets are to be evaluated at retarded time $t - |\mathbf{x} - \mathbf{y}|/c$. Quadrupole sources are distributed

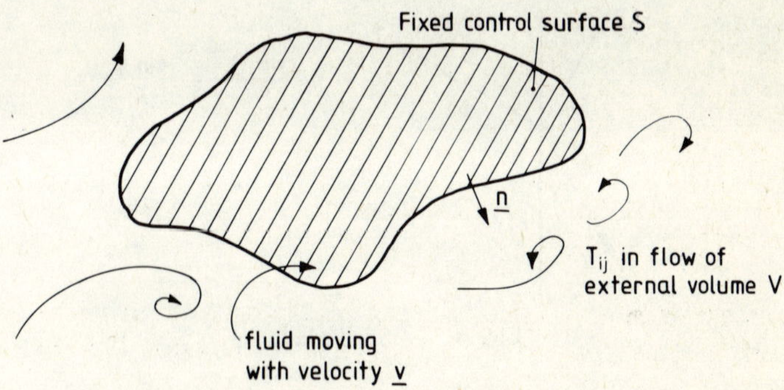

Fig. A.1 — Lighthill source geometry.

throughout fluid. Dipole sources are equivalent to body forces and momentum supply. Monopole sources seem to be equivalent to mass supply.

Exact solution of Lighthills equations in a region V, exterior to an impenetrable moving surface S

$$4\pi c^2 H \rho'(\mathbf{x}, t) = \frac{\partial^2}{\partial x_i\, \partial x_j} \int_V \left[\frac{J T_{ij}}{r|1 - M_r|}\right] d^3\boldsymbol{\eta} - \frac{\partial}{\partial x_i} \int_S \left[\frac{p_{ij} n_j K}{r|1 - M_r|}\right] dS(\boldsymbol{\eta}) + \frac{\partial}{\partial t} \int_S \left[\frac{\rho_0 \boldsymbol{v}.\boldsymbol{\eta} K}{r|1 - M_r|}\right] dS(\boldsymbol{\eta})$$

where H is unity in the region V and zero elsewhere. $\boldsymbol{\eta}$ is a co-ordinate system moving with velocity $\mathbf{u}(\mathbf{y}, t)$ in which the surface is at rest. M_r is the component of $\mathbf{u}/c$ in the direction of the observer, J is the ratio of volume elements in the fixed and moving co-ordinate systems, i.e. $d^3\mathbf{y} = J\, d^3\boldsymbol{\eta}$, and $\boldsymbol{K}$ is the ratio of area elements of the surface S in $\mathbf{y}$ and $\boldsymbol{\eta}$ spaces. If the moving co-ordinates are in rigid motion $J = K = 1$, while if the co-ordinates move with the material particles $J = \rho^*(\eta)/\rho$ where ρ^*, a reference density, is the density at $\boldsymbol{\eta}$ at the time when the fixed and moving co-ordinates coincide. r is again the distance between the observer and the source element at emission time and the square brackets denote that the function they enclose is to be evaluated at retarded time. Monopole sources are now seen to be due to volume and not mass supply at the surface.

SNELL'S LAW FOR OBLIQUE PLANE WAVES

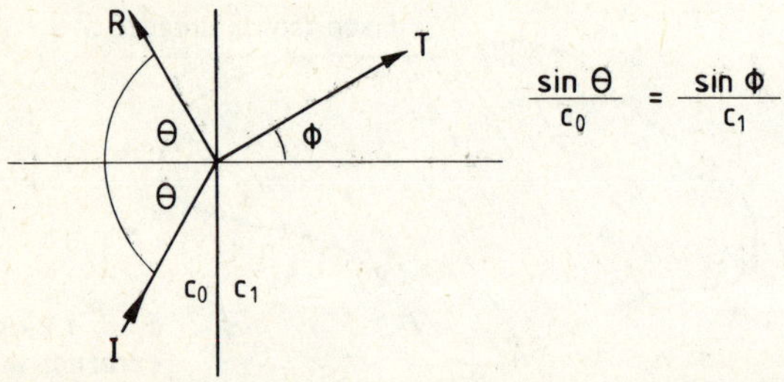

Fig. A.2 — Snell's law.

$$\frac{\sin\theta}{c_0} = \frac{\sin\varphi}{c_1}$$

Ray theory: for high frequency (short wavelength) waves in a stratified medium where ρ_0 and c vary with x

$$\frac{\sin\theta}{c} = \text{constant along a ray}$$

$$\frac{\overline{p'^2 A}}{\rho_0 c} = \text{constant along a ray tube.}$$

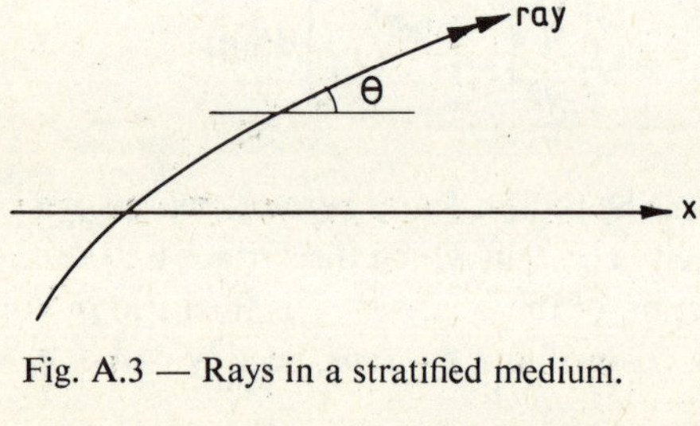

Fig. A.3 — Rays in a stratified medium.

DIFFUSE (OR REVERBERANT) SOUND FIELDS

The rate at which sound energy falls on one side of unit surface area, i.e. the intensity,

$$I = \tfrac{1}{4}ce.$$

Reverberation time of a room $T = 55.3(V/ac)$, where V is the volume of the room and a is the total sound absorption in the room.

RECIPROCITY

If $\varphi_1(\mathbf{x})$ is the source field at $\mathbf{x}$ due to a harmonic point source at $\mathbf{x}_1$ and $\varphi_2(\mathbf{x})$ is the field at $\mathbf{x}$ due to a harmonic point source at $\mathbf{x}_2$, i.e.

$$\frac{\partial^2 \varphi_1}{\partial t^2} - c^2\nabla^2\varphi_1 = \delta(\mathbf{x} - \mathbf{x}_1)\,\mathrm{e}^{i\omega t}$$

and

$$\frac{\partial^2 \varphi_2}{\partial t^2} - c^2\nabla^2\varphi_2 = \delta(\mathbf{x} - \mathbf{x}_2)\,\mathrm{e}^{i\omega t}$$

then $\varphi_1(\mathbf{x}_2) = \varphi_2(\mathbf{x}_1)$.

FOURIER TRANSFORMS

In one dimension, $\hat{p}(\omega)$ is the transform of $p(t)$ where

$$p(t) = \frac{1}{2\pi}\int_{-\infty}^{\infty} \hat{p}(\omega)\,\mathrm{e}^{i\omega t}\,\mathrm{d}\omega \quad \text{and} \quad \hat{p}(\omega) = \int_{-\infty}^{\infty} p(t)\,\mathrm{e}^{-i\omega t}\,\mathrm{d}t.$$

Some useful Fourier transforms:

$p(t)$	$\hat{p}(\omega)$
$\delta(t - \tau)$	$e^{-i\omega\tau}$
$\mathrm{e}^{i\omega_0 t}$	$2\pi\,\delta(\omega - \omega_0)$
$\cos\omega_0 t$	$\pi[\delta(\omega - \omega_0) + \delta(\omega + \omega_0)]$
$\sin\omega_0 t$	$i\pi[-\delta(\omega - \omega_0) + \delta(\omega + \omega_0)]$
$\mathrm{e}^{-\omega_0^2 t^2}$	$\dfrac{\sqrt{\pi}}{\omega_0}\,\mathrm{e}^{-\omega^2/4\omega_0^2}$
$\mathrm{e}^{-\omega_0\lvert t\rvert}$	$\dfrac{2\omega_0}{\omega_0^2 + \omega^2}$

$p(t)$	$\hat{p}(\omega)$
$\dfrac{d^n}{dt^n}\,p(t)$	$(i\omega)^n\hat{p}(\omega)$
$t^n p(t)$	$i^n\,\dfrac{d^n}{d\omega^n}\,\hat{p}(\omega)$
$\int_{-\infty}^{\infty} p(\tau)q(t-\tau)\,d\tau$	$\hat{p}(\omega)\hat{q}(\omega)$

In three-space dimensions, $\hat{p}(\mathbf{k}, \omega)$ is the transform of $p(\mathbf{x}, t)$ where

$$p(\mathbf{x}, t) = \frac{1}{(2\pi)^4}\int \hat{p}(\mathbf{k}, \omega)\, e^{i\mathbf{k}\cdot\mathbf{x}}\, e^{i\omega t}\, d^3\mathbf{k}\, d\omega$$

and

$$\hat{p}(\mathbf{k}, \omega) = \int p(\mathbf{x}, t)\, e^{-i\mathbf{k}\cdot\mathbf{x}}\, e^{-i\omega t}\, d^3\mathbf{x}\, dt.$$

The transform of the δ-function is unity and hence

$$\delta(\mathbf{x}) = \frac{1}{(2\pi)^3}\int e^{i\mathbf{k}\cdot\mathbf{x}}\, d^3\mathbf{k}.$$

CORRELATION FUNCTIONS AND SPECTRA

The autocorrelation function, $P(\tau)$, of $p(t)$ is defined by

$$P(\tau) = \overline{p(t)p(t+\tau)} = P(-\tau).$$

The power spectral density, or spectrum $\hat{P}(\omega)$, is defined by

$$\hat{P}(\omega) = \int_{-\infty}^{\infty} P(\tau)\, e^{-i\omega\tau}\, d\tau = 2\int_{0}^{\infty} P(\tau)\cos\omega\tau\, d\omega.$$

Then

$$P(\tau) = \frac{1}{\pi}\int_{0}^{\infty} \hat{P}(\omega)\cos\omega\tau\, d\omega \quad \text{and} \quad \overline{\hat{p}(\omega)\hat{p}(\omega')} = 2\pi\hat{P}(\omega)\,\delta(\omega+\omega').$$

The cross-correlation function for two signals $p(t)$ and $q(t)$ is defined by

$$C_{pq}(\tau) = \overline{p(t)q(t+\tau)}.$$

Hence

$$C_{pq}(\tau) = C_{qp}(-\tau).$$

The cross-spectral density function for two signals $p(t)$ and $q(t)$ is defined by

$$\hat{C}_{pq}(\omega) = \int_{-\infty}^{\infty} C_{pq}(\tau)\, e^{-i\omega\tau}\, d\tau.$$

Then

$$C_{pq}(\tau) = \frac{1}{2\pi} \int_{-\infty}^{\infty} \hat{C}_{pq}(\omega)\, e^{i\omega\tau}\, d\omega$$

and

$$\overline{\hat{p}(\omega)\hat{q}(\omega')} = 2\pi \hat{C}_{pq}(\omega')\, \delta(\omega + \omega').$$

The coherence $\gamma(\omega)$ between the two signals $p(t)$ and $q(t)$ is defined by

$$\gamma(\omega) = \frac{|\hat{C}_{pq}(\omega)|}{\sqrt{(\hat{P}(\omega))}\sqrt{(\hat{Q}(\omega))}}; \quad 0 < \gamma < 1$$

where $\hat{P}(\omega)$ and $\hat{Q}(\omega)$ are the power spectral densities of p and q respectively.

The cross-correlation function of $p(x, t)$ is defined by

$$\boldsymbol{P}(\Delta, \tau) = \overline{p(x, t)p(x + \Delta, t + \tau)} = P(-\Delta, -\tau).$$

Hence $P(0, \tau) = P(\tau)$.

The cross power spectral density, $\hat{P}(k, \omega)$, is defined by

$$\hat{P}(k, \omega) = \hat{P}(-k, -\omega) = \int_{-\infty}^{\infty}\int_{-\infty}^{\infty} P(\Delta, \tau)\, e^{-ik\Delta}\, e^{-i\omega\tau}\, d\Delta\, d\tau.$$

Then

$$\boldsymbol{P}(\Delta, \tau) = \frac{1}{(2\pi)^2} \int_{-\infty}^{\infty}\int_{-\infty}^{\infty} \hat{P}(k, \omega)\, e^{ik\Delta}\, e^{i\omega\tau}\, dk\, d\omega$$

and

$$\overline{\hat{p}(k, \omega)\hat{p}(k', \omega')} = (2\pi)^2 \hat{P}(k, \omega)\, \delta(k + k')\, \delta(\omega + \omega').$$

MOVING REFERENCE FRAMES

If the source distribution $q(\mathbf{x}, t)$ is specified with respect to a reference frame which is moving with respect to $\mathbf{x}$ at a constant velocity $\mathbf{U}$, then

$$q(\mathbf{x}, t) = q_m(\mathbf{x} - \mathbf{U}t, t)$$

and *the Fourier transforms* of q and q_m are related by

$$\hat{q}(\mathbf{k}, \omega) = \hat{q}_m(\mathbf{k}, \omega + \mathbf{U}\cdot\mathbf{k}).$$

The *cross-correlation functions* are related by

$$\boldsymbol{P}_m(\boldsymbol{\Delta}, \tau) = P(\boldsymbol{\Delta} + \mathbf{U}\tau, \tau)$$

and the *cross power spectral densities* are related by

$$\hat{P}_m(\mathbf{k}, \omega) = \hat{P}(\mathbf{k}, \omega - \mathbf{U}\cdot\mathbf{k}).$$

USEFUL MATHEMATICAL RELATIONS

In cylindrical co-ordinates (r, φ, z)

$$\text{grad}\, p' = \left(\frac{\partial p'}{\partial r}, \frac{1}{r}\frac{\partial p'}{\partial \varphi}, \frac{\partial p'}{\partial z}\right).$$

For $\mathbf{v} = (v_r, v_\phi, v_z)$

$$\text{div}\, \mathbf{v} = \frac{1}{r}\frac{\partial}{\partial r}(r v_r) + \frac{1}{r}\frac{\partial v_\phi}{\partial \varphi} + \frac{\partial v_z}{\partial z}$$

$$\nabla^2 p' = \frac{1}{r}\frac{\partial}{\partial r}\left(r\frac{\partial p'}{\partial r}\right) + \frac{1}{r^2}\frac{\partial^2 p'}{\partial \varphi^2} + \frac{\partial^2 p'}{\partial z^2}.$$

In spherical polar co-ordinates (r, θ, φ)

$$\text{grad}\, p' = \left(\frac{\partial p'}{\partial r}, \frac{1}{r}\frac{\partial p'}{\partial \theta}, \frac{1}{r\sin\theta}\frac{\partial p'}{\partial \varphi}\right).$$

For $\mathbf{v} = (v_r, v_\theta, v_\phi)$

$$\text{div}\, \mathbf{v} = \frac{1}{r^2}\frac{\partial}{\partial r}(r^2 v_r) + \frac{1}{r\sin\theta}\frac{\partial}{\partial \theta}(\sin\theta v_\theta) + \frac{1}{r\sin\theta}\frac{\partial v_\phi}{\partial \varphi}$$

$$\nabla^2 p' = \frac{1}{r^2}\frac{\partial}{\partial r}\left(r^2\frac{\partial p'}{\partial r}\right) + \frac{1}{r^2\sin\theta}\frac{\partial}{\partial \theta}\left(\sin\theta\frac{\partial p'}{\partial \theta}\right) + \frac{1}{r^2\sin^2\theta}\frac{\partial^2 p'}{\partial \varphi^2}.$$

Contracted notation in three dimensions

$$\delta_{ij} = 1 \quad \text{if } i = j$$

$$= 0 \quad \text{if } i \neq j$$

$$\delta_{ii} = 3$$

$$v_i = v_j\,\delta_{ij}$$

$$\frac{\partial x_i}{\partial x_j} = \delta_{ij}$$

$$\frac{\partial r}{\partial x_i} = \frac{x_i}{r}$$

$$\frac{\partial}{\partial x_i}\left(\frac{1}{r}\right) = \frac{-x_i}{r^3}$$

$$\nabla(\tfrac{1}{2}r^2) = \mathbf{x} \quad \text{i.e.}\ \frac{\partial}{\partial x_i}(\tfrac{1}{2}r^2) = x_i$$

$$\nabla \cdot \mathbf{v} = \frac{\partial v_i}{\partial x_i}$$

$$\nabla \cdot \mathbf{r} = \frac{\partial r_i}{\partial x_i} = 3$$

$$\nabla^2 = \frac{\partial^2}{\partial x_i \, \partial x_i}$$

$$\nabla^2 r = \frac{2}{r}$$

$$\nabla^2\left(\frac{1}{r}\right) = -4\pi\, \delta(\mathbf{x}).$$

Index